Conceptual Innovation in

American and Comparative Environmental Policy
Sheldon Kamieniecki and Michael E. Kraft, series editors

For a complete list of books in the series, please see the back of the book.

Conceptual Innovation in Environmental Policy

Edited by James Meadowcroft and Daniel J. Fiorino

The MIT Press
Cambridge, Massachusetts
London, England

This book was set in ITC Stone Sans Std and ITC Stone Serif Std by Toppan Best-set Premedia Limited. Printed and bound in the United States of America.

Library of Congress Cataloging-in-Publication Data

Names: Meadowcroft, James, editor. | Fiorino, Daniel J., editor.
Title: Conceptual innovation in environmental policy / edited by James Meadowcroft and Daniel J. Fiorino.
Description: Cambridge, MA : The MIT Press, 2017. | Series: American and comparative environmental policy | Includes bibliographical references and index.
Identifiers: LCCN 2016059714| ISBN 9780262036580 (hardcover : alk. paper) | ISBN 9780262534086 (pbk. : alk. paper)
Subjects: LCSH: Environmental policy–History. | Environmental protection–History. | Environmentalism–History.
Classification: LCC GE170 .C6418 2017 | DDC 363.7/0561–dc23 LC record available at https://lccn.loc.gov/2016059714

10 9 8 7 6 5 4 3 2 1

We dedicate this volume to our friend, colleague, and collaborator Judy A. Layzer. Judy was in equal measure a committed scholar and a dedicated environmentalist. Her intellect and enthusiasm brought tremendous benefits to this project. Judy had a deep understanding of environmental politics and policy. She also had a keen appreciation of irony and a wonderful sense of humor. Her death in 2015 represented a great loss to the environmental community. While we are sad she did not live to see the results of our collaboration appear in print, we feel privileged to have had the opportunity to work with her.

Contents

List of Tables and Figures

Tables

Figures

Box

Series Foreword

Environmental policy researchers have made significant progress in theorizing about and understanding the evolution of environmental concerns as well as the expanded role of government in addressing critical environmental problems, such as air and water pollution, toxic waste management, and climate change. Studies centering on political conflicts, the policy process, the choice and design of policy instruments, environmental movements, and business lobbying have demonstrated the importance of how conceptualizing environmental issues in certain ways can influence the policy approaches government takes (or does not take). A tendency toward risk aversion and "playing it safe" often results in scholars and policy makers conceptualizing environmental problems in standard and similar ways without attempting to rethink the underlying nature of such problems along with how best to ameliorate them.

The role of *conceptual innovation* in changing practices of environmental management has received relatively little attention. As the editors of this book point out, although there has been continuous debate about the meaning of specific ideas (such as *sustainability*, the *precautionary principle*, or *generational equity*), there has been little systematic analysis of the range and temporal development of the conceptual categories used in environmental controversies, linkages between innovative concepts and policy change, and the process of conceptual innovation itself. This edited volume addresses such gaps by focusing explicitly on conceptual innovation, and exploring the significance of adjustment to the categories in which scholars think and argue about environmental governance.

The editors (and no doubt the contributors as well) recognize that *conceptual innovation* sounds rather abstract and removed from practical concerns, and that potential readers may be inclined to wonder what this idea has to do with effectively addressing real-world environmental problems that confront local communities or preoccupy international decision makers. Anticipating this concern, the editors explain that conceptual categories are mechanisms through which scholars and others comprehend the world, vest it with meaning, reason about issues, argue over the path forward, and act. As such, conceptual categories allow students of environmental politics and policy to define problems, imagine solutions, and effect change.

The book employs a unique analytic strategy to understand the evolution of environmental governance. In doing so, the editors and contributors provide an in-depth understanding of the development of key concepts invoked in the environmental policy domain; highlight the temporal evolution of the conceptual field of environmental policy; shed new light on established discourses, arguments, and problems; and analyze the linkages between conceptual and policy change. The introductory chapter offers an overview of the volume, explains how the editors approach concepts and conceptual innovation in the environmental domain, outlines theoretical and methodological assumptions, and presents the material that is to follow. The second chapter provides a more detailed explanation of how concepts are used in environmental argument and examines the overall evolution of the "conceptual field" of environmental policy.

The following chapters present studies of individual concepts that play a prominent role in environmental policy making. There are eleven case studies in all, each of which reconstructs the origins and evolution of a specific concept (e.g., environmental risk, adaptive management, environmental justice, and sustainable development). Following the individual concept studies, the last chapter returns to the big picture, offering observations about patterns of change and the practical significance of conceptual innovation in the environmental policy arena.

Although this volume is intended for scholars with an interest in environmental governance, it is likely to appeal to a broader audience. In many respects, the book constitutes a rather-novel introduction to environmental policy, focusing not so much on regulation and policy instruments, parties and elections, or business lobbying and protest group activity, but more on key concepts that structure argument and action in the environmental policy sphere. Thus, it provides an alternative starting point for those without an extensive background in politics and policy who want to understand more fully the way environmental issues are presented in the policy field.

The book illustrates well our purpose in the MIT Press series in American and Comparative Environmental Policy. We encourage work that examines a broad range of environmental policy issues. We are particularly interested in volumes that incorporate interdisciplinary research, and emphasize the linkages between public policy and environmental problems and issues both within the United States and in cross-national settings. We welcome contributions that analyze the policy dimensions of relationships between humans and the environment from either a theoretical or empirical perspective.

At a time when environmental policies are increasingly seen as controversial, and new and alternative approaches are being implemented widely, we especially encourage studies that assess policy successes and failures, evaluate new institutional arrangements and policy tools, and clarify new directions for environmental politics and policy. The books in this series are written for a wide audience that includes academics, policy makers, environmental scientists and professionals, business and labor leaders, environmental activists, and students concerned with environmental issues. We hope they contribute to public understanding of environmental problems, issues, and policies of concern today, and also suggest promising actions for the future.

Sheldon Kamieniecki, University of California, Santa Cruz
Michael E. Kraft, University of Wisconsin–Green Bay
Coeditors, American and Comparative Environmental Policy Series

Acknowledgments

The authors are indebted to many people and institutions for supporting us in writing this book.

Core funding for research and collaboration was provided by a grant from the Social Sciences and Humanities Research Council of Canada. Thanks go also to the Canada Research Chairs program, which supported James Meadowcroft's research activities. We acknowledge the William K. Reilly Fund for Environmental Governance and Leadership and the School of Public Affairs at American University for providing financial support for the workshop in Washington, DC. And the Program Advisory Board of the Center for Environmental Policy supplied guidance in the early stages of the project.

At Carleton University, we wish to thank Mathew Retallack for contributing to tracking the usage of various concepts and supporting discussion at the project workshops. Kimberly Bittermann provided similar support at later stages of the enterprise. At American University, thanks go to Riordan Frost and Manjyot Bhan for their contributions. They helped with background research as well as the preparation and follow-up to the research workshops held in Ottawa and Washington. Riordan also played a major role in preparing the manuscript for publication and, along with Jennifer Hatch, organizing the reference materials.

This book would not have seen the light of day without the signal contributions of the authors of all the individual concept studies. This was a fruitful and enjoyable collaboration from start to finish. We wish to thank our authors for their scholarship, commitment, and insights.

Thanks also to Beth Clevenger, and before her to Clay Morgan, of the MIT Press for supporting this project and helping to bring it to fruition.

1 Conceptual Innovation and Environmental Policy

James Meadowcroft and Daniel J. Fiorino

This book is concerned with conceptual innovation and the development of environmental policy. It explores the evolution of the categories we use to think and argue about the environment, and their relationship to change in the practices of contemporary governance.

Over the past two decades, scholars have made considerable advances in understanding the evolution of modern environmental concerns and the expanded role of government in addressing environmental problems (Cohen 2006; Eisner 2007; VanNijnatten and Boardman 2009; Vig and Kraft 2015). Political conflicts, the policy process, the choice and design of policy instruments, environmental movements, and business responses have all been subject to analysis. With respect to ideas, attention has focused on the evolution of public attitudes and values (Dunlap 1991; Paehlke 1997), establishment of new international norms (Haas 1989; Lafferty 1996; Hoffmann 2005), and emergence of contrasting environmental discourses (Hajer 1997; Fischer and Hajer 1999; Dryzek 2005; Gabrielson et al. 2016). Modern perspectives have been illuminated by historical studies of alternatively situated understandings of human/nature interactions (Dunlap 1999; Andrews 1999; Radkau 2008). And there has been significant interest in "green" or "environmental" political theory (Barry 1999; Torgerson 1999; Eckersley 2004; Paehlke 2004).

Yet the role of *conceptual innovation* in changing practices of environmental management has received comparatively little attention. Although there has been continuous debate about the meaning of specific ideas (for example, *sustainability*, the *precautionary principle*, or the *green economy*), there has been little systematic study of the range and temporal development of the conceptual categories used in environmental argument, linkages between innovative concepts and policy change, and the process of conceptual innovation itself. This book begins to address such gaps by focusing explicitly on conceptual innovation, and examining the significance of adjustment to the categories with which we think and argue about environmental governance.

Each of the terms used in the book's title is therefore significant: *conceptual*—because concepts provide our starting point and basic unit of analysis; *environmental*—as this is the specific sphere of societal debate and action with which we are concerned; *policy*—because we are focused on concepts that play an important role in the policy realm and in the wider political interactions that shape policy making; *development*—as we track the evolution of particular concepts and the field as a whole over time; and *innovation*—as we are especially interested in novel understandings and their implications for practice.

We recognize that conceptual innovation sounds rather abstract and removed from practical concerns. And the reader may be inclined to wonder what all this has to do with solving real-life environmental problems that confront local communities or preoccupy international decision makers. As we hope to make clear in the following pages, the answer is: quite a bit. Conceptual categories are mechanisms through which we apprehend the world, vest it with meaning, reason about issues, argue over the path forward, and act. They allow us to define problems, imagine solutions, and effect change.

Since the birth of modern environmental policy in the late 1960s there has been a more or less continuous development of the categories invoked in the environmental domain (Meadowcroft 2012). Of course, conceptual change occurs in all policy fields. But in recent decades, it has been particularly marked in the environmental area as new problems have come to the fore and the perceived "reach" of environmental issues into diverse areas of social life has grown. *Biodiversity*, for example, now plays a central role in contemporary policy debates. Yet this concept was unheard of before the second half of the 1980s. New concepts have continued to emerge in the policy arena: consider the relatively recent arrival of *negative carbon emissions*, the *plastic gyre*, or *planetary boundaries*. And it should not be forgotten that concepts that we now more or less take for granted—such as *environmental impact assessment* or *the environment* itself—at one point also represented significant innovations in the policy sphere.

Most of the chapters in this volume present studies of individual concepts that have come to play a prominent role in the environmental policy domain. There are eleven case studies in all, each of which reconstructs the "career" of a particular concept. There are also three more general chapters that set out the framework of our analysis, assess broader conceptual developments in environmental policy, and draw lessons from the case studies.

This introduction offers an overview of the volume, explains how we approach concepts and conceptual innovation in the environmental sphere, outlines our theoretical and methodological assumptions, and presents the material that is to follow. The second chapter provides a more detailed discussion of how concepts are used in environmental argument and explores the overall evolution of the "conceptual field" of environmental policy. Following the individual concept studies, the final chapter returns to the big picture, offering observations about patterns of change and the practical significance of conceptual innovation in environmental policy.

Although this book is intended for an academic audience with a specialist interest in environmental governance, we believe it will appeal also to a broader constituency. In some respects, it constitutes an alternative introduction to environmental policy, focusing not so much on regulation and policy instruments, parties and elections, or business lobbying and protest group action (although all these make an appearance) but rather more on key ideas that structure argument and action in the environmental policy realm. Thus, it may supply an interesting starting point for those without an extensive background in politics and policy (for example, natural scientists) who want to make more sense of the way environmental issues are presented in the policy realm.

Concepts and Policy Argument

Concepts can be understood as "thought categories" through which we apprehend and, to some degree, constitute the world around us. They enable, but also constrain, reasoning and debate, and serve as building blocks for more elaborate arguments, theories, discourses, and ideologies. Ultimately they ground practical action in different spheres of human endeavor. Concepts are not just words. After all, one word may denote several concepts (consider *state* as a condition or international unit); and several words may denote a single concept (for example, *freedom* and *liberty* are usually understood as such equivalents). On the other hand, words and concepts are closely linked: typically we say someone possesses a concept when they are able to use the associated word in the relevant sense. And we should be aware of the risk of anachronism when we ascribe to historical figures or cultures conceptual understandings that they lacked specific words to denote. In the literature it is well established that conceptual innovation, linguistic shifts, and political change go hand in hand (Ball et al. 1989). New ideas require novel uses of language (new words or new meanings for old terms), and these are linked to changes in political practice. But the precise way these are related differs according to circumstance.

Concepts may be approached as distinctive "unit/ideas," but they must be understood in reference to other concepts. We clarify one idea by drawing on others, establishing parallels and contrasts, and assembling diverse elements into a distinctive pattern. Concepts acquire significance through their incorporation into broader processes of reasoning, argument, and practice. They are often found in "clusters": a group of related notions that are frequently invoked together. And they are employed in particular ways in specific discourses, forms of argument, and political or ideological perspectives. Concepts also have a history: they are born, are introduced in argument, and influence particular fields of endeavor (James and Steger 2014). They may become embedded in social practices, institutional structures, and legal forms. Ultimately, concepts can fade in significance as the ideas with which they are associated no longer attract interest. To take an example from the scientific domain, *phlogiston* was of interest in the sixteenth and seventeenth centuries, but has no place in contemporary scientific reasoning. Today, *dark matter* is a focus for lively scientific exchange, but whether this category will survive or be displaced by other theoretical constructs remains to be seen.

Our use of concepts is often routine and unreflective: we reach for terms and deploy concepts to reason, construct arguments, articulate claims, communicate beliefs and desires, and influence the conduct of others. On the other hand, we also make deliberate choices—selecting one concept and setting aside another, specifying that we are using a particular concept in this way and not that way, shifting received understandings by extending a category to cover new instances or adjusting its elements, or sometimes invoking a new term to express a novel meaning.

The investigation of concepts is sometimes understood to be the particular province of philosophical inquiry. Certainly conceptual analysis plays a central role in contemporary analytic philosophy, and perspectives on the essential character of concepts have long been intertwined with broader ontological and epistemological debates. The twentieth-century "linguistic turn" that affected not just philosophy but also the social sciences and humanities more generally brought out the importance of language for the constitution of social life (Rorty 1967). Discourse theorists emphasized linkages among language, dominant norms, and social hierarchies (Foucault 1977, 1984). Historians of political thought moved away from the idea of timeless debates with the classics to emphasize that ideas must be appreciated in their political and cultural contexts, and that encounters with historical texts should involve an effort to understand them in terms that would have been intelligible to their original authors. An explicit focus on concepts, historical context, and sociopolitical change has characterized the work of the school of "conceptual history," associated with Quentin Skinner (2009) and Reinhart Koselleck (2004), but to which many others have contributed (Williams 1982; Connolly 1983; Richter 1995; Freeden 1996). For our purposes, the key insights that can be drawn from these literatures are that language is not a neutral medium but one that reflects and reproduces particular perspectives, that conceptual categories are historical constructs that evolve over time, and that conceptual understandings and social practices are inextricably interconnected.

Since concepts are categories through which we think and communicate, it is not surprising that they have attracted interest from many disciplines, including philosophy, history, cultural studies, linguistics, psychology, mathematics, logic, neuroscience, information science, and artificial intelligence (Hjørland 2009). One strand of this work explores how individuals apprehend the world, acquire language, and manipulate concepts. It suggests that we do not grasp concepts by either compiling an exhaustive mental list of items that fall into a particular class or memorizing some formal rule against which future instances can be compared. Rather, it is more about appreciating patterns of similarity and difference. So we can understand that a rock could also be a table even though it lacks all the key features of a dictionary definition of a table as "a piece of furniture with a flat top and legs" (LoveToKnow Corporation 2015). In this book, however, we are concerned less with individual cognition than with concepts as socially shared thought objects—with the way they are taken up by multiple users and come to assume a particular place in the policy sphere.

All concepts are abstractions, but those that assume an important place in the political or policy sphere often have considerable internal complexity. Think of the idea of "public participation" that involves assumptions about who is participating, in what process, in what manner, and to what end, and that links to various empirical and evaluative dimensions of governance, including representation, responsibility, effectiveness, and so on. Some thinkers have invoked a special category of "essentially contested concepts" (Gallie 1964), yet for our purposes it is enough to note that political and policy concepts are subject to continuing contestation and reinterpretation. This is most obvious for high-level ideas with strong normative associations such as democracy, freedom, or justice, but it applies much more widely across the policy realm. Yet because they are complex and contested, it does not follow that these concepts are without meaning: typically the core range of uses is clear, and it is possible to rule out many potential understandings as implausible. And of course, the flexibility of these concepts is part of what makes them essential to political and policy discourse.

The concepts with which this book is concerned are those employed in the particular sphere of environmental governance. These are ideas such as *acid rain, adaptive management, biodiversity, carbon budget, common but differentiated responsibilities, dangerous climate change, environmental assessment, ecological footprint, the polluter pays principle, the precautionary principle, hazardous waste,* and so on. Like other issue domains, environmental policy relies on categories drawn from the general political arena, relating to basic values, institutions, and processes (democracy, parliament, government, the supreme court, the public interest, rights, and so on). But it is the *subject-specific conceptual repertoire,* delineating environmental problems and solutions, that is of primary concern in this volume. These are the ideas that establish the parameters and modalities of this particular field of governance activity, and that help frame the specific ways it is conducted.

The subject-specific repertoire of environmental policy/politics is linked to related areas such as law, economics, philosophy, and the natural sciences. The idea of sustainability, for example, could be invoked in any of these domains, although this might be done in somewhat-different ways. Although usage in these cognate domains will be considered in the discussion of the career of individual concepts, our core focus is politics and policy (rather than, say, legal or scientific discourse).

While concepts provide the starting point for this inquiry, our goal has not been to produce a conceptual dictionary of environmental thought. Instead, it is to explore the role innovation in these conceptual categories plays in reasoning and argument, and their significance for practical efforts to manage environmental problems.

Conceptual Innovation

Those who study processes of technological innovation examine the emergence and uptake of new technologies, from pioneering research efforts through prototype development, and scale up to the launch of marketable products and their subsequent societal adoption (Freeman 1974; Lundvall 2010). In an analogous way, we apply the expression *conceptual innovation in environmental policy* to refer to the broad process through which novel understandings win acceptance within the environmental policy realm. It involves more than the invention of a novel thought-construct (making the requisite ideational connections and perhaps coining an appropriate term) to include the way the new understanding is introduced into policy argument, gains adherents, and becomes linked in to policy design and implementation.

Our basic understanding of how this process operates runs as follows. Novel ideas are constantly being generated as individuals and groups struggle to apprehend and influence the world around them. Yet to enter the policy domain, some actor or set of actors must shape a new understanding into a *policy relevant construct* and explicitly introduce it into policy argument. If this move fails to resonate more broadly, matters may go no further. But if it finds a wider audience, the new conceptualization can become integrated into larger patterns of use and begin to shift the terms of debate.

Critical to extending the influence of a novel conceptualization is its institutional embedding within the environmental policy sphere. This can take a variety of forms: substantive uptake by prominent players within the policy community (environmental groups, business organizations, and expert advisory bodies); incorporation into the priorities of research councils and funding bodies; integration into political platforms or manifestos; adoption in government white papers, policy statements, plans, and programs; inclusion in the official remit of government agencies or ministries; and integration into legislation and perhaps even constitutional documents. To the extent that a concept is embodied in texts with legal force, its meanings and implications may ultimately be subject to judicial interpretation. Such institutional embedding, from "softer" manifestations (say, in the list of campaign priorities for an environmental nongovernmental organization) to "harder" forms (for example, in the legal responsibilities of a ministry), provide points of reference that can encourage further uptake and reinforce particular patterns of understanding. As a conceptual shift becomes more widely accepted, efforts may be made to specify its implications in particular and varied contexts, and it can be linked to the redesign of practical management activities. Throughout the process there will be continued argument over what the new idea actually signifies. Multiple interpretations may persist over the long term. And the forms that come to predominate may be somewhat different from those originally introduced into the policy realm. For a simple graphic representation, see figure 1.1.

Although this has been presented in relatively linear terms—from the launching of a policy-relevant conceptual shift, through diffusion and institutionalization, to practical impacts—in fact this process is more complex: the steps can be intertwined, practical impacts may flow from early diffusion, and differential institutional embedding can revise dominant understandings. Nevertheless, it is helpful to consider each of these interrelated elements of the innovation process.

From a policy perspective, the starting place is the articulation of the conceptual reconfiguration in policy-relevant arguments. Ideas may be drawn from diverse sources, within or without the policy sphere, broader social and political debate, and developments in the natural and social sciences and the humanities. Whatever the preceding genealogy, the idea must be articulated as part of an argument addressed to the public sphere, engaging with issues of societal interest and the realm of government. The "pitch" of a novel conceptualization can occur through general public argument (books, articles, or the press), via expert reports to policy makers (formal advisory bodies, consultants, or academics), in the findings of

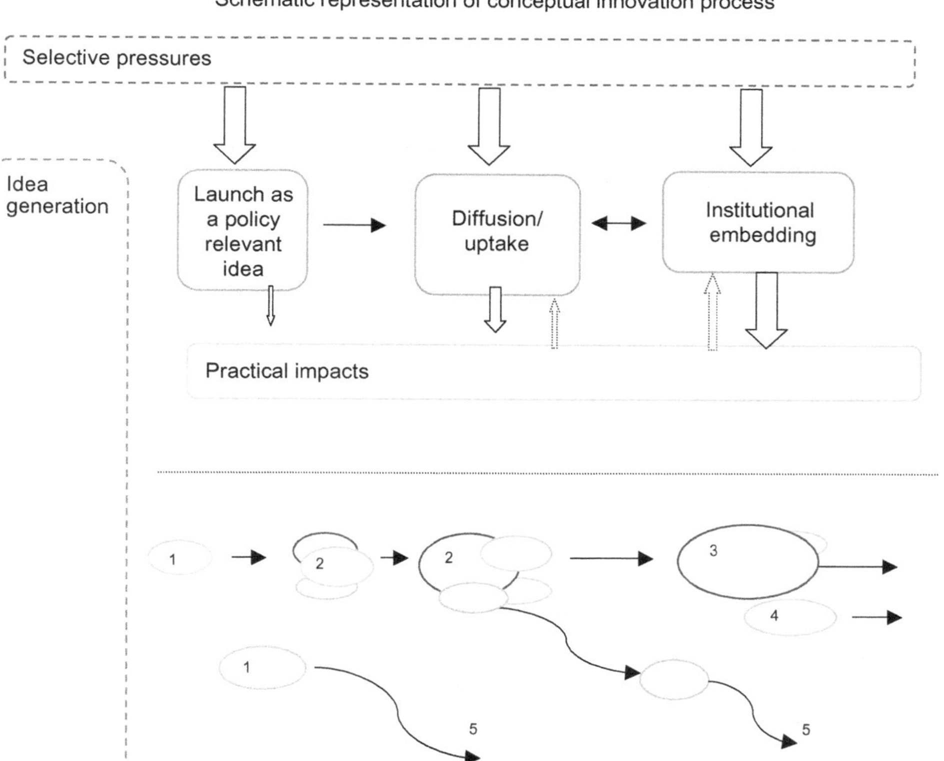

Figure 1.1
Schematic representation of conceptual innovation process.

public inquiries or commissions, or through communication from government agencies. Sometimes the conceptual shift will be explicitly flagged; on other occasions it may be introduced tangentially as the (new) normal way of approaching the subject matter. But in either case it must be sufficiently attractive to be taken up more widely. Sometimes such an innovation will spread rapidly; alternatively it may lie latent for many years, or may sink with barely a trace.

With respect to institutional embedding, a distinction should be made between uptake by organizations involved in policy argument and uptake by official bodies with formal responsibility for making and implementing policy (government agencies, ministries, and the legislature and executive). The latter groupings lie at the heart of the formal policy system. Similarly the solidity of the embedding matters: politicians (and their white papers) come and go, but legislation and regulatory procedures are more resistant to change. And there are issues of vertical and horizontal governance: embedding up and down the hierarchy (from local and regional institutions to central government), and across the breadth of government, not just in the environment ministry, but also in the ministries of energy, natural resources, economics, transport, and so on. Embedding at multiple levels or across ministries may be more resilient than embedding in one level or bureau only. International organizations can also provide a powerful mechanism for diffusion and continuity, although ideas lodged at an international level do not necessarily filter down to diverse domestic circumstances.

Sustainable development provides an illustration of the process sketched out above. Although by the early 1980s arguments about sustainability had been going on for more than a decade (and political and philosophical antecedents could be traced much further back), it is generally accepted that the report of the World Commission on Environment and Development (1987), *Our Common Future*, propelled the idea of sustainable development on to the international stage. The commission deliberately crafted the concept to advance the environment and development debate, and carefully defined the term in its report. So here was a concept that was "pitched" at the highest-possible level by a blue-ribbon UN panel. Even such a high-profile launch did not guarantee take-up: there were other UN commissions whose intellectual contributions are now barely remembered. But sustainable development did catch on. It was relatively rapidly institutionalized within the UN system during preparations for the 1992 Rio Earth Summit, and progressively adopted into national policy frameworks. Governments and international organizations devoted significant efforts to preparing sustainable development plans, strategies, and indicators. And today the idea remains very much alive in international discussion around implementation of the UN "sustainable development goals" for 2030. Over time, the dominant understanding of sustainable development has evolved; for example, neither the "three pillars" image (balancing economy, society, and environment) nor the emphasis on "participation" in environment and development decision making were explicit in the original World Commission on Environment and Development formulation, but both subsequently became prominent. And of course multiple and partially conflicting understandings of what sustainable development actually entails remain. This example also illustrates that conceptual innovation can occur at different scales: large shifts can give rise to an entirely new thought category (in this case, the invention of sustainable development), but smaller more incremental innovations (here, the three pillars image or emphasis on participation) can shift the bearing of an existing category.

So why do some conceptual innovations catch on while others fail? Why do some become integrated into the idiom of governance while others gain popularity among particular policy constituencies but remain locked out of the official sphere? This is an issue to which we will return throughout the volume. At the outset, though, we can suggest three elements that seem to characterize ideas that ultimately penetrate the core of the policy domain. First, the conceptual reconfiguration must address a perceived need: it identifies a new problem, diagnosis, or solution, or productively reorders established understandings. In functional terms we could say that it fills an ideational niche. Second, if a concept is to be taken up broadly and entrenched officially, it needs to be able to speak to multiple constituencies. Interestingly, this suggests that ambiguity can be an asset in the policy sphere. That a concept is open to varied interpretations widens its potential clientele and sphere of operations. It suggests its features can be rearranged into multiple configurations that generate diverse insights that appear significant to varied actors. Third, the reconceptualization should not be too alien to existing discursive patterns and dominant understandings of the way "the world works." In other words, the conceptual shift is more likely to find its way into the core of the policy system if it does not too directly and obviously challenge existing socioeconomic as well as political institutions and relationships. This is not to say that truly radical realignments are impossible; rather, these will then be accompanied with political/institutional changes of a comparable scale, and will still trade on existing understandings (resonating with some established categories to leverage significant change in others).

The Orientation of This Volume

The book deploys a distinctive analytic strategy to understand the evolution of environmental governance. The goal is to allow a fuller understanding of the development of key concepts invoked in the environmental policy domain; provide a deeper appreciation of the temporal evolution of the overall conceptual field of environmental policy; shed new light on established discourses, arguments, and problems (by illuminating them "from below," as it were, by approaching them from constitutive conceptual elements); and explore the linkages between conceptual and policy change.

At this point it is worth considering how our approach relates to complementary strands of scholarship. Reference has already been made to conceptual history. In a general sense, the work presented here could be considered a contribution to this field, but it is informed more by political science than historiography and has a more explicit focus on policy than most scholarship classed under this heading. Another obvious point of contact is with discourse analysis, which in the environmental policy field is often associated with the scholarship of Martin Hajer and John Dryzek (but consider also Karen Litfin). Hajer's (1997) best-known work examines acid rain policy in the United Kingdom and Netherlands, and chronicles the rise of the discourse of "ecological modernization." For his part, Dryzek (2005) has surveyed the whole field of environmental politics, identifying nine competing discourses. Despite their differences, both approaches take discourse as their fundamental analytic category. This volume, in contrast, starts from specific concepts, which may play different roles in varied discursive constructs. Thus, while Hajer discusses sustainable development as a characteristic "story line" of the dominant ecomodernist "discourse coalition," and Dryzek sees sustainable development as one among many competing environmental discourses, we approach it here as a specific concept and illustration of conceptual innovation deployed to structure thinking and argument in the environmental policy sphere. In other words, the point of departure and questions on which we focus are slightly different.

The literature on framing, particularly as used in media, communication, and policy studies, provides another point of reference (Rein and Schön 1993; Stone 2001). Framing studies typically examine a particular episode (perhaps a war, policy conflict, or electoral context), with the goal of discovering how issues are constructed in the public sphere by the media, political leaders, and other contending parties. Of course, one can also talk of how a concept is being framed, and one can consider the deployment of a particular concept (and/or a particular interpretation of that concept) as an exercise in framing. Finally, we should point to scholarship that has looked at the emergence and propagation of norms guiding expectations in the international environmental sphere. Steven Bernstein (2001), for example, has explored the rise of the "norm complex" of "liberal environmentalism" that reconciles environmental concerns with the liberal international economic order, and he considers sustainable development as a concept that has both legitimized this norm complex and masked the compromises that characterized its institutionalization. Again, we approach the issue from the bottom up, starting with individual concepts, rather than from the perspective of higher-level constructs such as norm complexes and discourses.

The chapters in the volume deal with the material in two basic ways. The case study chapters follow the career of individual concepts that are important in the environmental policy domain. Although the authors adopt slightly different narrative strategies, these chapters address a common set of questions that relate to the basic understanding of the target concept (what it implies, and how it is linked to adjacent or contrasting concepts), its origin (where it came from, who developed it, and in what context and when), the character of the associated ideational shift (what was new, and what problem was it intended to address), the way the novel understanding passed into political and policy argument (routes of transmission and institutional adoption), problems and internal tensions (varying perceptions, disputes, and controversies), and practical impacts (what difference it made).

Each case study chapter then concludes with an evaluative discussion that considers the significance of the concept and associated innovation for environmental policy. This revolves around answering two basic questions. First, *to what extent has the concept been important for argument and practice in the environmental policy domain*? This question is about spread and embedding. How popular has it been? How widely is it used? Has it been entrenched in regulation, legislation, treaties, and constitutions? This covers political/policy argument and practical deployment. Second, *to what degree has the concept proven productive or fruitful?* This question is about the extent to which it has contributed to clarifying thinking, introducing productive insights, resolving problems, and so on. The first question is more empirical, and the second is more normative. They link directly to the discussion of the future of the concept as well as its difficulties, internal contradictions, and so on.

The general chapters (1, 2, and 14) approach the issue in a more synthetic manner, exploring patterns of conceptual innovation across the environmental field. This involves an initial attempt to elucidate processes of conceptual innovation in environmental governance and map the evolution of the conceptual field. These general chapters draw on a variety of sources as well as seek to build on insights from the individual concept studies.

Selection of the concepts for the case studies began from a long list of about sixty, culled from the policy literature. Although hundreds of entries appear in environmental reference works (such as the US Environmental Protection Agency's (1997) *Terms of Environment: Glossary, Abbreviations, and Acronyms*), the overwhelming majority of these are scientific or technical terms (e.g., phenol, radioisotope, or osmosis), or relate to the text of particular laws and ordinances (exempted aquifer, CAS registration number, etc.). The long list was later narrowed to about twenty-five of the most widely used terms, and a group of environmental experts was consulted about their perspectives on the relative importance of these concepts for environmental policy past and future. In selecting the final group of concepts for detailed examination we considered a number of factors, including their *general importance* (that the concept was in fairly widespread use and appeared to play a prominent role in environmental argument); their *practical policy bearing* (that there was a link to politics, policy, and practice) so the idea was not *just* the province of environmental philosophers, natural or social scientists, or a particular activist group; *innovation/evolution*, that the concept emerged or underwent some significant change in the 1960–2014 period that covers the development of modern environmental policy; and their contribution to a good *overall mix* of concepts, to ensure a balance among established and emerging concepts, those with broad or narrow application, and those that relate to various areas of environmental policy.

As the chapter titles indicate, the final list was composed of environment, sustainable development, biodiversity, environmental assessment, critical loads, adaptive management, green economy, environmental risk, environmental security, environmental justice, and sustainable consumption.

The list includes two "macro" concepts that attempt to capture the whole domain with which we are concerned: environment and sustainable development. There are four environment-denominated concepts—that is to say, concepts that apply environment to modify a more general or established idea: environmental assessment, environmental risk, environmental justice, and environmental security. In each case the expression has come to constitute a distinct unit idea that signals more than a combination of two words. Some of these concepts came into use at the beginning of the period with which we are concerned (environment or environmental assessment), while others are newer arrivals (biodiversity or environmental justice). Some are used widely, notably environment, which also has the distinction of defining the policy sphere with which we are here concerned: environmental policy. Others such as critical loads or adaptive management have more restricted spheres of application, and are most often deployed by particular policy communities and in specific contexts.

There are many points of contact among the members of this group of concepts. For example, environment is frequently presented as one of the three pillars of sustainable development; environmental security is concerned with certain kinds of environmental risks; sustainable consumption is part of the green economy, and both these ideas are linked to sustainable development; and so on. And of course, these concepts link closely to many other environmental concepts that are not included on this short list: environmental assessment is one appraisal technique, but another is environmental cost-benefit analysis, biodiversity protection is often connected to resilience, and so forth.

There is no suggestion that the concepts on this list are in some sense exceptional or intrinsically more important than others that were not included. A somewhat-different group of concepts could reasonably have been selected. On the other hand, most researchers would probably agree that these are interesting and important ideas within the general field of environmental policy. A few absences require explanation. We did not include pollution or conservation because their usage was well established before the era of modern environmental policy on which this study focused. Certainly, they were critical in defining the emerging policy field—for environmental policy brought the two previously distinct areas of nature conservation and industrial pollution control into closer contact. And both ideas have continued to develop over intervening decades as, for instance, we have seen the progressive extension of the phenomena classed as environmental pollution. Some readers will feel the most obvious omission is climate change. Here we defer to the enormous scholarly effort that has already centered on clarifying this idea (Cowie 2007; Hulme 2009), and chose instead to highlight biodiversity—the second truly global environmental megaproblem that was addressed with a framework convention at the 1992 Rio Earth Summit. Other established concepts high on our priority list, but that in the end we were not able to cover, include environmental policy integration, decoupling, resilience, and the precautionary principle. There are also a number of ideas that are now beginning to make their mark, including planetary boundaries, the Anthropocene, sustainability science, and ecosystem services. Each of these ideas makes some appearance in this volume, but they are not subject to chapter-length treatment.

At this point it is worth considering some of the difficulties and limitations of the present study. In the first place, we set a challenging task for the authors of the concept studies. The selected concepts have been applied widely, in varying contexts, often over multiple decades. They have been invoked in political argument, linked to policy practice, and subjected to academic analysis. Some—like environment or sustainable development—are ubiquitous. Yet we aspire to tell their story in a comparatively brief chapter. So these discussions necessarily offer fairly high-level narratives that concentrate on key features and fault lines. Such broad-scale analysis is not the only possible way to approach the study of conceptual innovation in the environmental sphere. Instead, one could focus on a much narrower time period or institutional context (for example, how a concept was introduced in one government department, or how a group of concepts figured in a particular policy argument), but here we decided to explore the larger picture.

Second, we are interested in practical implications: with understanding whether, to what extent, and in what circumstances conceptual innovation really matters. Or are these changes in conceptual categories really more about fashionable buzzwords, rhetorical tricks, and symbolic politics, where the idiom of policy making and political argument shift, but the underlying relationships and outcomes remain largely unchanged? Clearly we think that changing the terms of discourse does matter, yet to what degree, and in what circumstances, remain issues for investigation. Certainly the kind of study undertaken here does not typically allow us to say in a simple and deterministic sense that change in the understanding of concept X led directly to specific policy outcome Y. We are simply not working with this form of linear causality. The relationship between ideational and policy change is more complex than that.

In the case of problem-identifying concepts (such as acid rain, climate change, and ozone disruption), there is an obvious link between problem definition and subsequent policy action to address the designated issue. But all sorts of ideas, interests, and institutions influence whether, when, and how the problem is actually handled. With management-oriented concepts, which denote a particular approach to dealing with environmental challenges (ecosystem management, the precautionary principle, and so on), there is sufficient interpretative flexibility to justify multiple policy outputs. Applying the approach in a particular case structures argument, but does not determine an exact policy prescription. The EU policy on genetically modified organisms and food is linked to an application of the precautionary principle. Yet there was a choice to apply this principle here, and in this particular way, when it is not applied, or is applied in different ways, in other areas of EU regulation. So exploring the practical linkages of conceptual innovation is not without pitfalls. Typically what we can state with confidence is that a shift in conceptual category X was associated with a particular (and often-diverse) set of policy experiences. And depending on the circumstances, one might want to say that the new understanding accompanied altered practices, that it facilitated the introduction of these practices (and the outcomes that ultimately flowed from them), or perhaps that it helped constitute the altered practice, because it only makes sense to those engaged in it in light of the conceptual shift.

Third, we approach the development of a more general picture of the evolution of the conceptual field of environmental politics in various ways. The case studies provide analysis of a number of important concepts, discuss some of the connections among them, and reference a number of additional concepts to which they are linked. The general chapters (especially chapter 2) try to point toward a more synthetic account by looking at families of concepts and tracing elements of the evolution over time. Yet we are far from being able to offer a full picture of the evolution of the conceptual field of environmental policy. For this a broader research effort will be required. The concluding chapter suggests some elements of that research agenda.

In other words, we understand the work presented here to be of an exploratory character, focused on an area that has not received much prior attention. Throughout, our goal has been to develop a broad initial investigation that can map out the terrain, develop interesting insights, suggest plausible interpretations, and point to areas for further work.

In the next chapter, we characterize the broad conceptual field of environmental policy and describe its evolution since its emergence as a distinct policy domain in the late 1960s.

References

Andrews, Richard. 1999. *Managing the Environment, Managing Ourselves*. New Haven, CT: Yale University Press.

Ball, Terence, James Farr, and Russell Hanson, eds. 1989. *Political Innovation and Conceptual Change*. Cambridge: Cambridge University Press.

Barry, John. 1999. *Rethinking Green Politics*. Thousand Oaks, CA: Sage Publications.

Bernstein, Steven. 2001. *The Compromise of Liberal Environmentalism*. New York: Columbia University Press.

Cohen, Steven. 2006. *Understanding Environmental Policy*. New York: Columbia University Press.

Connolly, William. 1983. *The Terms of Political Discourse*. Princeton, NJ: Princeton University Press.

Cowie, Jonathan. 2007. *Climate Change: Biological and Human Aspects*. Cambridge: Cambridge University Press.

Dryzek, John. 2005. *The Politics of the Earth: Environmental Discourses*. 2nd ed. Oxford: Oxford University Press.

Dunlap, Riley. 1991. "Trends in Public Opinion toward Environmental Issues: 1965–1990." *Society and Natural Resources* 4 (3): 285–312. doi:10.1080/08941929109380761.

Dunlap, Thomas. 1999. *Nature and the English Diaspora*. Cambridge: Cambridge University Press.

Eckersley, Robyn. 2004. *The Green State: Rethinking Democracy and Sovereignty*. Cambridge, MA: MIT Press.

Eisner, Marc. 2007. *Governing the Environment*. Boulder, CO: Lynne Rienner.

Fischer, Frank, and Maarten Hajer, eds. 1999. *Living with Nature: Environmental Politics as Cultural Discourse*. Oxford: Oxford University Press.

Foucault, Michel. 1977. *Discipline and Punish*. New York: Pantheon Books.

Foucault, Michel. 1984. *The Foucault Reader*. Ed. Paul Rabinow. New York: Pantheon Books.

Freeden, Michael. 1996. *Ideologies and Political Theory*. Oxford: Oxford University Press.

Freeman, Chris. 1974. *The Economics of Industrial Innovation*. London: Penguin Books.

Gabrielson, Tina, Cheryl Hall, John H. Meyer, and David Schlosberg. 2016. *The Oxford Handbook of Environmental Political Theory*. Oxford: Oxford University Press.

Gallie, Walter Bryce. 1964. "Essentially Contested Concepts." In *Philosophy and the Historical Understanding*, 157–191. London: Chatto and Windus.

Haas, Peter. 1989. "Do Regimes Matter? Epistemic Communities and Mediterranean Pollution." *International Organization* 43 (3): 377–403. doi:10.1017/s0020818300032975.

Hajer, Maarten. 1997. *The Politics of Environmental Discourse: Ecological Modernization and the Policy Process*. Oxford: Oxford University Press.

Hjørland, Birger. 2009. "Concept Theory." *Journal of the American Society for Information Science and Technology* 60 (8): 1519–1596. doi:10.1002/asi.21082.

Hoffmann, Matthew. 2005. *Ozone Depletion and Climate Change*. Albany: State University of New York Press.

Hulme, Mike. 2009. *Why We Disagree about Climate Change*. Cambridge: Cambridge University Press.

James, Paul, and Manfred Steger. 2014. "A Genealogy of Globalization: The Career of a Concept." *Globalizations* 11 (4): 417–434. doi:10.1080/14747731.2014.951186.

Koselleck, Reinhart. 2004. *Futures Past: On the Semantics of Historical Time*. Translated by Keith Tribe. New York: Columbia University Press.

Lafferty, William. 1996. "The Politics of Sustainable Development: Global Norms for National Implementation." *Environmental Politics* 5:185–208. doi:10.1080/09644019608414261.

LoveToKnow Corporation. 2015. "Table." *YourDictionary*. November 22, 2016, http://www.yourdictionary.com.

Lundvall, Bengt-Åke. 2010. *National Systems of Innovation: Towards a Theory of Innovation and Interactive Learning*. London: Anthem Press.

Meadowcroft, James. 2012. "Greening the State." In *Comparative Environmental Politics*, ed. Paul Steinberg and Stacy VanDeveer, 63–88. Cambridge, MA: MIT Press.

Paehlke, Robert. 1997. "Environmental Values and Public Policy." In *Environmental Policy in the 1990s*, ed. Norman Vig and Michael Kraft, 75–94. Washington, DC: CQ Press.

Paehlke, Robert. 2004. *Democracy's Dilemma: Environment, Social Equity, and the Global Economy*. Cambridge, MA: MIT Press.

Radkau, Joachim. 2008. *Nature and Power*. Cambridge: Cambridge University Press.

Rein, Martin, and Donald Schön. 1993. "Reframing Policy Discourse." In *The Argumentative Turn in Policy Analysis and Planning*, ed. Frank Fischer and John Forester, 145–166. Durham, NC: Duke University Press.

Richter, Melvin. 1995. *The History of Political and Social Concepts*. Oxford: Oxford University Press.

Rorty, Richard. 1967. *The Linguistic Turn: Essays in the Philosophical Method*. Chicago: University of Chicago Press.

Skinner, Quentin. 2009. "A Genealogy of the Modern State." *Proceedings of the British Academy* 162:324–370. doi:10.5871/bacad/9780197264584.003.0011.

Stone, Deborah. 2001. *Policy Paradox: The Art of Political Decision Making*. 3rd ed. New York: W. W. Norton and Company.

Torgerson, Douglas. 1999. *The Promise of Green Politics*. Durham, NC: Duke University Press.

US Environmental Protection Agency. 1997. *Terms of Environment: Glossary, Abbreviations, and Acronyms*. November 27, 2016, https://www.epa.gov/glossary.

VanNijnatten, Debora, and Robert Boardman, eds. 2009. *Canadian Environmental Policy and Politics*. Oxford: Oxford University Press.

Vig, Norman, and Michael Kraft. 2015. *Environmental Policy: New Directions for the Twenty-First Century*. Washington, DC: CQ Press.

Williams, Raymond. 1982. *Keywords: A Vocabulary of Culture and Society*, rev. ed. Oxford: Oxford University Press.

World Commission on Environment and Development. 1987. *Our Common Future*. Oxford: Oxford University Press.

2 The Conceptual Repertoire of Environmental Policy

James Meadowcroft and Daniel J. Fiorino

This chapter offers an overview of concepts employed in the environmental policy domain and reflects on the evolution of this conceptual repertoire. Its purpose is to provide a broad-brush portrait of the development of the conceptual field to complement the more focused and detailed examination of individual concepts presented in the chapters that follow. We use the expression *conceptual field* to denote the set of interconnected concepts that help establish the character of the environmental policy domain at a given point in time. The image of a field captures the idea of a plenitude of concepts related across multiple dimensions in argument and practice. Thinking about the general development of environmental concepts can help ground discussions of innovation in the environmental sphere. After all, the concepts in use at a specific conjuncture represent the product of previous rounds of innovation while also supplying the context within which further development takes place.

The discussion is organized into three parts. First comes a general look at environmental concepts. This is followed by an examination of concepts that play a particularly important role in structuring the policy realm. Finally, the chapter explores the temporal evolution of the field.

Contrasts and Connections

When one considers the contemporary environmental policy domain—including the full range of problems and responses linked to diverse economic and social activities—one is confronted with a bewildering array of concepts. Policy documents from national environmental ministries and the websites of international bodies such as the UN Environmental Program (UNEP), Organisation for Economic Co-operation and Development

(OECD), or European Environment Agency employ hundreds of distinct environment-related concepts. And many other ideas are invoked in broader societal debate. To make sense of this diversity, it is helpful to examine patterns of contrast and affinity among environmental concepts, and consider the varied roles they play.

To anchor the discussion, a sample of concepts employed in environmental argument is presented in table 2.1. The group has been selected to illustrate concepts of a quite-different character, with varied spheres of application.

One thing that emerges from a list of this type is the extent to which new environmental concepts are articulated by drawing together existing terms. A new thought category can, of course, be specified by repurposing an existing word; thus, *resilience* in the environmental domain draws on the long-established understanding of resilience as a capacity to recover from stress and endure. But more often, the conceptual lexicon is extended by way of conjoining terms. Sometime terms are fused into a single new word, as in *biodiversity* or *geoengineering*. Sometimes a longer phrase is required to capture the novel meaning, as with *extended producer responsibility* or the *polluter pays principle*. But most common is a simple coupling of terms, as in *sustainable tourism, planetary boundaries, nuclear waste,* or *emissions trading*. Indeed, by linking varied terms to a common partner, whole families of environmental concepts have been generated.

The largest such groupings are associated with the modifiers *environmental, ecological* (or simply *eco*), *sustainable*, and *green*. Table 2.2 lists examples of expressions in these groups. The root ideas of environment, ecology, sustainable, and green have quite-different histories and connotations. The

Table 2.1

A Selection of Environment-Related Concepts Used in Policy/Political Argument

Adaptive management	Ecotourism	Net zero housing
Biodegradable	Emissions trading	Nuclear waste
Biodiversity	Environmental audit	Organic farming
Carbon capture and storage	Environmental integration	Planetary boundaries
Carbon offset	Extended producer responsibility	Polluter pays principle
Circular economy	Flue gas desulphurization	Rewilding
Decoupling	Geoengineering	Sustainable tourism
Ecological footprint	Green growth	

Table 2.2
Groupings of Concepts Related to Four Core Environmental Terms

Environment	**Ecology**	Sustainable growth
Environmental assessment	Ecological fiscal reform	Sustainable livelihoods
Environmental audit	Ecological footprint	Sustainable production
Environmental crisis	Ecological interdependence	Sustainable use
Environmental education	Ecological justice	Sustainable waste management
Environmental ethics	Ecological rationality	Sustainable yields
Environmental fiscal reform		Sustainability
Environmental footprint	*Eco-*	Unsustainable
Environmental governance	Ecoaudit	
Environmental indicator	Ecoconscious	**Green**
Environmental justice	Ecoefficiency	Green buildings
Environmental law	Ecofriendly	Green business
Environmental liability	Ecojustice	Green consumer
Environmental management	Ecolabels	Green design
Environmental planning	Ecotaxes	Green economy
Environmental policy integration	Ecotourism	Green energy
Environmental protection		Green growth
Environmental restoration	**Sustainable**	Green infrastructure
Environmental risk	Sustainable business	Green investment
Environmental security	Sustainable cities	Green jobs
Environmental services	Sustainable communities	Green parties
Environmental space	Sustainable consumption	Green plan
Environmental taxes	Sustainable development	Green shift
Environmental valuation	Sustainable enterprise	Green taxes
	Sustainable fisheries	Green wash

environment refers to surroundings, and particularly natural surroundings threatened by human activities (for a more detailed discussion, see chapter 3). Ecology has been used to denote a particular branch of scientific inquiry, and later a philosophical or political perspective (McIntosh 1985). But in this context, it points more generally to the interdependence of organisms and broader natural systems. Sustainable literally means "can be continued," yet it has acquired an environmental resonance: environmental degradation may make certain practices unsustainable, and hence the importance of *sustainability* or *sustainable development*. Green has come to be used as a broad descriptor of things held to be more in harmony with nature or the environment.

Expressions within each broad family share a certain general resonance. But they embody distinct ideas, are invoked in alternative contexts, and are associated with varied controversies. *Environmental indicators* are measurements of an environment-related phenomenon that offer insight into broader conditions. The concept invites argument about what to measure and how it should be measured, and about the significance and interpretation of the results. *Environmental restoration* refers to the rehabilitation of areas or habitats that have been degraded or destroyed, particularly by human intervention (pollution, inappropriate development, invasive species, etc.). It implies discussion about what should or should not be restored, what such restoration actually entails (the desired environmental outcome), and the techniques employed to bring this about. *Environmental education* denotes the transfer of knowledge and understanding about the environment, and immediately raises issues concerning who is to be educated, who is to do the educating, and the character and content of the information and perspectives that are transmitted.

Sometimes users will employ expressions containing environmental, ecological/eco, sustainable, or green more or less interchangeably, thereby communicating an underlying concept in different terms. Alternatively, they may exploit the contrasting resonances to establish distinctions. *Environmental justice* and *ecojustice* can be taken as largely synonymous. But one can construe them so that ecojustice places more emphasis on justice toward the nonhuman natural world (animals, species, or ecosystems), while environmental justice stresses human inequities. *Ecolabeling, environmental labeling,* and *green labeling* appear as virtual synonyms. *Environmental footprint* and *ecological footprint* differ largely because the first term can be used relatively loosely to denote the aggregate impact of a particular product, practice, or community, while the second is linked to a specific methodology that assesses ecological burdens in relation to the land area required to support a particular way of life (Wackernagel and Rees 1996). *Green growth* has a clearer environmental resonance than *sustainable growth,* since the latter can be understood as "growth compatible with environmental sustainability" or simply "growth that goes on indefinitely." But both concepts are criticized by those who argue that growth is itself the core driver of expanding environmental destruction (Martinez-Alier et al. 2010).

Other environment-related concepts have also been used to spin out families of related ideas. Consider expressions that invoke either the older *pollution*, or relative newcomer *carbon*, listed in table 2.3 below.

These are somewhat more focused than the broad groups described above. Again we see that related expressions do quite different things. Pollution, and the various kinds of pollution (like air or water pollution), point to the release of damaging substances into the environment. The polluter pays principle is supposed to guide policy makers by indicating which societal groups should bear the cost of pollution control. *Pollution prevention pays* links pollution control with potential economic returns (inverting the more obvious claim that pollution prevention *costs*), and can be intended as an analytic statement about efficiency or as an injunction to motivate better environmental stewardship. A *pollution tax* is a particular policy instrument, which can be interpreted as a regulatory application of the polluter pays principle.

The carbon-related concepts are comparatively new, with the grouping only emerging strongly in the later 1990s as climate change became an internationally recognized policy problem. In this context, carbon serves as shorthand for *carbon dioxide*, or more specifically, *anthropogenic*

Table 2.3
Expressions Related to Pollution and Carbon

Pollution	Carbon
Air pollution	Carbon accounting
Carbon pollution	Carbon budget
Noise pollution	Carbon emissions
Polluted site	Carbon footprint
Polluter	Carbon neutral
Polluter pays principle	Carbon offset
Pollution control	Carbon pollution
Pollution index	Carbon tax
Pollution prevention	Carbon trading
Pollution prevention pays	Low-carbon economy
Pollution tax	Low-carbon society
Water pollution	Negative carbon emissions

carbon dioxide emissions driving climate change, but is frequently intended to embrace greenhouse gases (GHGs) more generally. So the carbon moniker opened space for a variety of new concepts and ways of thinking about the implications of the human release of GHGs along with potential management approaches.

A term that joins both the carbon and pollution lists—*carbon pollution*—provides a good example of how conceptual categories are adjusted and reframed. The novel use of carbon (to mean anthropogenic carbon dioxide emissions) has been coupled with the established idea of pollution (a harmful release to the environment) to denote a particular kind of pollution: GHG emissions that are driving climate change. Until comparatively recently, international policy debate generally presented climate change as a new and distinct environmental problem, rather than as an additional manifestation of the established issue of *air pollution*. Thus, carbon pollution does not appear in the UN Framework Convention on Climate Change. But the carbon pollution innovation has been especially salient in the United States, where it has opened up the possibility of invoking existing legal frameworks that deal with air pollution (for example, the Clean Air Act) to address GHG emissions, despite continuing opposition in Congress to dedicated climate change legislation.

The sample of environmental concepts presented in table 2.1 (as well as the family groupings in tables 2.2 and 2.3) displays the wide variation among different sorts of environmental concepts and their use in policy argument. In the first place, environmental *concepts do different types of things*: they denote particular physical phenomena, technological options, or societal practices; they identify problems and solutions and frame debate; and they articulate social critiques, management principles, or policy instruments. *Flue gas desulphurization* refers to a particular remedial technology for managing industrial emissions (for example, from coal burning power plants). *Ecotourism* points to a more responsible and environmentally sensitive recreational activity or business sector. *Extended producer responsibility* is an approach that assigns management duties for product impacts and disposal to those manufacturing and marketing goods. And of course, concepts can do more than one thing; thus, *acid rain* refers to a physical phenomenon that can be measured and its consequences assessed. But it is also a problem-framing concept that dramatized a complex issue and helped to mobilize social support for political action.

Second, *concepts vary with respect to their inherent complexity and the density of their external linkages* (connections to other ideas within and outside the environmental sphere). Internal complexity and external linkages may open the way to richer conversations, but they also entail multiple understandings and contestation. Compare an idea like *adaptive management* that has substantial internal complexity (integrating notions of resource management, experimental learning, adaptation, and stakeholder participation) and dense external linkages (to ideas such as conservation, sustainability, and ecosystems) with relatively straightforward concepts such as *mine waste* or *environmental audit*. Of course, these concepts can also open the door to certain kinds of complexity. There are different kinds of mine waste, associated environmental or social impacts, and remediation strategies. And environmental audits can be understood, designed, and implemented in many ways. Yet such concepts operate at a more bounded level of abstraction than more sweeping ideas such as adaptive management, ecosystem services, or sustainable development.

Third, *environmental concepts differ in the way they relate to normative commitments*. The vast majority of concepts in the tables above could be defined in relatively neutral terms that betray no explicit value judgment. As they are invoked in political and policy argument, however, they may acquire normative resonance. Policy is about the establishment of priorities for government action and is suffused with value choice. So in the policy domain, virtually any concept can become normatively charged. *Biodegradable* can be given a formal definition so that objective tests can determine whether the term can be applied legitimately to a given packaging material. But the concept only makes sense in the context of concern over environmental pollution (from plastics, for example), and in environmental policy argument it has an overwhelmingly positive connotation. Much the same is true of *renewable energy*. Of course, some environmental policy concepts are much more explicitly normative in orientation; the *precautionary principle* is formulated as a management rule that tells us how we *should* handle certain kinds of environmental issue. Other ideas are more ambiguous: thus, resilience sometimes appears as an objective, measurable property of ecosystems (or coupled socioecological systems), but in policy discourse it typically embodies a variety of normative assumptions (Olsson et al. 2015).

Environmental concepts also vary in their range of application. Some concepts are confined to a single policy subsphere (a fragment of the whole environmental policy terrain), while others have broad applicability. Thus, *carbon offset* denotes a compensatory GHG emissions reduction, and while offsets can be achieved in different ways (forestry, renewable energy projects, etc.), the idea is essentially part of the climate change policy subfield. On the other hand, the precautionary principle can be applied across a range of issue areas including chemicals management, resource-harvesting quotas, and climate change. There are also differences with respect to *the forms of social communication* in which different concepts are invoked. Pollution appears in all sorts of contexts from newspaper articles to expert reports, from cabinet memorandums to stakeholder consultations. But *decoupling* (of environmental impacts from economic growth) is largely confined to specialized policy documents and expert interactions.

Finally, these concepts differ in *the extent to which they are integrated into official policy argument and practice.* And this is closely related to the forms of institutional embedding discussed in chapter 1. So while an idea like *critical loads* is deeply intertwined with the scientific and policy decision processes in the Long-range Transboundary Air Pollution regime, the notion of *rewilding* (roughly, returning natural landscapes toward a wider state, often with the reintroduction of large, carnivorous species) currently has only a marginal purchase on conservation practices.

One way to illustrate these distinctions is to consider some of the issue areas into which the environmental policy field is typically divided. The waste area provides a good example, and a selection of the concepts related to this theme is presented in table 2.4. Some of these describe types of waste, which can be classed by origin (*construction waste, domestic waste,* and *municipal waste*) or particular characteristics (*toxic waste* and *radioactive waste*). Others relate to the handling or disposal of that waste: *composting, gasification,* and *landfill.* There are more general approaches or strategies for waste management: *recycle, reuse,* and *waste minimization.* There are analytic concepts that can be deployed to assess management options: *waste streams* and *life cycle analysis.* There are visions that suggest a comprehensive solution to waste issues: *cradle to cradle, circular economy,* and *zero-waste society.* And there are broad principles for dealing with environmental problems that find specific application in the waste area, such as the precautionary principle, polluter pays principle, or resource efficiency.

Table 2.4
Selected Concepts Associated with the Waste Issue Area

Anaerobic digestion	Industrial waste	Recycling society
Biodegradable	Landfill	Resource efficiency
Circular economy	Life cycle analysis	Resource recovery
Composting	Medical waste	Reuse
Construction waste	Mine waste	Sewage
Cradle to cradle	Municipal waste	Solid waste
Deposit return	Nuclear waste	Take-Back Scheme
Domestic waste	Packaging waste	Throwaway society
Ecoefficiency	Plastic waste	Toxic waste
Electronic waste	Polluter pays principle	Waste hierarchy
Energy from waste	Postconsumer waste	Waste management
Extended producer responsibility	Precautionary principle	Waste minimization
Hazardous waste	Product stewardship	Waste repository
Incineration	Radioactive waste	Waste streams
Industrial ecology	Recycling	Zero-waste society

An important organizing principle within the waste policy area is provided by the idea of a *waste hierarchy* that *ties together a series of other concepts* and suggests an appropriate lexical ordering of waste management approaches. When dealing with waste, the preferred option is to reduce, followed by reuse and recycle, then recovery (of materials or energy), and finally disposal. Underlying the approach is a notion of efficiency along with minimizing the waste of energy and raw materials. Thus, the waste hierarchy also links upward to wider ideas of environmental or resource efficiency.

If one takes *energy* as an environmental issue area, a similarly large array of concepts could be enumerated, relating to energy technologies and their use, environmental problems associated with energy production and consumption, and management approaches and policy measures. The energy and waste areas overlap: thus, energy is a consideration in the waste hierarchy, and waste management is of critical concern in the energy economy (consider radioactive waste associated with nuclear power or bottom ash from coal-fired power generation). On the other hand, the waste policy area is not reducible to energy and the energy/environment nexus is much broader than waste, so there are many distinct ideas in each field.

The energy/environment issue area provides an illustration of how alternative concepts can be used to influence problem framing. Consider the six different solution concepts presented in table 2.5, highlighting varied ways of understanding problems and solutions in the energy/environment domain. Each of these terms—alternative, clean, green, low carbon, renewable, and sustainable—is intended to evoke a positive resonance, but their implications may be quite different.

The widest and most rapidly expanding issue area in contemporary environmental policy is the climate change domain. The global and long-term nature of the problem, the centrality of fossil-fuel combustion and land use change to economic development, linkages to key sectors (such as industry,

Table 2.5
Six Ways of Characterizing Solutions to Energy/Environment Problems

Alternative energy	Alternative to dominant or mainstream energy. It can have the connotation of small scale and decentralized energy, but what is an "alternative" in a given context depends on what is incumbent.
Clean energy	A contrast to "dirty" energy. Coal is the principal target here, and the expression has been used by proponents of natural gas or nuclear. Of course, for some nuclear is also dirty, and gas releases carbon dioxide. Wind and solar can certainly claim to be clean, but their advocates usually emphasize their renewable character. There is also "clean coal," although it is not clear which technologies could actually deliver it.
Green energy	Energy that is kind to the environment. It is most obviously applied to new renewables (wind, solar, wave, etc.), but because of the ambiguity of "green," can be applied more broadly.
Low-carbon energy	Energy that does not release carbon dioxide (or GHGs more generally), and hence, it can include nuclear and carbon capture and storage–equipped fossil energy.
Renewable energy	Energy that is derived from natural flows that are continuously replenished, such as hydro, wind, wave, tidal, solar, and biomass.
Sustainable energy	Energy that contributes to sustainable development. Many sources can be justified, provided production and consumption furthers a sustainable societal development trajectory. Alternatively, energy produced from sources that can be sustained indefinitely, in which case it approaches renewable energy.

transport, and agriculture), and potential impacts on an array of other environmental and social problems contribute to the expansive nature of this field. A small selection of concepts from this issue area is presented in table 2.6 below.

This suggests some of the ways concepts link into political controversy. Most concepts appear as relatively neutral descriptors; consider, for example, *sea level rise*. The concept is anchored in a scientific understanding of natural processes (involving the thermal expansion of the oceans and locking up of less of the earth's water in the cryosphere), and its core meaning is relatively straightforward. Scientists may argue about how fast and how far the sea level will rise, given a particular atmospheric concentration of GHGs. In the policy world, this gives rise to disputes about how concerned societies should be about sea level rise and what should be done. For those unwilling to confront the risks of human-induced climate change, the answer may be to avoid talking about it at all—witness the North Carolina legislature's 2012 law forbidding state agencies from including anticipated (climate change–induced) sea level rise in state maps used for coastal planning. The idea of a *carbon tax* (a policy instrument involving a charge on GHG emissions) is also relatively straightforward. This time the meaning can be anchored in economics (or law, when legislation is adopted). But there is endless potential for argument about whether such a tax is desirable

Table 2.6
Selected Concepts Related to the Issue Area of Climate Change

Adaptation	Common but differentiated responsibilities	Greenhouse gas
Carbon budget	Consumption emissions	Intergenerational justice
Carbon capture and storage	Contract and converge	Land use change
Carbon dioxide equivalent	Dangerous climate change	National GHG inventories
Carbon divestment	Deforestation	Negative carbon emissions
Carbon emissions	Emissions scenarios	Offsets
Carbon leakage	Emissions trading	Production emissions
Carbon sinks	Energy efficiency	Reforestation
Carbon tax	Extreme weather events	Renewable energy
Climate emergency	Fossil energy	Sea level rise
Climate equity	Fugitive emissions	Sufficiency
Climate mitigation	Geoengineering	Two-degree climate target
Climate refugees	Global warming	

or undesirable, whether it should be imposed upstream or downstream, what should be done with the revenues, and so on. The *2-degree climate target* has a basis in scientific assessments. But it is an explicitly political goal, which commanded formal international agreement, although the 2015 Paris climate summit has now floated a still more ambitious 1.5-degree target (while sidestepping the tortuous nature of the emissions trajectories that could actually deliver such an outcome). *Geoengineering* gives rise to disputes over what the concept actually covers (have we already been conducting geoengineering since the dawn of the fossil-fueled era, or at least since we understood the implications of GHG emissions?), whether societies should contemplate *solar radiation management* or *carbon removal* projects, or even pursue research into these alternatives.

To this point the discussion has stressed the variety of environmental concepts, while pointing simultaneously to certain patterns of affinity. There are the broad families of concepts defined by conjoined terms that employ the same core term (environmental, sustainable, carbon, etc.). As we saw with the waste and energy examples discussed above, there are also clusters of concepts defined by relevance to a particular subfield or issue that contain both topic-specific concepts as well as particular applications of more general concepts. Another form of linkage involves logical ties among concepts. In some cases, one concept is logically entwined with another. Thus, when resilience is invoked as a property of *ecological systems* or *socioecological systems*, the understanding of such systems becomes critical to making the idea of resilience intelligible. Or consider the waste hierarchy mentioned above: the idea only makes sense in relation to the series of approaches to handling waste that it orders. In other instances, concepts can be understood as related to some underlying integrative idea. The notion of *environmental limits* provides a good illustration here, as concepts such as limits to growth, maximum sustainable yield, ecological footprint, critical loads, or planetary boundaries can all be presented as embodying the idea that there is a limit to the burdens humans can place on natural systems without causing damage with significant repercussions.

Critical Organizing Concepts

So far we have explored the variety of environmental policy concepts, identified particular patterns of affinity, and noted some of the ways new expressions are generated. At this point it is important to emphasize that certain environmental policy concepts play a particularly important role in structuring discourse and practice, in shaping the policy field and serving as a focus for reflection, analysis, controversy, and action. Such ideas form critical interconnections, assuming an influential role in specific patterns of discursive and practical intervention, and tying together various areas, themes, and actors. These may be concepts that identify critical problems or solutions, explicitly or implicitly linking the two, and marking out a sphere of influence over other ideas and practices. Such concepts often have clear normative content (like sustainable development or the precautionary principle); they may bridge science and policy (consider adaptive management or critical loads), and/or operate at a higher aggregative level (pollution or the green economy). Frequently they are among the most internally complex and externally interlinked ideas. These concepts typically serve as focal points for explicit contestation. Such contestation may involve open argument about the meaning of the concept, its value as an analytic category, and its biases and blind spots. But concepts that have achieved substantial institutional embedding may ultimately become so integrated into the everyday landscape of policy argument that they are more or less taken for granted, even if they continue to structure and focus controversy over policy implications.

There is no closed list of such critical concepts. And to some extent it is a question of degree rather than of absolutes. Different concepts may come to the fore in relation to specific issue areas, discursive contexts, and moments in time. Basic properties of a concept (its bearing, scope, internal complexity, and external linkages) influence its potential to play such a role. But whether or not it actually comes to do so depends on ideational and practical developments as well as the political and policy interactions among actors pursing diverse goals. In other words, such concepts do not acquire such a role merely because they are logically satisfying, but because they serve an important purpose in discursive interactions.

When considering the environmental domain, it is convenient to distinguish several categories of such organizing concepts. First, there are what can be described as *meta-concepts*: these are the overarching ideas that offer structure to the entire environmental domain. So far two concepts unambiguously play such a role. On the one hand, there is the environment, which defines this policy domain and distinguishes it from other areas of policy concern. On the other hand, there is sustainable development or sustainability, which provide an image of a social development trajectory where key environmental problems threatening continued human flourishing (even existence) can be addressed. Sustainability looks beyond environmental policy narrowly conceived to link up with broader issues of economic and social welfare. Each of these metaconcepts is given more detailed treatment later in this volume (chapters 3 and 4).

Second, there are *problem or issue concepts* that identify key topics for policy attention. In the early decades of the modern environmental era, pollution was the most important such concept, and the "war on pollution" legitimized the institutionalization of environmental policy as a distinct realm. Acid rain and ozone depletion were later important in articulating the international dimensions of environmental threats, while today climate change and biodiversity (loss) are the most sweeping environmental problem concepts.

Third, there are *mesolevel analytic or management concepts* that deal with particular dimensions of policy, offer alternative lens to structure understanding, or indicate principles that should be applied when dealing with environmental problems. This is where ideas such as environmental policy integration, environmental security, ecosystem services, or the precautionary principle fit in. Many of these concepts are less well known outside expert circles and policy communities. Thus, the concept of common but differentiated responsibilities is rarely found in the mass media, even when the press covers debates about the burdens that richer and poorer states should be willing to assume to abate GHG emissions. Yet the concept plays a crucial part in structuring argument in international forums attempting to address climate change. Concepts in this group tend to be somewhat more abstract than the problem concepts referred to above or the more instrumental concepts considered next. Box 2.1 illustrates a selection of mesolevel analytic or management concepts. A number of others are subject to analysis in the chapters in this volume.

Finally, there are concepts denoting core *policy approaches and instruments* that are expected to bear the burden of managing environmental problems. Examples here include environmental regulation, environmental assessment, emission and effluent trading, or feed-in tariffs. These can become the focus of ongoing political and policy battles. Consider the continuing arguments about the effectiveness and design of carbon pricing

Box 2.1
Selected Mesolevel Analytic or Management Concepts

Polluter Pays Principle

The costs of preventing pollution should be borne by the polluter, although there may be exceptions or special circumstances, especially during a transitional period. Championed by the OECD, in part to discourage national subsidies that would interfere with trade, it remains a staple of major environmental declarations. It is consistent with encouraging the "internalization of externalities." It is often interpreted to mean that the consumer pays (as the ultimate polluter). In practice, governments may support polluting industries (to keep them competitive) by subsidizing the acquisition of remediation technology or assuming ultimate liability (for example, cleaning up toxic sites once underregulated companies have ceased operations). So the taxpayer pays (O'Connor 1997).

Balancing Economy, Environment, and Society: The Three Pillars of Sustainability

Decision makers should balance economic, environmental, and social considerations when developing policies, plans, programs, and projects. What this balancing actually entails, and how much one element can be sacrificed if gains are made on the others, remains an issue of debate. According to critics, environmental goals are typically sacrificed to economic and social objectives. Some contend sustainable development is about solutions that advance all these objectives simultaneously and not trading one off against the others (Robinson 2004; Meadowcroft 2013).

Environmental Policy Integration

Environmental protection is a task of the whole government. It should be integrated into key sector ministries (transport, agriculture, energy production, manufacturing, etc.) and considered at all levels (local, regional, and national). By integrating environmental consideration early in the decision cycle, a more sustainable development trajectory can be secured. Such integration is compatible with a sector-based environmental policy and emphasizes engagement with relevant stakeholders. Yet academic studies suggest that environmental policy integration is often more symbolic than real and sometimes results in the recapture of environmental policy by sectoral interests (Jordan and Lenschow 2010).

Box 2.1 (continued)

Precautionary Approach/Principle

When confronted with potential threats of serious or irreversible damage to the environment, appropriate measures should be taken to mitigate risks even in the absence of definitive scientific evidence of harm. The precautionary approach or principle has been invoked in various international environmental agreements, including the 1992 Rio Declaration on Environment and Development. It has been applied to areas such as chemical release, species extinction, the release of genetically modified organisms, and climate change. There are continuing arguments over the evidence of potential harm required for its invocation, the extent to which the "burden of proof" should be shifted to those proposing a novel activity, and the degree to which cost-benefit calculus should be applied. Environmentalists sometimes talk about inadequate or inconsistent application, while critics contend it can slow technological innovation that can provide societal benefits, including environmental benefits (O'Riordan and Cameron 1994).

Common but Differentiated Responsibilities

States share a responsibility to address global environmental issues, but contributions should take account of different responsibilities for generating environmental pressures and capacities to address them. This norm has been incorporated into international agreements including the 1992 Rio Declaration and UN Framework Convention on Climate Change where the rich, developed parties accepted greater responsibility for historic GHG emissions and the capacity to take action (including to assist developing countries). While generally accepted as an equity principle of international law, in practice arguments revolve around the degree of action and assistance required by different parties in particular contexts.

Ecosystem Services

The concept tracks essential contributions that natural systems make to human welfare. These contributions are often understood to comprise *supporting services*, *provisioning services*, *regulating services*, and *cultural services*. An economic value of such services can be established by a variety of techniques, and incorporated in the system of national accounts or used to underpin *payment for ecosystem services*. Critics contend that the approach neglects noninstrumental reasons for appreciating nature, and suggest that such schemes may have negative consequences for social equity and undermine nonmonetary motivations for environmentally friendly behavior (Primmer et al. 2015; Kull et al. 2015).

mechanisms such as the EU Emissions Trading System. But even when such approaches become settled routines and attract less attention in high-level debate (consider environmental assessment), their continued implementation structures the policy landscape and impacts future policy choice.

Needless to say, these categories are not mutually exclusive. Biodiversity, for example, can be understood as an analytic and management construct as well as a problem-defining concept. And there is plenty of room for debate about whether a specific concept actually assumes the roles identified here. There is variation among national contexts and international forums, so that a concept may have salience in one venue that it lacks elsewhere. And the situation is dynamic. Resilience is an idea that has come into policy prominence on both sides of the Atlantic over the past decade. It has strong champions (Folke 2006), yet also critics (Olsson et al. 2015). But just how important has it become for contemporary policy argument? Clearly there are many concepts that have the potential to assume an important role, but have yet to acquire adequate institutional support. Thus, the notion of a circular economy has elicited discussion among specialists and activists, and is now being pursed in EU structures. Yet one would hesitate to class it among the most established mesolevel -analytic and management concepts.

In this regard, it is important to underscore again the extent to which the relative penetration and dominant interpretation of these organizing concepts results from political struggle. Conceptual entrepreneurs—individuals (politicians, scientists, and public figures) or organizations (research institutes, blue-ribbon panels, national environment agencies, and international organizations such as the UNEP or OECD)—actively promote innovative ideas that articulate specific values, interpretations, and interests. And to the extent that these take root in particular manifestations, they shape subsequent rounds of interaction. Just as physical infrastructure shapes future policy debate, so too does conceptual "infrastructure" (particular understandings of concepts embedded in institutions, programs, foundational documents, laws, etc.). As we will see in chapter 4, for example, the idea of sustainable development articulated by northern governments often strips the concept of its intragenerational equity dimension, collapsing it back to something like "improved quality of life." And in chapter 5, we will learn that it was a tamer and more circumscribed conception of environmental assessment that ultimately became established in the policy world.

The notion of planetary boundaries provides an instance of such a struggle unfolding in the 2010s. The concept posits critical thresholds related to nine key environmental pressures that human societies must not cross if they are to continue to enjoy the relatively benign circumstances in which human civilization flourished over the past ten thousand years. For some, the idea (and related notions such as *safe operating space*) supplies a scientific foundation for political judgments about the management of global environmental issues (Rockström et al. 2009). For others, the idea is conceptually muddled (confusing different forms of boundary) or even represents a misguided attempt to introduce neo-Malthusian biases into policy process (Nordhaus et al. 2012). It achieved some recognition in *Living Well, within the Limits of Our Planet* (European Commission 2013), but failed to win a place in the UN's *Sustainable Development Goals* (UN Department of Economic and Social Affairs 2015).

Evolution of the Conceptual Field

Reference has already been made to the emergence or eclipse of individual concepts as well as the changing conceptual repertoire of environmental policy. The final section of this chapter will examine broad developments over the decades since modern environmental policy emerged at the end of the 1960s.

One way to introduce movement over time is to consider the relative uptake of individual concepts. Figures 2.1a–b and 2.2a–b present data on the frequency with which eight selected concepts appeared in two printed sources: the *OECD Observer* and the EU Environmental Action Programs (EAPs). Issued by the OECD, the *Observer* provides news of the organization's activities and has a continuous publication record over the period with which we are concerned. The EAPs are official documents establishing the European Union's environmental priorities and have been issued since 1973.

Pollution, sustainable development, biodiversity, and climate change are tracked in figures 2.1a–b. The OECD data show the proportion of articles in the environment-related coverage where the specified concept appeared, while the EAP data show frequency of usage in each of the seven EAPs. Thus, between 1963 and 1995, pollution typically appeared in more than half the environment-related articles in the *OECD Observer* (tabulated in multiyear

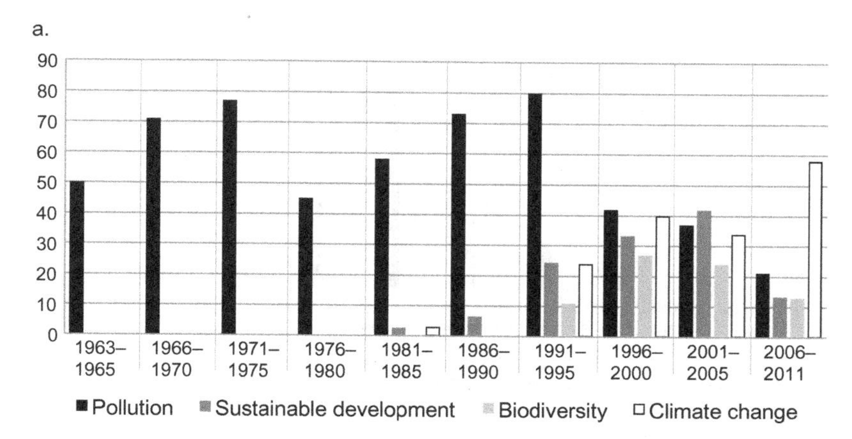

Figure 2.1a

Proportion of articles dealing with environmental themes in the *OECD Observer* in which pollution, sustainable development, biodiversity, and climate change appeared at least once.

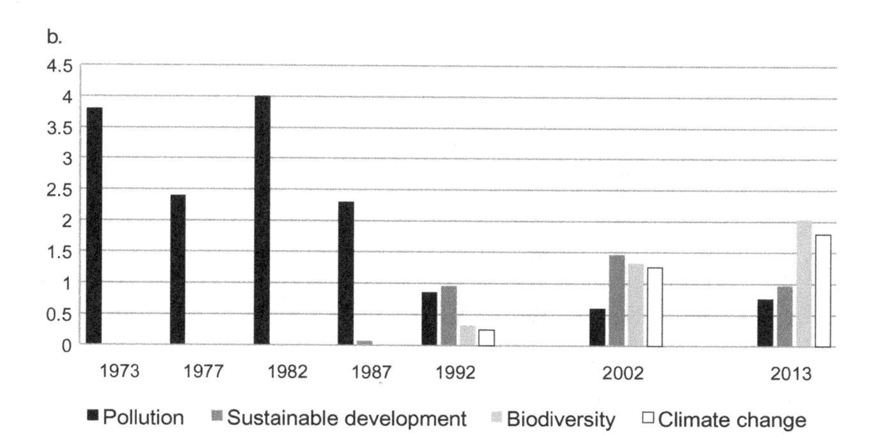

Figure 2.1b

Appearance (per page) of the concepts pollution, sustainable development, biodiversity, and climate change in the seven EU Environmental Action Programs.

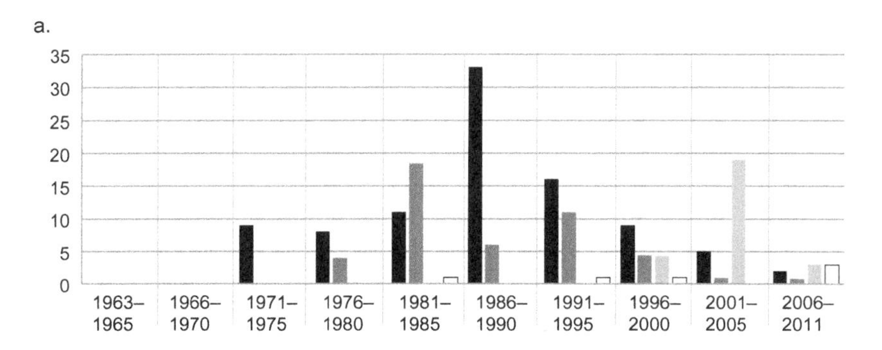

Figure 2.2a

Proportion of articles dealing with environmental themes in the *OECD Observer* in which polluter pays principle, environmental assessment, decoupling, and resilience appeared at least once.

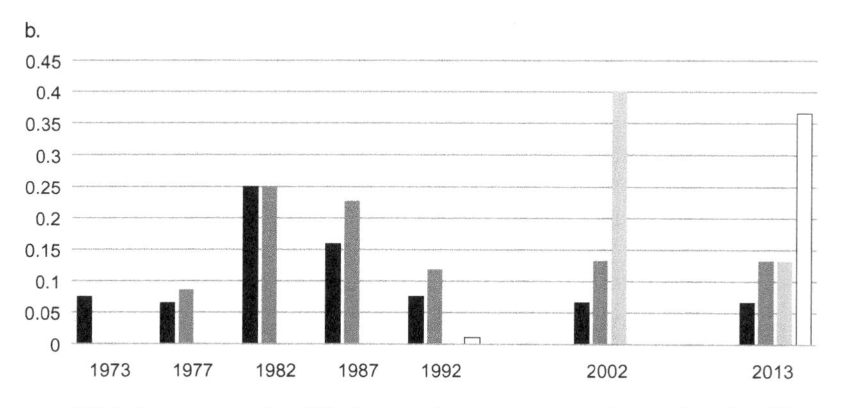

Figure 2.2b

Appearance (per page) of the concepts polluter pays principle, environmental assessment, decoupling, and resilience in the seven EU Environmental Action Programs.

periods) with rates of 75 percent in the early 1970s and early 1990s. This concept was similarly prominent in the EAPS in the 1970s and 1980s, averaging 2.5 to 4 appearances a page. In both sets of documents, the relative prominence of pollution declined over the course of the 1990s and beyond (down to just 20 percent of *OECD Observer* articles in the late 2000s, and less than one reference per page in the seventh EAP). In both sets of documents, sustainable development gained prominence after the late 1980s, with a peak in the early 2000s. Biodiversity received attention from the early 1990s on; although this continued to rise in the EAPs, it seemed to decline in the *Observer*. Climate change first appeared in the *OECD Observer* in the early 1980s, but grew in prominence in both publications from the early 1990s on. Similar data for the concepts of the polluter pays principle, environmental assessment (or impact assessment), decoupling (of economic growth from environmental impacts), and resilience are presented in figures 2.2a–b. They show interest in the polluter pays principle and environmental assessment rose and then fell over time. Decoupling peaked in the early 2000s, while resilience seemed to be receiving more attention at the end of the period.

One must take care when interpreting these sorts of data. Inevitably, the counts reflect institutional biases associated with the OECD and the European Union, as well as the character of these specific publications. Moreover, the nature of the *Observer* and EAPs changed over time. It is most convincing with respect to the first arrival of concepts within OECD- or EU-level debate. Pollution was already established as a policy problem before the modern environmental era; environmental assessment enters the intergovernmental world from the late 1970s (though it originated in the United States almost decade earlier); biodiversity takes off in the early 1990s; and so on. The changing frequencies may also say something about the ideas and problem areas that are viewed as most salient at a given point in time. A decline in reference to the concept could indicate that the idea has fallen out of favor. But it might simply be a consequence of the increased attention given to other issues and concepts. And it may signal that an idea has been incorporated into policy routines and no longer garners the attention of intergovernmental bodies. Environmental assessment is probably an example of this last phenomenon. The proportion of articles in which it appeared dropped off from the mid-1990s, yet we know that in practice, such assessments continue to be performed in huge numbers around the world. Thus, the concept remains important in the environmental policy domain, even if it no longer occupies much attention in high-level discussions at the OECD or European Union. Decoupling, in contrast, is probably an illustration of the first process: the OECD actively promoted decoupling at the turn of the millennium, featuring it in the *OECD Environmental Strategy for the First Decade of the 21st Century* (OECD 2001). But the idea did not catch on readily in national and international policy circles, and its appearance dropped off as the new century advanced. Although the polluter pays principle continues to be cited in major environmental declarations, it seems to be playing a diminishing role in practical policy discussion.

Finally, let us consider the evolution of the field as a whole. Prior to the 1960s, *environmental policy* did not exist as a distinct domain. Only with the emergence of the modern concept of the environment (discussed in more detail in chapter 3) did the constitution of such a sphere of government activity make sense. Figure 2.3 offers a graphic representation of the way in which environmental policy emerged through a reconceptualization of the policy world, partially absorbing areas that previously had been defined in different terms.

To capture key developments over the following decades, we will offer three idealized portraits of the conceptual architecture of environment policy spaced at twenty-year intervals. These are not intended to be exhaustive but rather to capture important elements of relatively high-level policy argument, particularly as manifest in leading industrial states and international policy institutions such as the OECD and European Union.

By the early 1970s, environmental policy had emerged as a distinct area of national government activity with dedicated ministries or agencies, framework laws and regulations, and an increasing level of international coordination. Environment was the key metacategory that anchored the field. The reference was often to "man's environment" or "the human environment." The explicit goal was the "protection and improvement of the environment," which included both "natural" and "man-made" dimensions. Pollution, understood as the release of noxious substances harmful to humans and the environment, appeared as the main problem category. Pollution was already an established concept, with existing regulatory and institutional anchors in leading jurisdictions. But its control and prevention were catapulted to national political attention, to be monitored and

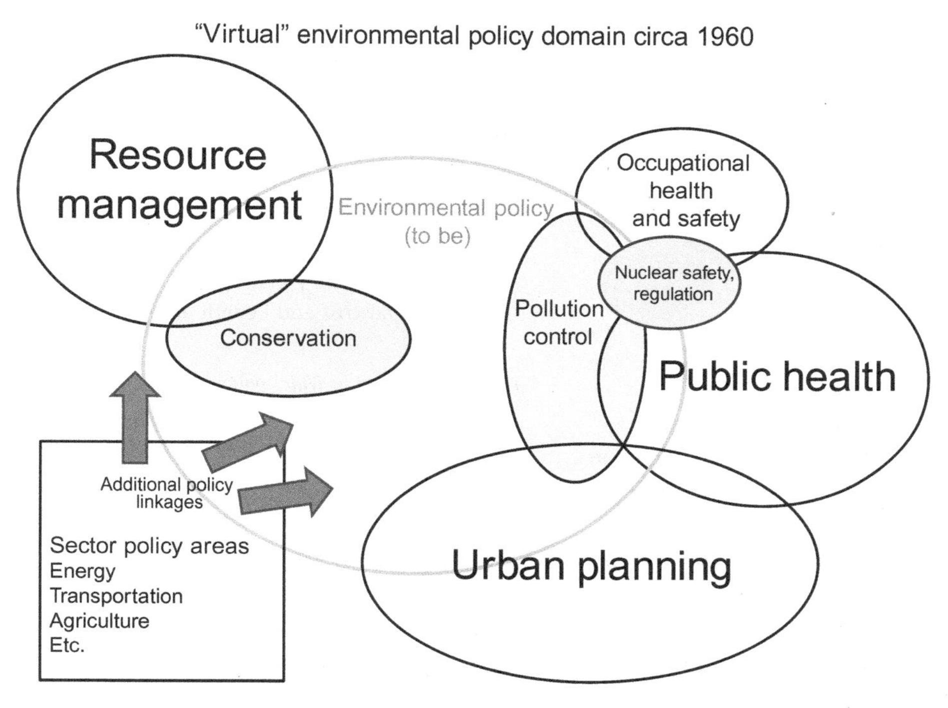

Figure 2.3
The emergence of the environmental policy domain.

controlled by central bureaucracies (Enloe 1975). There were four main pollution-related issue areas: air pollution, water pollution, chemicals, and waste. The concern was with risks to human health, living conditions, quality of life, and property as well as to wildlife and ecological systems. Other problems were defined by the urban environment, threats to the countryside, wildlife and endangered species, and the overexploitation of resources. *Conservation* was an established management concept. Ideas imported from general policy debates included planning and rational policy design as well as the careful estimate of costs and benefits. The polluter pays principle was an early mesolevel management concept promoted by the OECD. (For a summary of the conceptual field in the early 1970s, see table 2.7.)

At this point, pollution and other environmental damage was understood as an undesirable side effect of industrialization, economic growth, human expansion, urbanization, and growing technological capacities (Jänicke and Weidner 1997; Hanf and Jansen 1998). There was also some

Table 2.7
Developments in the Conceptual Field of Environmental Policy by the Early 1970s

Metaconcepts	*Mesolevel analytic and management concepts*
Environment	Polluter pays principle
Core problem or issue concepts	*Policy approaches and instruments*
Pollution	Environmental regulations, standards
Air pollution	Environmental planning
Water pollution	Environmental cost and benefit assessment
Chemicals	
Waste	*Concepts invoked in wider policy argument*
Urban environment	Limits to growth, environmental crisis, ecology,
Nature protection, conservation	ecological balance, overpopulation, resource scarcity, environmentalism

concern with *resource scarcity*, especially in relation to water, energy, and minerals. And there were worries about the costs of pollution control and, particularly from international bodies such as the OECD, the potential adverse impacts of regulation on international trade. More critical or challenging ideas such as "limits to growth," "ecological crisis," and "population control" were often invoked by societal actors in broader public debate.

Two decades later, by the early 1990s, the environmental policy domain was firmly established in developed countries. There was a general appreciation in government that environmental problems, and continuing argument over what to do about them, had become a permanent fixture of political life (Durant et al. 2004). Despite regulatory progress, the range of problems continued to expand, and the existing suite of policy measures was often perceived to be ineffective and/or was the object of increased resistance from business. Following the publication of the Brundtland Report in 1987, sustainable development emerged as a second overarching metaconcept. Pressing environmental issues had come to include acid rain, ozone depletion, *tropical deforestation*, and *nonpoint source water pollution*. And climate change and biodiversity marked out two new domains of truly global significance. From the mid-1980s, increased emphasis had been paid to anticipating and preventing environmental problems. It was argued that "pollution prevention pays," and tighter regulation could encourage innovation and competitive advantage. By the early 1990s, a series of innovative mesolevel analytic and managements concepts had been brought forward, including the precautionary approach, environmental policy integration, three pillars of sustainable development (balancing environmental, economic, and social goals), and common but differentiated responsibilities. The notion of critical loads had helped unblock negotiations over the regulation of long-range, transboundary air pollution. Attention was also focused on diversifying the portfolio of policy instruments, and *negotiated instruments, market-based instruments* or *economic instruments,* and *environmental information* rose in prominence (Jordan et al. 2003; Fiorino 2006).

Environmental assessment was increasingly integrated into planning and decision routines across the developed world. And more emphasis was placed on involving societal stakeholders in environmental policy making and implementation processes. (See table 2.8.)

By the early 2010s, twenty years after the Rio Earth Summit, and following a deep recession, global environmental problems appeared more persistent than ever. The *green economy* was launched by several international organizations as an additional overall framing concept, although its ability to maintain this status over the long haul remains to be established. Climate change emerged strongly as *the overarching environmental policy problem*—understood as the issue with the largest potential to provoke long-term socioeconomic and political disruption, and exacerbate other environmental problems (such as biodiversity loss). *Adaptation* was attracting more attention as the impacts of climate change became increasingly evident. Energy had returned as an important environmental issue area with discussion of renewable energy, energy efficiency, and low-carbon energy. Mesolevel analytic and management ideas broadened further with more attention given to concepts such as resilience, ecosystem services, natural capital, low-carbon development, and enhanced resource efficiency.

The increasing significance of the climate change problem area and its reframing in more transformative terms became evident as the 2010s advanced. As the discussion of solutions began to move from incremental emission reductions to the elimination of GHG emissions and phase out of fossil energy dependence, innovative concepts such as *carbon budgets, unburnable carbon,* and *carbon divestment* started to acquire political salience. The underlying idea of limits implicit in much environmental policy debate was made increasing explicit with the two-degree climate target, along with the emergence of novel concepts such as planetary boundaries or *the*

Table 2.8
Developments in the Conceptual Field of Environmental Policy by the Early 1990s

Metaconcepts	*Mesolevel analytic and management concepts*
Environment	Environmental risk
Sustainable development, sustainability	Precautionary approach/principle
	Three pillars of sustainability: balancing economy, environment, society
Core problem or issue concepts	
Acid rain	Common but differentiated responsibilities
Ozone depletion	Critical loads
Pollution	Environmental policy integration
Climate change	Integrated pollution control
Biodiversity	Adaptive management
	Intergenerational equity
	Policy approaches and instruments
	Environmental assessment
	Market-based environmental policy instruments
	Environmental management systems
	Environmental audit
	Environmental indicators, sustainability indicators
	State of the environment reporting
	Sustainable development plans and strategies
	Ecolabeling
	Concepts invoked in wider policy debate
	Ecoefficiency, triple bottom line

Anthropocene (which posits a fundamentally altered relationship between human kind and planetary processes). Another consequence has been to strengthen the link between environmental issues and core governmental concerns with economic development welfare and security (Meadowcroft 2012; Duit et al. 2016). (See table 2.9.)

What do these idealized snapshots suggest about the overall evolution of the conceptual landscape of environmental policy? First and most obviously, the field has become broader and more complex: as decades pass, there are more concepts relating to more issues and a wider portfolio of policy response. For the most part, new ideas have simply been added on

Table 2.9
Elements of the Conceptual Field of Environmental Policy by the Early 2010s

Metaconcepts	*Mesolevel analytic and management concepts*
Environment	Resilience
Sustainable development, sustainability	Ecosystem services
(Green economy?)	Natural capital
	Low-carbon society, economy, transition, development
Core problem or issue concepts	Environmental security
Climate change	Environmental equity, environmental justice
Biodiversity and nature conservation	Sustainable production and/or consumption
Renewable energy, energy efficiency	
Pollution	*Policy approaches and instruments*
	Environmental information
	Life cycle analysis
	Carbon pricing
	Emissions trading
	Green procurement, sustainable procurement
	Feed-in tariffs, renewable portfolio standards
	Carbon budgets
	Concepts invoked in wider policy debate
	Decoupling, the Anthropocene, planetary boundaries, circular economy, climate crisis, climate justice, climate refugees, degrowth

top of older concepts, with older ideas continuing to play a role in everyday argument and policy implementation even if newer ideas often predominate in higher-level discussion.

Second, the field appears to reflect the increasing societal entanglement of environmental problems: concepts increasingly relate to issues at multiple scales (local, regional, national, international, and global) and implicate diverse economic sectors (resource extraction, agriculture, manufacturing, transport, retail, finance, and trade). Environmental concepts bridge into a wider range of societal domains (economy, employment, welfare policy, security, households, and so on), reflecting an understanding that problems and solutions are both interlocked with broader societal practices. For example, sustainable development and green economy link environment, economy, and equity; *integration* explicitly posits a drawing together of environment and sector decision making; resilience is about socioecological interdependence (embracing economic and social structures as well as environmental performance); and *sustainable consumption* potentially challenges existing ways of living, not just existing techniques of production.

Third and in a related sense, the conceptual field seems to reflect a widening appreciation of the scale of social adjustment (and associated policy interventions) required to come to terms with environmental problems. This is most clear with respect to climate change, where *mitigation* and *GHG emissions reductions* are increasingly accompanied by discussion of a *low-carbon society*, the *transition to a low-carbon economy*, *energy transitions*, and *climate adaptation*. But it is also evident in concepts such the green economy (offering an image of a revamped economic system that respects ecological limits) or ecosystem services (which provides a framework to conceptualize the overall dependence of human societies on environmental systems). And yet each of these concepts is contested, and one of the main areas of contention is precisely what the idea really implies about the depth and direction of the requisite transformation of existing institutions. Needless to say, the increasing prominence of concepts that point toward a more profound transformation of existing institutions and practices does *not* imply that policies have actually been put in place to realize such a transformation, or even that ideas that acquire increased salience in the environmental policy field hold sway across the governmental apparatus as a whole.

Fourth, environmental policy concepts have been marked by the broader political and economic contexts in which they have emerged. Interest in *voluntary instruments* as well as market-based ones since the early 1990s reflected the general mood of increased skepticism toward the state and a renewed faith in markets, although of course measures such as emissions trading were in fact predicated on a critique of *market failure* (i.e., a failure to internalize environmental externalities) and could only be implemented through government action. And the turn toward a wider portfolio of instruments also followed from an appreciation within the environmental policy realm that the growing complexity of environmental issues required more diversified instruments and greater engagement with social actors, rather than simply an endless expansion of the regulatory rule book. *Globalization* was another major feature shaping conceptual development, particularly from the later 1990s. Moreover, in this discussion we have concentrated on environmental concepts, but in policy argument such concepts are always interwoven with other, more general political and policy ideas (for example, *new public management, evidence-based policy making, value for money, austerity*, and so on). This more general climate influences which environmental concepts gain traction and how they are interpreted.

Finally, although the conceptual field has seen truly remarkable change, there has also been substantial continuity. With institutional embedding—especially in organizational and legal form—important environmental policy concepts endure over decades. Furthermore, many ideas that later came to acquire great salience (for instance, *integrating policy* or applying economic instruments) were already present in embryo in the earliest international policy discussions of the late 1960s and 1970s. And yet other ideas with equally deep roots (such as the problematization of *population, consumption*, and *growth*) have consistently remained at the margin of official environmental policy discourse.

This chapter has offered an overview of concepts invoked in environmental policy argument and analyzed key developments in the evolution of the general conceptual field. The discussion has illustrated how new ideas have captured attention in the environmental policy arena, and has provided examples of the ways in which innovative concepts have identified new issues and reframed arguments. So far, however, not much has been said about how new ideas actually come to the fore, achieve institutional embedding, and evolve over time as different understandings compete. We have not focused on the struggles among and implications of different understandings of these concepts. Nor has their practical impact on policy outputs and outcomes been closely examined. Tracing these dimensions requires a much more detailed examination of the career of individual concepts. And this is the challenge that will be taken up in the following chapters.

References

Duit, Andreas, Peter Feindt, and James Meadowcroft. 2016. "Greening Leviathan: The Rise of the Environmental State?" *Environmental Politics* 25 (1): 1–23. doi:10.1080/09644016.2015.1085218.

Durant, Robert, Daniel J. Fiorino, and Rosemary O' Leary. 2004. *Environmental Governance Reconsidered: Challenges, Choices, and Opportunities*. Cambridge, MA: MIT Press.

Enloe, Cynthia. 1975. *The Politics of Pollution in a Comparative Perspective: Ecology and Power in Four Nations*. New York: David McKay Company.

European Commission. 2013. *Living Well, within the Limits of Our Planet*. Accessed November 28, 2016, http://ec.europa.eu/environment/pubs/pdf/factsheets/7eap/en.pdf.

Fiorino, Daniel J. 2006. *The New Environmental Regulation*. Cambridge, MA: MIT Press.

Folke, Carl. 2006. "Resilience: The Emergence of a Perspective for Social-Ecological Systems Analyses." *Global Environmental Change* 16 (3): 253–267. doi:10.1016/j.gloenvcha.2006.04.002.

Hanf, Kenneth, and Alf-Inge Jansen, eds. 1998. *Governance and Environment in Western Europe: Politics, Policy, and Administration*. Abingdon, UK: Routledge.

Jänicke, Martin, and Helmut Weidner, eds. 1997. *National Environmental Policies: A Comparative Study of Capacity Building*. Berlin: Springer.

Jordan, Andrew, and Andrea Lenschow. 2010. "Environmental Policy Integration: A State of the Art Review." *Environmental Policy and Governance* 20 (3): 147–158. doi:10.1002/eet.539.

Jordan, Andrew, Rüdiger Wurzel, and Anthony Zito. 2003. *New Instruments of Environmental Governance? National Experiences and Prospects*. London: Frank Cass Publishers.

Kull, Christian, Xavier Arnauls de Sartre, and Minica Castro-Larranaga. 2015. "The Political Ecology of Ecosystem Services." *Geoforum* 61:122–134. doi:10.1016/j.geoforum.2015.03.004.

Martinez-Alier, Joan, Unai Pascual, Franck-Dominique Vivien, and Edwin Zaccai. 2010. "Sustainable De-growth: Mapping the Context, Criticisms, and Future Prospects of an Emergent Paradigm." *Ecological Economics* 69:1741–1747. doi:10.1016/j.ecolecon.2010.04.017.

McIntosh, Robert. 1985. *The Background of Ecology: Concept and Theory*. Cambridge: Cambridge University Press.

Meadowcroft, James. 2012. "Greening the State." In *Comparative Environmental Politics*, ed. Paul Steinberg and Stacy VanDeveer, 63–88. Cambridge, MA: MIT Press.

Meadowcroft, James. 2013. "Reaching the Limits? Developed Country Engagement with Sustainable Development in a Challenging Conjuncture." *Environment and Planning C: Government and Policy* 31 (6): 988–1002. doi:10.1068/c1338j.

Nordhaus, Ted, Michael Shellenberger, and Linus Blomqvist. 2012. "The Planetary Boundaries Hypothesis: A Review of the Evidence." Breakthrough Institute. Accessed November 28, 2016, http://thebreakthrough.org/blog/Planetary%20Boundaries%20 web.pdf.

O'Connor, Martin. 1997. "The Internalisation of Environmental Costs: Implementing the Polluter Pays Principle in the European Union." *International Journal of Environment and Pollution* 7 (4): 450–482.

OECD. 2001. *OECD Environmental Strategy for the First Decade of the 21st Century.* Accessed November 28, 2016, https://www.oecd.org/env/indicators-modelling-outlooks/1863539.pdf.

Olsson, Lennart, Anne Jerneck, Henrik Thoren, Johannes Persson, and David O'Byrne. 2015. "Why Resilience Is Unappealing to Social Science: Theoretical and Empirical Investigations of the Scientific Use of Resilience." *Science Advances* 1 (4): e1400217. doi:10.1126/sciadv.1400217.

O'Riordan, Tim, and James Cameron, eds. 1994. *Interpreting the Precautionary Principle.* London: Earthscan.

Primmer, Eeva, Pekka Jokinen, Malgorzata Blicharska, David N. Barton, Bob Bugter, and Marion Potschin. 2015. "Governance of Ecosystem Services: A Framework for Empirical Analysis." *Ecosystem Services* 16:158–166. doi:10.1016/j.ecoser.2015.05.002.

Robinson, John. 2004. "Squaring the Circle? Some Thoughts on the Idea of Sustainable Development." *Ecological Economics* 48 (4): 369–384. doi:10.1016/j.ecolecon.2003.10.017.

Rockström, Johan, Will Steffen, Kevin Noone, Åsa Persson, F. Stuart Chapin III, Eric Lambin, Timothy M. Lenton, et al. 2009. "Planetary Boundaries: Exploring the Safe Operating Space for Humanity." *Ecology and Society* 14 (2): article 32.

UN Department of Economic and Social Affairs. 2015. *Sustainable Development Goals.* Accessed November 28, 2016, https://sustainabledevelopment.un.org/?menu=1300.

Wackernagel, Mathis, and William Rees. 1996. *Our Ecological Footprint: Reducing Human Impact on the Earth.* Gabriola Island, BC: New Society Publishers.

3 The Birth of the Environment and the Evolution of Environmental Governance

James Meadowcroft

Today the notion of the environment is so ubiquitous that it is hard to imagine a world without it. Yet sixty years ago, the environment had not yet emerged as an explicit focus for political argument or policy design. It is not just that many of the issues that we would today class as *environmental*—such as acid deposition, the eutrophication of lakes and estuaries, the impact of ozone-depleting chemicals, the release of endocrine disruptors, or the risks of climate change—had yet to make it onto national political agendas, or that the specialized institutions of modern environmental governance (including dedicated ministries and agencies, scientific advisory bodies, and national air and water pollution laws and programs) had still to be established. Rather, the very idea of *the environment* as a category for articulating societal concern and structuring political debate around the damage industrial civilization is doing to natural systems had not achieved widespread currency.

This chapter explores the emergence of the modern concept of the environment, its evolution in political controversy, and its linkages to governance and policy.

The Discovery/Invention of the Environment

In 1971, the *Oxford English Dictionary* offered the following entry for the word environment:

1. The action of environing; the state of being environed. ...
2. *concr[etely]*. That which environs; the objects or the region surrounding anything. ... *esp[ecially]*. The conditions under which any person or thing lives or is developed; the sum-total of influences which modify and determine the development of life and character.

In this context, to *environ* means "to form a ring round, surround, encircle." So environment refers to surroundings, to the situation or circumstances within which a person or thing was located and developed. This broad understanding of environment as context or setting can be applied in many different ways when, for example, one refers to a challenging "business environment," a "classroom environment" conducive to learning, or unit survival in the modern "battlefield environment." We may talk about an environment hostile to innovation or the importance of the social environment for adolescent development.

But the distinctive and quite-modern usage that concerns us here is the invocation of the environment in political and policy discourse to refer to the physical surroundings, especially the natural surroundings, that provide the context for human development, which are being adversely impacted by our activity and have become an object of serious concern. The current *Oxford Online Dictionary* captures something of this new understanding with its second stated usage (the first recapitulates the more general meaning of surroundings cited above): "*the environment*: the natural world, as a whole or in a particular geographical area, especially as affected by human activity." And the *Longman Dictionary of Contemporary English* suggests "the environment: the air, water and land on Earth which can be harmed by man's activities."

A number of features of this understanding of the environment should be emphasized at the outset. First, note the definite article and the absence of any close specification of *whose* surroundings are in question. It is implicit that *the* environment is a distinct thing, a totality of biophysical processes and phenomena. And a little reflection makes it clear that human civilization provides the reference point from which this environment is delimited. *The* environment is ultimately *our* environment, the context within which human societies live and develop. And yet the absence of a *specific* reference to humans as constituting the subject with respect to which this environment is defined (in contrast to the now less common expressions *man's environment* or *the human environment*, about which more will be said below) means that the environment acquires a certain distance from humanity, appearing as a more or less independent entity that can be an object of interest and action.

Second, the notion of the environment is "scalable": it can be applied locally, to a wider region, or to the globe as whole. Concern with the environment may relate to a chemical discharge to a local stream, the preservation of wilderness areas, the movement of air pollutants across continents, or global biodiversity loss. And since the environment evolves through time, it can be considered as it is now, was once, or will be in the future.

Third, there is considerable ambiguity with respect to what might be described as the "naturalness" of the phenomena that the environment denotes. Although the primary referent is the natural world—the natural systems and processes that surround and support human existence—this is a nature that has to varying degrees been altered by human intervention. Landscapes and ecological communities have been influenced by human action for millennia (Goudie 1986). Thus, protecting the environment may imply conserving an environment as it has been transformed by human activity (think of efforts to preserve the English countryside). In cities, moreover, the environment that directly surrounds the inhabitants is still more obviously in large part a human creation, and terms like the *built environment* or *man-made environment* are often applied to denote these elements of our surroundings.

Fourth, this notion of the environment embodies an implicit problematization of human interactions with our surroundings. There is a suggestion that the environment requires our attention; that it is under pressure from human activities and vulnerable to misuse. So in contemporary political or policy usage, the concept already transmits a posture of concern, pointing to difficulties in the relationship between humanity and the global biosphere.

Finally, while discussion of the environment may involve appeals to the intrinsic worth of nature or a duty of care toward the nonhuman natural world, emphasis is typically placed on the impacts on human beings and their societies. Air pollution, the dumping of toxic wastes, or the careless exploitation of resources will damage ecosystems, but above all they represent threats to human health and welfare. It is because the environment is ultimately *our* environment that we must take particular care.

Construction of a Concept

How did this notion of "the environment in jeopardy" emerge? By the early twentieth century, the word environment was firmly established in English-language usage to refer to surroundings or circumstances, especially with regard to influences, interactions, and interdependencies with the surrounded entity. Trevor Pearce (2010) has emphasized the role played by the nineteenth-century philosopher Herbert Spencer in conceptualizing "organism-environment" interaction and popularizing the term environment to capture the totality of contextual circumstances within which an organism developed. In the biological sciences, environment was associated with debates about evolution and the emerging scientific discipline of ecology. Through the first half of the twentieth century, environment was used widely in the natural sciences, but also increasingly in the social sciences. Key pairings were organism-environment (biological sciences), society-environment (sociology), organization-environment (business and organization theory), and system-environment (engineering, cybernetics, and systems theory). In political terms, an early application related to the "nature" versus "nurture" controversy: whether heredity (lineage) or environment (physical surroundings, family upbringing, moral teaching, etc.) had a greater influence on human character, intelligence, criminality, poverty, and so on.

Movement to the modern policy-relevant idea of the environment involved three steps. The first was increased attention to man's environment in the sense of the *physical* surroundings in which humans actually lived. It could include elements near and far, built and natural, and the resources on which society drew. But it definitely excluded as part of man's environment what we might now describe as the "social context for human existence" or social milieu (family, culture, religion, politics, and so on)—those interhuman connections that could also be understood as part of an individual's surroundings. The second and most critical step was to see this physical environment—especially the natural environment—as somehow imperiled by human action, and understand that this in turn constituted a threat to humanity. It was not just that man and his environment were intertwined: that the environment shapes humankind (as individuals and a society), and that humans had progressively transformed their environment to make it more amenable (clearing forests, planting crops, modifying watercourses, domesticating animals, and building cities). The character of this transformation also was leading to dangerous consequences for humanity. Hence, contemporary concern with the environment embodies some notion of *reflexivity*, of feedback with negative implications as human powers to transform nature bend back to harm us—a development that can itself only be apprehended by reflection and remedied by social intervention to alter the offending practices (Voß, Bauknecht, and Kemp 2006; Smith and Stirling 2008). The third step was to focus increasingly on this vulnerable natural environment, to drop the explicit reference to man or human, and add the definite article. So the environment became in the first instance the threatened natural surroundings that provide the context for human development, but that are now being damaged by our actions, with consequent costs and risks.

While authors understood environment in many different ways, these changes are clearly manifest in the evolution of political argument from the late 1950s through the early 1970s. The passage toward this threatened environment can be illustrated by three works now considered cornerstones of the evolution of modern environmental consciousness. *A Sand County Almanac* by Aldo Leopold (1949) contains barely a mention of environment. Much of the discussion centered on personal observations of particular natural phenomena, but when the author speaks more broadly, it is the categories of *nature, land,* and *ecology* that frame his argument. In the famous passage defining a *land ethic,* Leopold refers to a "third element in human environment": an "ethic dealing with man's relation to land and to the animals and plants which grow upon it" (203). But the first two elements of man's environment are understood to be the "relations between individuals," and the link "between the individual and society," so it is clear that environment is not being deployed in the sense discussed above.

Consider next Rachel Carson's (1962) *Silent Spring,* which is typically cited as having played a pivotal role in bringing environmental issues before the US public. In fact, the environment appears perhaps a dozen times in the entire 259-page work, although it does so at particularly important junctures where the author is summarizing her overall argument. Much more prominent are references to the grave threat to "nature" and "our Earth." At various points, Carson refers to "man's assaults upon the environment" (16), "the contamination of man's total environment" (18), and "the tide of chemicals born of the Industrial Age [that] has arisen to engulf our environment" (168). Here we are close to modern usage, although the talk of man's total environment and our environment shows the author believed particular emphasis was required to establish exactly which environment was at issue.

Finally, if we move forward to the famous report on *The Limits to Growth* (Meadows et al. 1972, 21, 85, 126), although the environment appears infrequently, it is consistently invoked in this modern sense, with reference being made to the problem of "a deteriorating environment," the "total pollution load on the environment," and "the natural absorptive capacity of the environment."

It is also worth considering a remarkable article, "Environment: A New Focus for Public Policy?" written by US political scientist Lynton Caldwell (1963), who later served as an adviser during the drafting of the National Environmental Policy Act (1969) (NEPA). At the core of Caldwell's argument was a plea for a more integrative and comprehensive approach to decisions with environmental implications. Caldwell noted that the US government was "continually trying to solve problems and to salvage the wreckage caused by misguided, heedless, or inadvertent environmental change" (136), and he decried "our national tendency ... to deal with environmental problems segmentally" (134).

At various points Caldwell spoke of "human environment," "environment as a generic concept," "the total environment," "artificial environments," and the "altered immediate natural environment." The specific issues to which he linked environment were urban life, resource management, changes to rural landscapes with farm mechanization, and the large-scale transformation of river basins. Pollution and nature conservation were not explicitly mentioned. Caldwell called for the development of a new field, "environmental administration," which would "deal with environments comprehensively, as environments, in contrast to focusing upon their component parts" (136). Indeed, he went further to suggest "the need for a generalizing concept of environmental development that will provide a common denominator among differing values and interests" (138). Yet his article privileges environment or total environment (rather than *the* environment), often understanding it in an inclusive manner sometimes including society itself: "environment is not only the complex interrelating reality surrounding us; it includes us" (133).

The NEPA spoke of encouraging a "productive and enjoyable harmony between man and his environment," and highlighted the need to "prevent or eliminate damage to the environment." Reference is made to preserving "important historic, cultural and natural aspects of our national heritage," and assessing the state of the "major natural, manmade, or altered environmental classes of the Nation." Yet the main thrust of the text is clearly threats to the natural environment with their implications for resources and human welfare. Around the same time, the United Kingdom's Royal Commission on Environmental Pollution was established with a mandate to "advise on matters, both national and international, concerning the pollution of the environment; on the adequacy of research in this field; and the future possibilities of danger to the environment." Its first report spoke of "our physical environment," "preserving" the environment, safeguarding "the natural environment," and improving "the quality of the environment" through "education, legislation, financial measures and international agreements" (Royal Commission on Environmental Pollution 1971).

Major UN conferences provide another point of reference. The Declaration of the UN Conference on the Human Environment held in Stockholm in 1972 opens with the words "man is both creature and molder of his environment." It calls for "the protection and improvement of the human environment," arguing "both aspects of man's environment, the natural and the man-made," are critical (UN Environment Program 1972). Four forms of "man-made harm" to the environment are cited in the declaration:

> dangerous levels of pollution in water, air, earth and living beings; major and undesirable disturbances to the ecological balance of the biosphere; destruction and depletion of irreplaceable resources; and gross deficiencies, harmful to the physical, mental and social health of man, in the man-made environment, particularly in the living and working environment.

This integration of quite-different kinds of concern—with pollution and ecological balance, resource depletion, and living and working conditions—within the ambit of environment was critical to building the political coalition that made the conference possible.

What is surprising to the early twenty-first-century reader is not the differentiation between developing countries ("where most of the environmental problems are caused by underdevelopment") and industrialized countries (where "environmental problems are generally related to industrialization and technological development"), or the strong affirmation of state sovereignty over resources. Rather, it is the overwhelming faith in future efforts to transform nature and build an environment "more suited" to humans, and the emphasis on rational planning and control. It reflects the different political/ideological context (where planning and state control were part of the default response to societal problems) along with the need to balance political preoccupations of East and West as well as developed and developing countries. Using language that now sounds archaic, the Stockholm Declaration affirmed that "along with social progress and the advance of production, science and technology, the capability of man to improve the environment increases with each passing day." And a little later in the context of a discussion of population pressures, it insists, "Of all things in the world people are the most precious" for they "propel social progress, create social wealth, develop science and technology and, through their hard work, continuously transform the human environment" (ibid.).

By the time of the 1992 Rio Declaration issued by the UN Conference on Environment and Development, reference to man's environment had disappeared. Throughout the twenty-seven principles, the talk is of "the environment" and particularly "the global environment"—which clearly relates in the first instance to the threatened natural environment (United Nations 1992). And instead of appearing positive, continuous environmental change has acquired a more negative resonance.

Considerable attention has been devoted to the question of why the idea of threatened nature on which we depend acquired increased cultural resonance during the 1960s. In 1970, Neil Young captured this sentiment perfectly in his song "After the Gold Rush": "Look at Mother Nature on the run in the 1970s." Concern with the destructive effects of industrial transformation (on nature and human communities) goes back well into the nineteenth century (Andrews 1999), and arguments about relations with the nonhuman natural world were known to antiquity. The issue was a staple of nineteenth-century romantic and utopian theorists, and later came the efforts of the nature and resource conservation movements as well as the campaigns for urban sanitation and clean air. During the period of post–World War II reconstruction, and as growth in both rich and poor countries accelerated through the 1950s, there was increased talk of resource scarcities and the challenge of feeding a rapidly expanding world population. A revitalized conservation movement called for state planning and international cooperation to manage potential resource constraints (Linnér 2002). But in the late 1960s and 1970s, these currents blossomed into a much wider perception of the destructive impacts of industrial society that linked up with more general countercultural and protest movements along with a certain loss of faith in established institutions (government, big corporations, the military, and the unquestioned authority of science). However the story is told, it must include elements such as the increasing scale of human impacts on nature and the lived environment (agricultural chemicals, oil spills, industrial hazards, deteriorating air and water quality, and nuclear fallout); the growth of "consumer society," which provided material affluence but also unwanted side effects and unmet needs; and an increased perception of the interconnectedness of the world (images of the earth from space, but also ICBMs and the Cold War).

Of course, the environment was not the only concept that focused concern on these issues. As we have seen, "nature" and "the Earth" were already favored by those worried about the encroachments of industrial civilization. In its opening paragraph, the NEPA referred to the *biosphere* and *ecological systems* as well as the environment. And terms like *our planet, the ecosphere,* or the ecology were also in vogue. Features that made environment particularly serviceable included the concept's:

- impeccable pedigree in the natural and social sciences, planning, occupational health, and engineering. Particularly in the later 1950s and early 1960s, there was an explosion of interest in environment across multiple scientific domains (witness, for example, the creation of both the UK Natural Environment Research Council and US Environmental Sciences Service Administration in 1965) (Hare 1969).
- general currency to denote "surroundings" in many different contexts, so that it was not excessively technical and instead easily accessible to nonscientists such as politicians, officials, and journalists.
- potential to span a vast range of phenomena relating to air, water, and land; pollution, resource management, and nature protection; and the natural but also the build environment—so groups and individuals concerned with many different issues could find some purchase in the concept.
- scalability (from local to global).
- ambiguity because it was open textured with multiple meanings and related in various ways to the points listed above.
- anthropic resonance—explicit in our environment, and implicit in the environment, but not so strong as to preclude an opening to less anthropocentric readings as *the* environment is increasingly abstracted from humanity.

In contrast, the planet, the earth, and the biosphere were comprehensive but not scalable. Biosphere, ecosphere, ecology, or *ecosystem* all had less obvious popular resonance, nor were they in wide use across multiple scientific and engineering disciplines. Moreover, for most of these terms the initial anthropocentric reference was weaker. It is hard to think of *our* nature, and while we can talk of *our* earth, planet, or biosphere, the connection is less immediate than with *the* environment that surrounds and sustains us.

Practical Implications

What were the implications in the political/policy field of this emerging understanding of the environment as surroundings (especially natural surroundings) that were imperiled by human action? Most obviously, this unifying abstraction allowed activists, officials, scientists, and citizens concerned with specific problems (such as pesticides, urban smog, or wildlife protection) to relate their preoccupations to a common notion of human pressure on the natural world on which we depend. Disparate problems at multiple scales could be understood as causally interconnected. Local struggles (to protect this lake, clean up a waste site in this town, or conserve that species of bird) could be understood not as parochial responses to idiosyncratic concerns but rather as part of a broad effort to save the environment as a whole.

Once the environment was conceptualized as a *critically important but vulnerable* entity, it made sense to take systematic action to protect that environment and the human interests to which it was tied. Faced with such a comprehensive and complex issue—with implications for human health, economic prosperity, and long-term national welfare—action by governments to alter the societal practices giving rise to environmental degradation appeared essential. Of course, governments at various levels had already done much in this area—taking measures to limit air and water emissions, enacting early health and sanitary ordinances, regulating resource extraction (fishing, logging, and mining), planning urban development, protecting wildlife or valued landscapes, establishing national parks, and so on (Andrews 1999). But with the emergence of a threatened environment critical to human well-being, the necessity for a more vigorous institutional response came to the fore. And this is exactly what happened at the end of the 1960s and early 1970s as countries across the Organisation for Economic Co-operation and Development (OECD) moved to create environmental ministries and agencies, set up expert advisory councils on the environment, and introduce national air and water pollution legislation and/or framework environmental laws (Janicke and Weidner 1997; Hanf and Jansen 1998).

In this way, environmental policy acquired the status of a distinct policy domain, with dedicated officials, organizations, and budgets along with associated administrative practices, analytic approaches, and policy tools. Although politicians and officials initially regarded it as a second-tier function (in contrast to core areas such as economic management, foreign affairs, defense, welfare provision, and so on), and the new ministries and agencies lacked clout in internal bureaucratic wrangling, its institutional embedding within the apparatus of governmental power mattered. This emerging environmental policy domain provided intellectual (and also increasingly organizational) cohesion across a wide range of previously disparate and disjointed government activities. Most obviously this related to government units, personnel, and functions that were reorganized into the new environmental departments. But it also provided conceptual and intellectual links that extended across a much wider political/administrative space, linking departments concerned with resource management (fisheries, forests, and mineral extraction), urban affairs, land use planning, transport, public health, worker health and safety, nuclear regulation, and so forth.

The birth of environmental policy was above all the birth of *national* environmental policy. Local and regional governments had long been involved in pollution abatement and other efforts to manage impacts that were now being classed as environmental. But the environmental framing pointed to the pervasive and acute character of the threat, and the importance of national governments stepping in to supply an overarching policy framework within which other levels of government might act. Yet even as national governments established a distinct environmental policy domain, they simultaneously prepared the way for its internationalization. By the end of the 1960s, it was clear that environmental problems confronted all industrialized societies, and that it was only a matter of time before they gathered pace in less developed countries (OECD 2002). Pollution of the high seas and the trade in endangered species had been early areas for international efforts. The potential trade implications of increased environmental regulation attracted the attention of the OECD from its earliest discussions of the environment (Long 2000). By the time of the Stockholm conference in 1972, the environment had been identified as a potential focus for East-West cooperation across the Cold War divide, and major lines of tension between the perspectives of developed and developing states were becoming apparent.

The formal institutionalization of environmental policy also brought a significant boost for those campaigning around environmental issues who had long demanded more vigorous action. It was not just that they now had a point of contact and potential advocate within government, or that policy and bureaucratic routines provided enhanced entry points for intervention. It was also that they received a powerful boost of legitimacy. After all, if governments had come to acknowledge that environmental threats were real, and thus that the critics had been right all along, who could say that the environmentalists were not also correct about a whole list of other emerging challenges?

Finally, with this institutionalization of environmental protection, the environment became established as a more or less permanent site of political contention in modern states. It came to serve as a focus for different sorts of arguments about the nature of progress and welfare, social economic and political priorities, and the meaning of equity and societal security.

Environmental Problems and Environmental Policy

Disputes over the environment involve different sorts of claims about the burdens humans impose on their natural surroundings, the reasons for concern over such impacts, the social forces generating these changes, and the formulation of potential solutions. Discussion may relate to individual cases (poor water quality in a lake), broad classes of problem (particulate air pollution, genetically modified food, or overfishing), or the environmental issue or crisis writ large. With respect to *problem identification*, there is continuous political and policy argument—extending from local communities to the global level. Critical here is the identification of a worrying phenomenon and establishment of a causal link to some human practice. *Reasons for concern* involve notions of harm to human well-being (health or livelihoods) and/or to the nonhuman natural world (species, ecosystems, or the biosphere). Arguments are framed in prudential or other regarding terms:

if we fail to protect the environment, we will suffer and/or the well-being of others (in other locations, future generations, or nonhuman nature) will be harmed. They may involve utilitarian or deontological reasoning. The *causes of environmental pressure* can be considered from the perspective of proximate causality (release of a toxin from this facility), intermediate drivers (rising demand for mobility), or underlying processes (industrialization or population growth). They can be framed in terms of individual responsibility (thoughtless consumers, rapacious developers, or poorly educated farmers), faulty governance mechanisms (lax regulations or poorly defined property rights), or broader system failures (human nature, perverse values, capitalism, or the anarchic international system). Ideas (consumerist mentality or the instrumental legacy of the Enlightenment), institutions (free trade or the market), and interests (oil companies or farmers) are all called on to play their part. *Policy prescriptions* relate to the diagnosis of the problem, but also to the relative significance of issues, the magnitude and distribution of the costs and trade-offs associated with proposed solutions, and the merits of different policy instruments.

From the perspective of the political and policy system, environmental problems only come into being when someone with a capacity to influence others says, "Hey, there is a problem here." Of course, the underlying physical processes (and the social relationships that generate the physical impacts) may have been going on for years, but it is when politically relevant actors take them up that they emerge explicitly as a societal problem. As long as wolves are mainly understood as a threat to livestock, their eradication is not a problem but rather a solution. But if the decline of tree cover in a national park can be linked to overgrazing, or there are public complaints about excess deer populations, or ecologists and nature lovers bemoan the loss of wild species, it can achieve recognition as an environmental problem.

In the early 1970s, environmental policy was mainly focused on the pollution of air and water from industry and agriculture. Soon waste disposal and the clean up of contaminated sites were added. Toxic metals (lead, cadmium, and mercury), PCBs (and persistent organic pollutants more generally), acid rain, and ozone depletion further expanded the list. Climate change and biodiversity loss later emerged as critical global issues. The basis for the scientific understanding of many of these issues may have been laid decades before, but over time they percolated into the public arena, as environmental activists and civil servants increasingly took note. Four factors appear to underlie the extension of the field. First, there is the growth of physical impacts driven, on the one hand, by the increasing scale of established activities (volume or geographic extent), which reaches a point where negative effects become manifest (more fossil-fueled combustion, more deforestation, etc.), and on the other hand, by the introduction of novel technologies, which gives rise to new problems (e-waste following the telecommunications boom or health risks from nanotechnology). Second, there is a gradual advance in scientific knowledge (and the increasing diffusion and accessibility of this knowledge) that identifies previously unappreciated effects and causal linkages (acidification of lakes or ozone depletion). Third, changes in societal values and perceptions mean that previously accepted practices are now deemed to be problematic. And finally, the institutionalization of environmental concern, in government structures and legal systems, scientific organizations, and environmental groups, led to continuing expansion of the range of problematized practices.

Particularly from the late 1980s onward, environmental policy makers talked of dealing with the sources (rather than just the symptoms) of environmental degradation. The idea was to displace polluting products and processes by making production and consumption more environment friendly as opposed to simply applying "end-pipe" controls. Critical here was the notion of *integration*, which was applied in "integrated pollution control" and "integrated water management,' but also much more generally in the notion of integrating the environment into decision making in key economic sectors and across government. To some extent, the integrative ideal was present from the outset (consider Caldwell, the NEPA, and the goal of environmental assessment). Now it was made a more explicit objective. The point was to design away—through improved technology and policy—environmental problems from the outset. Sustainable development represented the most ambitious integrative project, emphasizing the link between environment and development decision making. In more recent years, integrative ideas have focused particularly on the climate change problem area.

At this point we reach the ironic situation where successful environmental policy is no longer really environmental policy, or at least not just environmental policy (Meadowcroft 2012a). While specialized institutions are required to care for the environment (and such specialization is a hallmark of the modern administrative state), it is only if environmental policy penetrates into mainstream economic and social decision making that it can move beyond a ghetto that is always reactive, patching up problems that follow mainstream development decisions. Yet difficulties with integration over several decades raise questions about whether the idea cuts deeply enough to articulate reform to established economic and governance arrangements.

Four Enduring Themes in Environmental Argument

To appreciate policy argument over the environment, it is helpful to consider four critical and interrelated themes: valuing the environment, environmental limits, economy-environment linkages, and the distribution of environmental harms and goods.

Valuing the environment relates to the various ways the environment matters: which aspects do we value and why? Among environmental theorists there are continuing debates over "anthropocentric" and "ecocentric" approaches to valuation, and whether we should be concerned with protecting the environment because of its utility to humanity or its intrinsic worth. Should we preserve lions because they deserve independent moral consideration (or are elements in an ecosystem that merits such consideration) or because humans benefit from their continued existence (for example, as an attraction in zoos or to provide game park income from tourism or hunting)?

Valuing environmental elements can involve a moral/ethical, experiential/aesthetic, religious/spiritual, or economic perspective. And it implies choices about the appropriate units of care: individuals, species, ecosystems, landscapes, the oceans, the planet, and so on (Dobson 1995). Economists emphasize economic valuation of environmental goods and services (which often lack market prices), so their worth can more accurately be reflected in the economic calculus that guides decision making. The concept of *ecosystem services* tracks the various ways environment contributes to economic activity and social well-being. Yet some environmentalists criticize the tendency to reduce social choice to an economic logic. In fact, human individuals and groups experience the environment (or elements of it) in a vast array of different and overlapping ways that relate to where they live, how their earn their living, the cultural context, and so forth. But societies are continuously transforming the environment. Typically it is when some dimension that has previously been taken for granted is called into question that groups and individuals think more consciously about its significance to their lives. And thinking about these threatened environmental dimensions (and the human social consequences) brings to the fore questions about how our society operates and how we see ourselves.

The idea of *environmental limits* is central to arguments about the environment. The core notion is one of biophysical limits, of constraints that the environment imposes on human activity. For example, beyond a certain point dumping sewage into a river will make the water undrinkable; if we harvest too many fish from a lake, there will be none to catch next year. Limits are first experienced locally (exceeding pollution thresholds or exhausting resources), but today we are increasingly concerned with regional limits (for instance, to water use) and global limits (the greenhouse gas emissions that are generating climate change). Limits can be discussed in terms of carrying capacity, critical loads, threshold effects, and tipping points. And much of environmental policy can be understood in terms of imposing limits to human activity to avoid crossing environmental thresholds. But environmental limits are often difficult to define, appearing somewhat "plastic": there is uncertainty about complex causal mechanisms and ultimate impacts, and their assessment depends on normative assumptions. By deploying new technologies, changing social values, or redefining the terms of the argument, seemingly hard limits can appear to evaporate. The *Limits to Growth* report (Meadows et al. 1972) with its assertion that exponential growth of human population and economic activity could not continue indefinitely is seen as a paradigmatic limits argument. But limits are equally present in *Our Common Future* (World Commission on Environment and Development 1987), which, while encouraging a change in "the quality" of economic growth, also insisted that development trajectories had to acknowledge the reality of "ultimate" ecological limits. And they appear more recently in the discussion of "planetary boundaries" (Rockström et al. 2009).

In addition to the core idea of biophysical constraints, limits take other forms in environmental argument. There is concern with the *limited capacity of humans to understand and successfully manipulate nature* (Meadowcroft 2012b). The barrier here is more "internal" than "external," relating to our ability to appreciate nature's complexity and comprehensively to manage socioecological interactions. It indicates the virtues of prudence and precaution when contemplating grand, rationalistic schemes to reorder nature or human society. Limits can point to the *inability of current materialistically oriented development models to fulfill human needs*. Material affluence has brought us much, but at a certain point further growth in the economy produces diminishing real welfare gains. By turning away from consumerist values (with the environmental burdens they entail), we may open the way toward more healthy and satisfying lives. And then there is the critical notion of *limits as self-imposed restraint*: as fully conscious beings we can take stock of the situation and deliberately choose to curtail human demands on the environment. Whatever the motivation (to protect our health, leave a bountiful environment for our children, or protect other species), the point is to take responsibility for our acts and exercise restraint.

A third critical area is *economy-environment linkages*, which have always been at the core of environmental argument (Dryzek 2013). The relationship between the environment (and environmental policy) and the economy (and economic policy) can be thought about as a four-way impact grid. For starters, economic activity (in specific instances and at the macro level) can harm the environment (and so affect human welfare) as it draws in raw materials, transforms nature, and releases wastes. As economies grow, the range and scale of environmental pressures expand as natural ecosystems are increasingly altered by human use. On the other hand, economic prosperity can open the way to enhanced environmental protection and rising environmental quality (cleaner air and water, say). With basic material needs satisfied, other social priorities emerge; increasing resources are available for environmental protection; structural economic change may reduce environmental pressures (from smokestack industries), while technological innovation can raise materials and energy efficiency as well as sometimes reduce absolute environmental burdens. Third, environmental protection also imposes costs on producers and the economy as whole. Pollution control and clean up are expensive, emissions charges and taxes raise costs, proliferating regulations stifle business, and conservation measures put certain resources (for example, in protected areas) out of reach, increasing costs and driving activity elsewhere. Fourth, in contrast, the environment provides a foundation for economic activities, and environmental problems can impose costs (on receptor industries) and act as a drag on economic performance. Environmental quality also promotes a healthy and productive workforce, and measures to protect the environment can promote economic development: conserving potentially renewable resources in danger of being eroded, ensuring provision of environmental services that may attract investment, encouraging materials and energy efficiencies, and stimulating innovation that creates green products and services that can open up new markets, create jobs, and so on.

These relationships are subject to endless argument, such as the long-running controversies over the "environmental Kuznets curve" (which links improved environmental quality to economic growth) and the potency of innovation-forcing, environmental-regulatory initiatives (Ambec et al. 2011). In fact, all these linkages hold true to differing degrees in differing circumstances. And it is precisely those circumstances that matter.

At the cusp of contemporary debate is the extent to which it is possible to "decouple" the existing social economy from environmental burdens. The issue is brought into focus with climate change and the greenhouse gas intensity of GDP. A relative decoupling of these variables (where emissions per unit of output fall) is inadequate to meet the challenge as total emissions still can rise. What is required is an "absolute decoupling" where emissions intensity improves year on year (and substantially faster than total emissions are pushed upward by economic and population growth) to bring absolute levels down over time to a few percentage points of the current emissions. Of course climate change is only one problem. So far, absolute decoupling of environmental impacts from economic growth has only been achieved for certain impacts, in certain countries, for particular periods (for example, sulfur oxide emissions in OECD [2002] countries since the 1980s). Achieving this across a range of major environmental issues for all countries over the long haul looks to be a tall order. And this leads to broader arguments about the appropriate scale of the social economy, the relationship between social welfare and continuing economic growth, and the compatibility of existing economic institutions and practices with a flourishing natural environment (Victor 2008; Stiglitz et al. 2009).

A fourth recurring theme concerns the *distribution of environmental harms and goods*. Every environmental issue is also a distributive issue that can be understood as a struggle over the allocation of different sorts of harms and goods among existing groups and individuals, across human generations, and with respect to the nonhuman natural world. Typically environmental problems and solutions involve crosscutting distributional dimensions generating multiple harms and goods, impacting different types of groups at different spatial and temporal scales. Coal combustion can produce cheap electricity that encourages economic competiveness, but releases innumerable pollutants (particulates, mercury, sulfur dioxide, carbon dioxide, etc.) that impact varied constituencies spread across space and time. Any solution will disturb existing entitlements (property rights) and so provokes resistance. One may talk of aggregate social costs and benefits, but the distribution of these costs and benefits is also important—for the efficiency, effectiveness, and legitimacy of policy outcomes. The theory of collective action shows why it is more difficult for widespread yet diffuse interests (asthma sufferers or nature lovers) to coordinate their efforts as compared to concentrated producer interests (fossil fuel producers or utilities) (Olson 1965). And of course nature and future generations have no explicit representation of their interests in current deliberations, although proxies may argue their corner (nongovernmental organizations, environmental commissioners, etc.)

While the arguments over responsibility, vulnerability, and capacity to pay at the international climate talks, or campaigns for "environmental justice" in the United States explicitly raise distributional issues, they are implicit in every claim against environmental degradation, every act of pushback by those associated with problem generation, and every discussion of remedial approaches and policy instruments.

The Future of the Environment

This chapter has argued that the conception of the environment as threatened nature is a relatively recent addition to the conceptual lexicon. The innovation consisted of taking a term that had been applied generally to denote surroundings, and used more specifically in the natural (and to some degree, the social) sciences to frame subject-environment interactions, to denote the physical environment within which humans make their home, emphasizing the threat to nature from human activity and consequent societal risks. This concept was then used to justify political and policy action to protect the environment, link specific issues to a broader societal challenge, unify discrete impacts, define management frames, and ultimately interrogate the current development trajectory.

It seems strange to talk of the "invention" or "discovery" of *the environment.* Surely the existence of our biophysical surroundings—the natural processes and ecosystems within which human societies are embedded—is independent of our perceptions. Moreover, social and natural scientists have long appreciated the dependence of societies on their surroundings, while recognition of human impacts goes back millennia. Yet by changing the way these things were conceptualized, something new did come into the world: the environment emerged as a political as well as policy and legal/administrative entity that is understood to underpin social life, is threatened by human carelessness, and is a legitimate object of political argument and governmental care. This notion arose at a particular moment in time. And its consequences are still being worked through.

To ask about the future of the environment is to ask about the future of humankind, the planet on which it evolved, and the categories through which we understand them. Now that the environment is with us, it is hard to see how it would fade away. No matter how successful we are at managing environment problems, it is difficult to imagine a world where the types of issues we now class as environmental will have disappeared. Certainly, political dilemmas and arguments around the four themes discussed in the last section appear to be with us for the long haul: what and why we value environmental elements, the negotiation of environmental limits, the economy-environment linkage, and the distribution of environmental costs and benefits. In a more obvious sense, the sheer weight of human numbers and scale of existing technologies mean that "threatened nature" and its risks for humanity are unlikely to go away anytime soon. And even if we resolve the problems that have already been identified (no small feat for issues such as climate change), the emergence of new technologies that reach down ever deeper into nature's treasure trove—especially through nanoscale and bioengineering, with the possibility of deliberately altering the human genome (as well as the biosphere at large)—are certain to raise new quandaries.

All this is not to say that we have to understand these things in terms of the environment concept. Concepts come and go, and just as issues are aggregated and integrated through the notion of the environment, so they might conveniently be disaggregated and reaggregated under alternative frames. But to anticipate what such categories might be is no easy feat.

What can be done in closing is to reflect briefly on the problems, weaknesses, or inadequacies that may be discernible in the concept of the environment as its history has played out to date. There are three obvious points on the policy side. First, environmental policy has since its inception remained essentially reactive, struggling to manage problems that are already well established. Second, efforts at policy integration remain elusive, and environmental goals are routinely trumped by other social objectives. And third, the underlining increase in the absolute scale of global environmental pressures has not been slowed. Even if substantial progress is made in decarbonizing energy supplies to address climate change, there are many other issues (including the relentless losses of biodiversity) waiting in the wings.

There are two major issues on the conceptual side, and unsurprisingly these "weaknesses" are the flip side of the dimensions that make the idea of the environment so useful. First, there is the reification of the environment as a thing, as a coherent entity about which general claims can be made, and that needs to be "protected," "managed," "governed," and so on. Of course, the great advantage of this move is that we now have a single abstract entity that can frame discussion, serving as a proxy for innumerable discrete processes and phenomena. It is precisely this aggregative, synthetic, or integrative potential that gives the environment its power as a device to structure argument and analysis. Yet there is always a risk that we forget that what falls under the environmental umbrella is not really a single thing. Rather, it is composed of many interrelated processes operating at different spatial and temporal scales, which are experienced in different ways by different actors (human and nonhuman) at different times and places. So while we can hope to find solutions to particular "environmental problems," there cannot really be a solution to *the* environmental problem writ large. And talking about the environment in such general and abstract terms may even distract us from identifying the focused and context-specific approaches required to address particular difficulties. It can conjure up a misplaced "concreteness," a seemingly obvious whole, that can actually never be engaged with in its totality.

Second, there is the underlying binary fissure that is marked out when one starts with the environment, on the one hand, and that which it implicitly surrounds (humanity, society, or civilization), on the other hand. Into one basket goes Homo sapiens, and into another goes the rest of nature. But our biological and social existence is intertwined with ecological processes—from the complex biotas that flourish on and in our bodies (whose relation with physical and psychological health is only now beginning to be understood), through the ecosystems that provide the resources that underpin distinct ways of social life (with their technologies, economic and social relations, and cultural constructions), to the global "life-support" systems. Moreover, the biosphere has been transformed, consciously and unconsciously, by human actions that stretch back millennia but have accelerated over the past two centuries. Some environmentalists now talk of the "death of nature" or the end of the idea of a "wild" nature independent of humanity (McKibben 1990). While scholars, starting with geologists and geographers, yet extending increasingly to those in the social sciences and humanities, explore the idea of the Anthropocene, the era when humans have become the major determinant of the future evolution of the biosphere (Lovbrand et al. 2009; Dalby 2015). Although environment and that which it surrounds have long been understood as interdependent—indeed, at its most powerful the concept serves to emphasize the *relationships* that link living beings and their surroundings—in practice the environment sometimes serves to abstract away from this intimate connection, providing for campaigning or managerial efforts that pull attention away from the coevolution of societies and ecologies.

References

Ambec, Stefan, Mark A. Cohen, Stewart Elgie, and Paul Lanoie. 2011. "The Porter Hypothesis at 20: Can Environmental Regulation Enhance Innovation and Competitiveness?" *Review of Environmental Economics and Policy* 7 (1): 2–22. doi:10.2139/ssrn.1682001.

Andrews, Richard N. L. 1999. *Managing the Environment, Managing Ourselves*. New Haven, CT: Yale University Press.

Caldwell, Lynton. 1963. "Environment: A New Focus for Public Policy?" *Public Administration Review* 23 (3): 132–139. doi:10.2307/973837.

Carson, Rachel. 1962. *Silent Spring*. Boston: Houghton Mifflin.

Dalby, Simon. 2015. "Framing the Anthropocene: The Good the Bad and the Ugly." *The Anthropocene Review* 3 (1): 33–51.

Dobson, Andrew. 1995. *Green Political Thought*. 2nd ed. Abingdon, UK: Routledge.

Dryzek, John. 2013. *The Politics of the Earth.* 3rd ed. Oxford, UK: Oxford University Press.

Goudie, Andrew. 1986. *The Human Impact on the Natural Environment.* 2nd ed. Oxford: Blackwell.

Hanf, Kenneth, and Alf-Inge Jansen, eds. 1998. *Governance and Environment in Western Europe: Politics, Policy, and Administration.* Harlow, UK: Longman.

Hare, Kenneth. 1969. "Environment: Resuscitation of an Idea." *Area* 1 (4): 52–55.

Janicke, Martin, and Helmut Weidner, eds. 1997. *National Environmental Policies: A Comparative Study of Capacity Building.* Berlin: Springer.

Leopold, Aldo. 1949. *A Sand County Almanac: And Sketches Here and There.* Oxford: Oxford University Press.

Linnér, Björn-Ola. 2002. *The Return of Malthus.* Isle of Harris, UK: White Horse Press.

Long, Bill. 2000. *International Environmental Issues and the OECD, 1950–2000: An Historical Perspective.* Paris: OECD.

Lovbrand, Eva, Johannes Stripple, and Bo Wiman. 2009. "Earth System Governmentality: Reflections on Science in the Anthropocene." *Global Environmental Change* 19 (1): 7–13. doi:10.1016/j.gloenvcha.2008.10.002.

McKibben, Bill. 1990. *The End of Nature.* New York: Viking Press.

Meadowcroft, James. 2012a. "Greening the State." In *Comparative Environmental Politics*, ed. Paul Steinberg and Stacy VanDeveer, 63–88. Cambridge, MA: MIT Press.

Meadowcroft, James. 2012b. "Pushing the Boundaries: Governance for Sustainable Development and a Politics of Limits." In *Governance, Democracy and Sustainable Development: Moving beyond the Impasse*, ed. James Meadowcroft, Oluf Langhelle, and Audun Rudd, 272–296. Cheltenham, UK: Edward Elgar.

Meadows, Donella, Dennis Meadows, Jørgen Randers, and William Behrens III. 1972. *The Limits to Growth: A Report for the Club of Rome's Project on the Predicament of Mankind.* New York: Universe Books.

National Environmental Policy Act. 1969. Pub. L. 91–190, 42 U.S.C. § 4231–4347.

Olson, Mancur. 1965. *The Logic of Collective Action: Public Goods and the Theory of Groups.* Cambridge, MA: Harvard University Press.

Organisation for Economic Co-operation and Development(OECD). 2002. *Indicators to Measure the Decoupling of Environmental Pressure from Economic Growth.* Paris: OECD.

Pearce, Trevor. 2010. "From 'Circumstances' to 'Environment': Herbert Spencer and the Origins of the Idea of Organism-Environment Interaction." *Studies in History and*

Philosophy of Biological and Biomedical Sciences 41 (3): 241–252. doi:10.1016/j.shpsc.2010.07.003.

Rockström, Johan, Will Steffen, Kevin Noone, Åsa Persson, F. Stuart Chapin III, Eric F. Lambin, Timothy M. Lenton, et al. 2009. "A Safe Operating Space for Humanity." *Nature* 461:472–475. doi:10.1038/461472a.

Royal Commission on Environmental Pollution. 1971. *First Report*. London: Her Majesty's Stationery Office.

Smith, Adrian, and Andy Stirling. 2008. "Social-Ecological Resilience and Socio-Technical Transitions: Critical Issues for Sustainability Governance." STEPS Working Paper 8. Brighton: STEPS Centre.

Stiglitz, Joseph, Amartya Sen, and Jean-Paul Fitoussi. 2009. *Report by the Commission on the Measurement of Economic Performance and Social Progress*. Paris: Government of the French Republic.

UN Environment Program. 1972. Declaration of the UN Conference on the Human Environment. Nairobi: UN Environment Program.

United Nations. 1992. Rio Declaration on Environment and Development. Report of the UN Conference on Environment and Development. Rio de Janeiro: United Nations.

Victor, Peter. 2008. *Managing without Growth: Slower by Design, Not Disaster*. Cheltenham, UK: Edward Elgar.

Voß, Jan-Peter, Dierk Bauknecht, and René Kemp, eds. 2006. *Reflexive Governance for Sustainable Development*. Cheltenham, UK: Edward Elgar.

World Commission on Environment and Development. 1987. *Our Common Future*. Oxford: Oxford University Press.

4 Environmental Impact Assessment: Can Procedural Innovation Improve Environmental Outcomes?

Richard N. L. Andrews

An environmental impact assessment (EIA) is a procedure used to anticipate and document in advance the environmental impacts, effects, and consequences of proposed actions by governments as well as by many major international institutions, banks, insurers, and other businesses. It has been widely adopted, and ideally it is used to choose an alternative that is environmentally preferable, prevent or mitigate adverse effects, or at least anticipate potential risks and controversies. It also provides public access to far more information about proposed actions and their impacts than was previously available, and in some organizations it is also used to identify and compare alternatives.

The EIA originated in a requirement of the United States' National Environmental Policy Act (NEPA) in 1970. The NEPA required for the first time that in every recommendation or report on proposed government actions that might "significantly affect the quality of the human environment," the responsible agency must include a "detailed statement" discussing the environmental impacts of the proposed action, any adverse environmental effects, alternatives to the proposed action, relationships between short-term uses of the environment and its long-term productivity, and any irreversible and irretrievable commitments of resources that would be involved. Over the following decades, the governments of more than a hundred other countries plus more than half the United States' state governments—as well as international lending institutions such as the World Bank—adopted their own versions of EIA requirements, and many large business corporations have done so as well.

EIAs have figured prominently in many controversial decisions affecting the environment. Examples include the Trans-Alaska Pipeline and the more recent Canada-US Keystone XL pipeline proposal in the United States, the drying up of the Aral Sea in Central Asia, the Three Gorges Dam in China, and countless other proposed projects around the world. As a conceptual innovation, the EIA idea also has inspired many other types of impact assessments, such as inflation, small business, social, health, "strategic," technology, and life cycle impact assessments.

The authors of the EIA concept intended it not as an end in itself but instead as a means to assure that government actions would help "to create and maintain conditions under which man and nature can exist in productive harmony, and fulfill the social, economic, and other requirements of present and future generations" (NEPA 1970, sec. 101): in effect, to plan for what today would be called "sustainable development." Their fundamental purpose was not simply to document the environmental impacts of a single proposed action: it was to promote comparison of the impacts of alternative actions, so that those with the most positive impacts—or at least the least negative ones—would be chosen. This intent has been widely overlooked and marginalized in its implementation, but it was explicit and central in the vision of the concept's authors.

Over time, the EIA requirement has been implemented far more narrowly as merely a project-specific procedural requirement. Even in that role, it has caused some projects not to be built, and others to be significantly redesigned to mitigate their environmental impacts. It has vastly increased public access to information about proposed government actions along with documentation of their environmental impacts and alternatives, and it has greatly increased opportunities for the public to voice concerns during project planning and decision processes. It also has forced governments and businesses to identify and evaluate alternatives—including alternatives proposed by different agencies and members of the public—and hire experts on environmental impacts.

At the same time, however, EIA has come under recurrent attacks by critics arguing that it is ineffectual, its assessment documents are too long and complex, and its process is unduly burdensome and simply delays rather than improves decisions. Some of these criticisms are legitimate, but others appear to blame the EIA for other causes of project delays; still others seem to reflect continuing political resistance by some project advocates to the environmental policy goals and public decision processes that the EIA represents. In short, the EIA has clearly been a major conceptual innovation in environmental decision making, and remains so, but its influence is still contested, and its full conceptual vision has been far from fully achieved.

The EIA Concept: Origins

Unlike many other conceptual innovations in environmental policy, EIA emerged from its birth as a legislative requirement, in a statute passed by overwhelming majorities in the US Congress. The president signed it into law in a televised public ceremony on the first day of a new decade in which he declared the 1970s the "Decade of the Environment."

The concept of the EIA had antecedents in long-standing US public concerns over the effects of human activities—natural resource extraction, landscape transformation, urbanization, and industrialization, and especially those resulting from government actions—on seemingly pristine natural landscapes (Andrews 2006). These activities were promoted by government agencies, most of which were organized by specific sectors and statutory missions, and staffed by professionals trained only for those missions: water management by civil engineers, transportation by highway engineers, the US Forest Service by foresters, the US Department of Agriculture by agronomists, and so forth. In many agencies there was no requirement even for consideration of alternatives: the agency simply developed project proposals to fulfill a statutory mandate, or approved or disapproved requests for grants or permits. When alternatives were considered at all, they were limited to the agency's mission and traditional expertise, with no meaningful consideration of other options that might achieve the agency's mission with less damaging environmental effects.

Administrative agencies also had little responsibility to coordinate potential conflicts with other agencies' missions. Each reported to different legislative committees, and each typically coordinated with its state-level counterparts and beneficiary constituencies, but not necessarily with other agencies whose responsibilities might be affected by their actions. Some agencies' decisions required cost-benefit analysis, yet in practice these analyses considered only direct expenditures, not broader social and environmental costs. The requirement to maximize net benefits or minimize costs also discouraged the consideration of environmental impacts that did not add to their calculations of economic benefits.

Finally, public access to information about proposed government actions was severely limited. Agencies were required to notify anyone who had a direct economic interest in a proposed action, but broader public notification was typically limited to an announcement in the local newspaper and, in some cases, a public hearing. Public access to interagency comments also was limited: these were considered private communications, even by the landmark 1966 Freedom of Information Act, supposedly to protect the agencies' candor in communicating with one another.

Concerns about these practices became acute in the 1960s, with a growing stream of news stories about the unforeseen consequences of new technologies and of uncoordinated actions by government agencies: radioactive fallout from nuclear weapons testing, ecological destruction by aerial pesticide spraying (as captured in Rachel Carson's *Silent Spring* in 1962), dams and highways proposed in parks, clear-cutting of National Forests, and many other proposals driven by narrowly defined agency missions.

Around the world, other countries and international institutions also lacked any mandate to collect systematic information on the environmental consequences of their proposals as well as any significant formal rights for public access to such information or voice in the decision-making process—let alone a right to challenge agency proposals in the courts. Some countries have long been more accepting of coordinated planning than the United States, but none had procedures for assuring systematic consideration of environmental impacts and alternatives.

The EIA as an "Action-Forcing Mechanism"

In 1963, public administration scholar Lynton Caldwell authored a seminal article proposing "environment" as a new focus for public policy. Caldwell suggested that many problems that were becoming evident resulted from the failure to consider the environment as a complex whole in designing government actions. As Caldwell (1963, 138) put it,

> Many of the worst environmental errors are direct or indirect results of segmental public decision making, of failing to perceive specific environmental situations in comprehensive environmental terms. A policy focus on environment in its fullest practicable sense would make more likely the consideration of all the major elements relevant to an environment-affecting decision. ... In contrast to some of our more primitive current methods of environmental decision making, ... [this task] would indeed be complex. But methods oversimplified in relation to the problems are not likely to produce satisfactory results.

In brief, Caldwell proposed that governments embrace environmental policy as an integrative approach to problems whose simple "solutions" have more complex consequences. Caldwell saw natural systems and ecological communities as essential foundations for all societies and their economies, and vitally interconnected with them: a precursor in all but name to the more recent concept of sustainable development. He did not explicitly state the concept of EIA in this article, yet he clearly implied it in his articulation of the need to make the more comprehensive reality of the environment more explicit and integrate this understanding into the tasks of government agencies.

US legislators adopted these ideas in an unusual, bipartisan legislative statement of principles to be included in a national environmental policy. Priorities and choices among alternatives must be planned and managed at the highest level of government; alternatives must be actively generated and widely discussed; actions should proceed only after an ecological analysis and projection of their probable effects; the social as well as natural sciences should be used to inform environmental management; and decisions on new applications of technology must consider their potential for unintended and unwanted consequences (US Congress 1968, 15–16).

In 1969, Senator Henry Jackson introduced legislation based on these principles, and hired Caldwell as a consultant to help draft it. Caldwell in turn pressed to include in the law not only a statement of policy but also specific "action-forcing" provisions to ensure that its policy statement would be implemented by the agencies (US Congress 1969a, 116; Caldwell 1976, 74). The most explicit of these provisions became the NEPA's requirement for a "detailed statement" of environmental impacts, effects, alternatives, and other considerations: the environmental impact statement (EIS), or in effect, the EIA requirement (NEPA 1970, sec. 102[2][c]).

The purpose of this requirement, Caldwell emphasized, was not to create a new administrative procedure for its own sake but rather to use an enforceable documentation requirement to assure action-by-action implementation of the law's substantive statement of environmental policy—in effect, to supplement or modify each agency's statutory mission so as to include an environmental policy mandate (US Congress 1969b, 7–9; Caldwell 1976, 74; Caldwell 1997, 25–26, 30). The NEPA was adopted by overwhelming bipartisan majorities, and signed by President Richard Nixon on January 1, 1970 (United States 1970).

The NEPA's (1970, sec. 101[a]) policy statement declared that

> it is the continuing policy of the Federal Government ... to use all practicable means and measures, including financial and technical assistance, in a manner calculated to foster and promote the general welfare, to create and maintain conditions under which man and nature can exist in productive harmony, and fulfill the social, economic, and other requirements of present and future generations of Americans.

To effectuate this policy, the NEPA (sec. 101[b]) went on to establish six environmental policy goals:

> to use all practicable means, consistent with other essential considerations of national policy, to improve and coordinate Federal plans, functions, programs, and resources to the end that the Nation may—

- fulfill the responsibilities of each generation as trustees of the environment for future generations;
- assure for all Americans safe, healthy, productive and [a]esthetically and culturally pleasing surroundings;
- attain the widest range of beneficial uses of the environment without degradation, risk to health and safety, or other undesirable and unintended consequences;
- preserve important historic, cultural, and natural aspects of our national heritage, and maintain wherever possible an environment that supports diversity and variety of individual choice;
- achieve a balance between population and resource use which will permit high standards of living and a wide sharing of life's amenities; and
- enhance the quality of renewable resources and approach the maximum attainable recycling of depletable resources.

Finally, to assure implementation of these goals, section 102 of the law directed that

> (2) all agencies of the Federal Government shall ... (C) include in every recommendation or report on proposals for legislation and other major Federal actions significantly affecting the quality of the human environment, a *detailed statement* by the responsible official on (i) the *environmental impact* of the proposed action, (ii) *any adverse environmental effects* which cannot be avoided should the proposal be implemented, (iii) *alternatives* to the proposed action, (iv) the relationship between *local short-term uses of man's environment and the maintenance and enhancement of long-term productivity*, and (v) any *irreversible and irretrievable commitments of resources* which would be involved in the proposed action should it be implemented. (emphasis added)

Several implications of this language are worth highlighting. First, the law applied not merely to large infrastructure projects but instead to all "major federal actions," including, for instance, land and water management practices, applications for federal grants or permits, and other actions that require federal approval. Second, it required documentation of environmental impacts, adverse effects, and other *consequences* of a proposed action; no longer could an agency justify its proposals simply by listing its economic costs and benefits. Third, it required identification of *alternatives*, not merely the impacts of a *single* proposed action. Finally, it required consultation with "any Federal agency which has jurisdiction by law or special expertise" with respect to any environmental impact involved, and that "copies of such statement and the comments and views of the appropriate Federal, State, and local agencies ... shall be made available to the President, the Council on Environmental Quality [CEQ] and to the public ... and shall accompany the proposal through the existing agency review processes."

In short, the NEPA created a requirement for documentation not simply of the normal justifications for a proposed action but rather the environmental impacts, adverse effects, and consequences, both of a proposed action and its alternatives, and disclosure of this information in advance to other agencies and the public. Its purpose was to ensure that actions were consistent with national environmental policy goals, and its means was a procedure for documentation and public disclosure that was intended to make the EIA mechanism self-enforcing.

Implementation, Implications, and Consequences

The NEPA thus created the legal mandate for the EIA, but its details were spelled out in guidelines and regulations issued by the US Council on Environmental Quality, which also was established by the NEPA. Each EIS was to "utilize a systematic interdisciplinary approach which will insure the integrated use of the natural and social sciences and the environmental design arts." In particular, alternative actions that would minimize adverse impact were to be explored. The phrase "significantly affecting the quality of the human environment" was to be construed "with a view to the overall, cumulative impact of the action proposed (and of further actions contemplated)." An EIS was always required if the environmental impacts of an action might be controversial. Importantly, the guidelines directed that a *draft* EIS—in effect, the underlying EIA—be made public "early enough in the agency review process before an action is taken in order to permit meaningful consideration of the environmental issues involved" (US CEQ 1970).

By 1978, some 70 US agencies had produced more than 11,000 EISs, of which 1,052 were challenged in the courts and 217 proposed actions were stopped by court orders (Clark 1997, 19). Early EISs were criticized as being too long, unreadable, and often unused; completed too late to be meaningful; and frequently inconsistent in treating important environmental considerations. In response, the CEQ issued regulations in 1978 that sought to make EISs more succinct: they were to be "analytical rather than encyclopedic," stress issues that were useful to decision makers and the public, reduce emphasis on background material, be written in plain language, and integrate the EIA with other environmental review procedures. A "scoping" procedure was to be added early in the process of project development, through which anyone interested could propose additional impacts and alternatives for consideration. The EIAs were to be completed in multiple "tiers" as an action moved from a broadly conceived program to more site-specific projects (for example, from selection of a highway corridor to approval of site-specific projects).

The 1978 regulations also defined for the first time what an "assessment" was. An EIA, they stated, is "a concise public document that (1) provides sufficient evidence and analysis for determining whether to prepare an EIS or a finding of no significant impact (FONSI), (2) aids agency compliance with NEPA when no EIS is required, and (3) facilitates preparation of an EIS when one is necessary" (US CEQ 1978). They did not specify any standard format, but they recommended a length of ten to fifteen pages including a brief discussion of the need for the proposal, alternatives, impacts of the proposal and alternatives, and a list of agencies and individuals consulted. If the impacts were significant enough to warrant a full "detailed statement" (an EIS), a more detailed EIA would then be necessary. In short, the EIA concept took two forms: an initial, preliminary analysis to determine whether or not a formal EIS was required, and a starting point for the more detailed assessment of impacts and alternatives that was required in a full EIS.

Most important, the regulations described the *alternatives* section as "the heart of the EIS" (US CEQ 1978). The purpose of the EIA concept was not simply to document what would happen once a preconceived action was approved but rather to require consideration of alternatives, and therefore to encourage environmentally preferable actions (US CEQ 1981).

Finally, the regulations required that at the end of its decision process, the agency must prepare a public record of decision, including a statement of the final decision, all alternatives considered, and whether all practicable means to minimize environmental harm had been adopted (and if not, why not). This record provided a final public document, reviewable by the courts, specifying how the information in the EIA was considered in reaching the agency's decision.

Even with these guidance documents, however, translating the EIA concept into practice required answers to many more specific questions. What was a "major Federal action"? What counted as an "impact" or "adverse effect"? What impacts should be considered "significant"? What range of alternatives should be considered? All these questions and others were left to case-by-case answers by the agencies.

The courts also heavily influenced the EIA's implementation in the United States. The earliest decisions confirmed the right of citizen groups to demand implementation of the law; others then required that reluctant agencies actually produce the statements (Anderson 1973, 77–79). Subsequent decisions addressed the adequacy of the EIA's content.

Some agencies, for instance, tried to treat EISs as merely paperwork exercises as opposed to serious assessments of impacts and alternatives. The initial EIS for a haul road to build the proposed Trans-Alaska Pipeline, for example, was just eight pages long, and described the action's "irreversible and irretrievable commitments of resources" as merely the gravel, roadbed right-of-way, and "use of 13 square miles, or 1/22,300 of this total area to establish the first all-weather secondary road connecting arctic Alaska with Yukon." One legal scholar compared this to analyzing "the loss of the *Mona Lisa* as 'a disposition of a few ounces of oil paint and 620 square inches of canvas (used)'" (Sax 1971, 98). A judge ruled that the EIS was inadequate, and subsequent decisions confirmed that the courts had the authority to decide whether the agencies' compliance with the NEPA was adequate or not, even if they had conformed to the letter of the law. In the case of the Trans-Alaska Pipeline, for instance, court rulings required the agency to consider alternatives, including an alternate route across Canada. The agency then produced a six-volume revised EIS. Lengthy as it was, the EIS itself was based on no original research, relied entirely on data provided by the pipeline company, and consisted largely of vague generalizations rather than rigorous predictions (Lindstrom and Smith 2008, 86–90).

Ultimately, however, the courts declined to decide whether an agency's eventual decision complied with the NEPA's policy statement. The courts never reviewed the revised EIS for the Trans-Alaska Pipeline: Congress intervened during the 1973 energy crisis and narrowly approved the pipeline by legislation. In 1978, the US Supreme Court overruled a different decision that would have required the Atomic Energy Commission to consider energy conservation as an alternative to a new nuclear power plant, stating that the "NEPA does set forth significant substantive goals for the Nation, but its mandate to the agencies is essentially procedural."[1] Two years later, it held specifically that once an agency has complied with the procedural requirement to produce an EIS documenting its "consideration" of environmental issues, the courts could not intrude further to question the environmental merits of the agency's actual decision.[2] In 1989, the Court went on to say that "other statutes may impose substantive environmental obligations on federal agencies, but NEPA merely prohibits uninformed, rather than unwise, agency action."[3]

These decisions broke the connection the NEPA's authors had intended: that the EIA serve as an "action-forcing mechanism" to ensure implementation of the NEPA's substantive environmental policy. Instead, the courts ruled that it would be treated as merely a procedural requirement to document impacts and alternatives as well as to inform the public. Despite this narrowing of its meaning, though, the EIA requirement even as a procedure has continued to play a crucial role in US decision proposals affecting the environment, and has been widely adopted worldwide as well.

International Diffusion

Beyond the United States, more than a hundred countries have adopted the EIA, including most developed and many developing nations as well as many international banks and businesses. The European Union, UN Environment Program, World Bank, and Organisation for Economic Co-operation and Development all adopted or recommended EIA procedures

(Sadler 1996, 25). In 1995, the World Business Council for Sustainable Development endorsed EIA as a practice to "assist companies in their quest for continuous improvement by identifying ways of maximizing profits through reducing waste and liabilities, raising productivity and demonstrating a company's sense of duty towards its customers and neighbors" (ibid., 33). Nearly eighty transnational investment banks also adopted the EIA as a due diligence screen for identifying the environmental and social impacts of investment proposals (Equator Principles Association 2013).

EIA also has been incorporated into international environmental agreements. It was identified as a key implementing mechanism at the UN Conference on Environment and Development in 1992 (Rio Declaration on the Environment and Development, principle 17), and several international conventions included EIA requirements. In recent years, it also has been used by the European Commission (2013a, 2013b) to address transboundary environmental impacts and to integrate climate change and biodiversity considerations into development project decisions.

In the course of this international diffusion, important variations have emerged.[4] In a few countries, such as Canada, the EIA was linked from the start to a substantive statement of environmental policy goals, like the original US concept, but in most it emerged solely as a procedural requirement. In many countries, the EIA was limited to large physical infrastructure projects, and did not include other government actions such as grant and permit approvals. In the original EIA concept, the requirement to consider alternatives was identified as the "heart of the EIS," and the World Bank, other development banks, and some countries also adopted this requirement (World Bank 1991). Other countries, however, did not. The EIA law of the People's Republic of China, for instance, "encourages relevant units, specialists and the public to participate in the evaluation of environmental effects in an appropriate manner" (Article 5), but it was largely limited to governmental plans and construction projects, and it required only a report on the environmental effects of a proposed project that had already been fully planned rather than the consideration of alternatives.[5]

Another major difference among international EIA practices is that the judicially enforceable public rights of access to information and to participation in decision processes, which were integral elements of the original EIA innovation, do not exist in all countries, and some countries deliberately avoided creating new opportunities for public challenge (Wood 1997,

100). On the other hand, some countries created more formal interagency approval processes than the original version of the EIA, enhancing that aspect of its "action-forcing mechanism." In China, for instance, the EIA is prepared by the agency initiating an action, but it must then be examined by a group of specialists representing other relevant departments that then submit opinions in writing; if the group's conclusions are not adopted, the approving authority must file a written explanation for the record.

Some countries even charged a separate agency to conduct the EIA. The original EIA concept required the agency initiating a proposed action to conduct the EIA in order to promote the development of expertise in each agency to consider the full range of environmental impacts. In Australia, however, the decision to conduct an EIA and the actual study are carried out by an independent agency reporting to the minister of the environment, not by the agency proposing the action. Canada's EIA studies were conducted by the independent Canadian Environmental Assessment Agency, and were then reviewed by panels of independent members appointed by the minister of the environment (Wood 1997, 101; Sadler 1996, 25). The use of an independent agency may well produce higher-quality EIAs; an important unanswered question is whether or not better statements by a separate agency lead to environmentally superior actions.

The EIA has undergone a number of positive refinements as it has continued to develop. One is the creation of formal "scoping" procedures to involve a wide range of stakeholders during early stages of project planning, allowing the public to participate in identifying impacts and alternatives for fuller analysis. Another is the identification of additional kinds of impacts that EIAs are explicitly expected to consider, such as pollution prevention, energy efficiency, cumulative impacts, environmental justice, climate change, transboundary impacts, and others.

In some cases, EIA has also been applied in strategic environmental assessments (SEAs) to address not just site-specific impacts but also more systemic, long-term, and cumulative effects of broad policies and entire programs (Partidario 1996). US EIA regulations require "programmatic" EIAs and "tiering" of them to include programs as well as their site-specific implementing actions, and there have been successful examples of programmatic EIAs for such actions as geothermal leasing, solar energy installations, and agricultural fruit fly control programs, among others (Stewart 2014). The World Bank (2011) also encourages preparation of sectoral and regional EIAs, and in 2011 issued detailed guidance on using SEAs for policy and program reform. The Organisation for Economic Co-operation and Development, UN Development Program, UN Economic Commission for Europe, and European Community also adopted policies encouraging programmatic assessments. In general, though, SEAs are less frequently conducted, and many practical questions remain about what procedures, methods, and institutional frameworks are appropriate for them (Partidario 2012; Noble 2009; Song and Glasson 2010). In December 2014, the US CEQ (2014) issued new guidelines encouraging greater use of programmatic EIAs when appropriate, and providing advice as to how to do so. Whether this will lead to increased or more effective use of them remains uncertain.

Finally, a few countries' EIA procedures require postdecision follow-up studies to monitor actual impacts, ensure the effectiveness of mitigation measures, and improve future impact assessments. Under Korea's EIA procedure, for instance, the Ministry of Environment has the responsibility to ensure implementation of the conclusions of the EIA, and in the case of at least one major project, a post-management investigation commission was charged with monthly monitoring and inspection of the implementation of environmental mitigation measures (Cha et al. 2011). Post-decision review and follow-up were not mentioned in the initial EIA concept, nor have they been common practice in most countries. Where they exist, they represent a positive extension of the EIA concept.

Beneath these variations, EIAs around the world share a set of common unifying principles, distilled and disseminated by the International Association for Impact Assessment (1999, 2009). The EIA continues to produce prior documentation of the effects of development proposals, although it does not by itself drive a more fundamental process for integrating planning toward sustainability or other environmental goals, or assure that the environmentally optimal action will be chosen. It requires the identification of effects on the natural environment as well as related social effects, not simply economic benefits and costs, or other project justifications. In some but not all applications, it requires identification and comparison of alternatives. And to varying degrees, it provides a mechanism that did not previously exist for assuring public opportunities to comment on and challenge proposed actions of governments, businesses, international development banks, and other institutions.

Significance of the EIA

Over the four decades since it was conceptualized, the EIA has become deeply embedded in both the language of environmental policy debates and procedures for project planning and decision making, worldwide as well as in the United States. How significant has the EIA been, then, as a conceptual innovation in environmental policy? Ideally, one would like to see a direct causal relationship between the findings of an EIA and the cancellation of environmentally damaging proposals, environmental improvement of proposed activities, more far-reaching indirect benefits such as broader policy and institutional improvements, and positive shifts in environmental attitudes and behavior more generally (Sadler 1996, 53). Has the EIA's documentation of environmental impacts and alternatives, and increased public disclosure of them, actually produced better decisions and actions? And has it offered any other benefits?

In some of the most controversial cases, the EIA has led to more elaborate mitigation measures and even some different project outcomes than would otherwise have been chosen. In the case of the Trans-Alaska Pipeline, for example, executives of the pipeline company later acknowledged that the EIA process resulted in a far safer and better-constructed pipeline than had originally been planned. Even so, the EIA seriously underestimated the potential risks of major oil spills: it predicted a worst-case spill of 140,000 gallons, whereas the *Exxon Valdez* disaster in 1989 spilled nearly 11 million gallons (see also Morgan 2012).

More generally, the EIA has had at least six consequences of enduring significance.

First, it has greatly expanded the documentation of both the environmental conditions that may be affected by proposed actions and the potential consequences of those actions. Almost any major development project today, and many other major actions of governments, development banks, and other institutions, is framed not simply in terms of its economic benefits and costs but also its environmental impacts. Far more information about proposed projects and their environmental impacts is made publicly available in advance, fueling more active public discussion, challenges in controversial cases, and often proposals for mitigation measures and even for better alternatives that the agency had not previously considered. No longer can government projects—and in some jurisdictions, government grants and permit approvals—be justified merely by an agency's mission and cost-benefit analysis.

Second, in many cases the EIA has compelled documentation of alternatives to a proposed action, and comparison of their environmental pros and cons, not merely the presentation of a single action proposal—information that previously was neither required nor produced. The requirement to consider alternatives has led in some situations to different actions being chosen. The magnitude and significance of this consequence is frequently overlooked or taken for granted, but in those EIA regimes where it is required, it has been at least as important as the documentation of environmental impacts, and arguably even more so.

Third, EIA has greatly increased the transparency of information on environmental impacts and alternatives of proposed actions, both to the public and independent experts, and even to other agencies and levels of government. In some countries, it also has opened this information to independent scrutiny by the courts, or review panels of other agencies or independent experts. It has thus dramatically expanded public access to information, and has made proposed actions more accountable to independent evaluation and controversy.

Fourth, the EIA has produced many adjustments of proposed actions to mitigate impacts, even if those adjustments are often localized rather than producing more fundamental changes in policies and programs. Some major proposals have actually been stopped, others modified by mitigation measures, and an unknowable number not advanced in the first place due to anticipation of controversies.

Fifth, the EIA has significantly increased environmental awareness and understanding among those who participate in it. It has substantially increased expert capacity for identifying environmental impacts, sometimes in government agencies but also in business firms using EIA, and particularly in the consulting firms that do much of the EIA and project design work (Sadler 1996; Jay et al. 2007, 94).

Finally, the EIA has inspired the proliferation of many other kinds of impact assessments as well in place of the previous near-universal reliance merely on mission- and economics-based justifications for proposed actions.

A worldwide survey of practitioners in 1996 found that in the view of the majority, the EIA is at least moderately successful in identifying appropriate mitigation measures and providing clear information to decision makers on the potential consequences of proposals. It is unsuccessful or only marginally successful in making verifiable predictions, specifying the significance of residual impacts, and offering advice to decision makers on alternatives. Nevertheless, it does create a learning process, providing important benefits beyond informing decision makers, such as the promotion of greater awareness of environmental and social concerns, upgrading of professional capabilities, and promoting public involvement in decision making (Sadler 1996, iii).

Criticisms

Despite these positive effects, EIA implementation has been controversial. It has rarely, if ever, been abolished once instituted, but there have been frequent attempts to weaken and marginalize it as well as limit its applicability.

One criticism is the charge that it has been ineffective. Law professor Oliver Houck (2009, 10648), for instance, has argued that the "NEPA is missing the point. It is producing lots of little statements on highway segments, timber sales, and other foregone conclusions; it isn't even present, much less effective, when the major decisions on a national energy policy and a national transportation policy are made." Critics such as these contend that to be effective, the EIA must be far more deeply integrated into agencies' overall planning processes—as its original authors in fact intended—rather than merely attached as additional paperwork to hundreds of site-specific projects implementing those plans. EIA expert Ray Clark (2014), similarly, has asserted that the EIA staff should be considered policy staff, attached to the agency's senior executive or manager's office, and shaping fundamental policy and program alternatives as opposed to simply the implementing actions. This of course would require active interest and consistent support by top management—a commitment whose absence has often been the real reason for limits to the EIA's effectiveness.

The lack of serious interest and support from top management is itself thus a second important concern. For the EIA to work effectively, people must want to do it well and use it to make better decisions. When they do not, it is often ignored and reduced to a burdensome, overly comprehensive paperwork requirement simply to preempt potential lawsuits. One study found that the EIA does exert some influence on development decisions, but that it was common for EIA findings to be marginalized in favor of other considerations. As its authors concluded, "Decision-makers ... were operating primarily within the parameters of the planning procedures to which they were accustomed and ... [the] EIA was generally seen as external to those procedures. ... So [the] EIA was resulting in some modification of projects, though not usually of a major kind: the overall types and scales of development were unaffected" (Jay et al. 2007, quoting Wood and Jones 1997).

Third, in many cases the EIA process has been contracted out to consulting firms as merely a documentation requirement rather than conducted within government as a planning process. The authors of the EIA concept clearly intended that it force action by all responsible agencies to diversify their staff expertise in environmental impacts, and integrate their expertise more fully into government planning and decision-making processes affecting the environment. Initially this happened, but at least in some major countries, government budgets and staff have been reduced over time, and more and more government functions have been contracted out to consulting firms—often with close business ties to project sponsors—with government employees increasingly serving only as contract managers rather than as substantive experts themselves (Clark 2014). As a result, many EIAs in practice are conducted as paperwork exercises by low-bid consulting firms, without serious review or quality control of their content unless challenged (Wright et al. 2013). And even with good quality control, consulting firms may have expertise in predicting and documenting particular types of environmental impacts, but they generally lack incentive and even authority to weigh environmental values, generate new and different alternatives, and integrate that information fully into government policy, program, and planning decisions.

Fourth, over time, many actions have been exempted from EIA requirements, either by statute, by "categorical exemptions" for whole classes of actions, or by "findings of no significant impact" (FONSIs) in particular cases. In the United States, for instance, some major actions that clearly had significant impacts were exempted by statute, such as the Trans-Alaska Pipeline during the 1973 Arab oil embargo, the REAL ID Act authorizing the secretary of the Department of Homeland Security to waive all laws preventing the construction of a security fence along the US-Mexico border, and US congressional attempts (subsequently vetoed by President Barack Obama) to override the EIA review and approve the Canada-US Keystone XL pipeline by statute. A recent review also found that some 95 percent of US EIA analyses were to justify categorical exemptions, less than 5 percent were environmental assessments, and less than 1 percent were fully documented EISs (US Government Accountability Office 2014, 8). Similarly, in the name of "streamlining," the Canadian government in 2012 radically rewrote its EIA statute, restricting its applicability to a limited list of specific project types and a narrowed list of specific types of impacts (not including, for example, climate change, or strategic assessment of policies and programs); it also eliminated language requiring discussion of alternatives and cumulative impacts, devolved most EIA requirements to lower levels of government, and made many of the remaining decisions about the scope of the EIA discretionary (Gibson 2012).

Finally, the EIA has come under increasing attack by claims that it takes too long, imposes excessive paperwork burdens, delays worthwhile projects, and produces overly long and turgid documents, rather than concise and useful decision assessments. In the United States, a wide range of government actions have now been exempted from full EIA requirements as categorical exclusions, but those full assessments that are still being written take on average at least two years, and sometimes as long as seven years or more. Many of these assessments, though, still were not models of good analyses leading to better decisions. And even many categorical exclusion justifications were becoming longer and slower. At the least, such assessments represented a failure to follow CEQ regulations on limiting the EIAs to succinct decision documents, and instead a practice of "lawyering them up" to encyclopedic levels to preempt potential lawsuits. Many project delays actually resulted from incomplete funding, local opposition, low priority, or noncompliance with other laws and regulations, not from the EIA process, but the EIA was often blamed (Trnka and Ellis 2014). Complaints against delay and paperwork burdens arose partly from a backlash against environmental advocates using delay as a weapon against projects they opposed, but partly also from project advocates using claims of delay and paperwork burdens to evade the questions that the NEPA intended the EIA to raise. The reality is that government actions affecting the environment remain contested, and while the EIA makes these contested issues and alternatives more visible, and hence more open to debate, the EIA alone cannot assure a better outcome than the political process produces.

Limitations as a Concept

In addition to these implementation issues, the EIA has some important limitations even as a concept.

First, while it was conceptualized as an "action-forcing mechanism" to motivate more integrated and proactive planning, in itself it remains an ad hoc, action-by-action procedure operating within the existing structures of agency missions rather than across them. Even with requirements for early stage "scoping" processes, the EIA process is normally triggered by a proposal originating from a particular agency mission and client constituency, as opposed to an overarching vision of a national environmental policy or sustainable development goal. If one's goal is to integrate overall policies and programs more effectively toward environmental goals, as Caldwell and the other originators of the EIA concept intended, then the EIA must be integrated into policy-making processes at a strategic level, not just in more specific project proposals once they have already begun to take shape. And there must be serious commitment and active support for this from senior policy makers and managers. The political will to do this, however, is often absent, and strategic environmental assessments have so far been rare. Despite more than forty-five years of EIAs, human pressures—climate change, fisheries deterioration, freshwater availability, and many others—still seriously impact both local and global environmental conditions, and serious commitment by senior policy makers and managers to solving these problems has often been lacking.

Second, the EIA rarely includes post-decision monitoring and mitigation, and did not even include these functions in its initial conceptualization: it functions mainly as an element of ex ante justification. Without follow-up, though, the potential to both assure achievement of environmental policy goals and improve the methods of the EIA—through the related concept of adaptive management, for instance—is seriously constrained.

Finally, the EIA continues to pose inherent theoretical and methodological challenges. Many actions produce complex mixtures of consequences, which are also influenced by other factors that cannot all be foreseen. Prediction of their environmental impacts, let alone their significance and how they would differ from other alternatives, is thus inherently difficult. And articulating their significance requires specification of appropriate spatial and temporal scales of ecological changes as well as environmental impacts (Nash 2014).

The Future of the EIA

The EIA has not, then, fully accomplished the visionary aspirations of its authors. The goal that governments would "use all practicable means and measures ... to create and maintain conditions under which man and nature can exist in productive harmony," and that an EIA requirement would compel all government agencies to conform their actions to this policy, clearly has not been achieved. Other political and economic priorities continue to exercise heavy influence. As a result, the EIA has been implemented primarily as a procedural requirement, not as a substantive mandate to affirmatively plan toward an environmentally sustainable economy, or adopt the most environmentally beneficial or sustainable alternatives.

Yet even as a procedural requirement, the EIA has had important and continuing influence, both on environmental policy outcomes and on how we even conceptualize environmental policy. It has stopped some proposals, and led to modification and mitigation of others. More broadly, it has made environmental considerations a permanent element of project planning and justification, and has greatly expanded the access of the public and even other relevant agencies, as well as environmental scientists, to information and decisions about proposed actions by governments, international funding institutions, and more than a few businesses. The limits to the EIA's effectiveness bear witness simply to the enduring influence of immediate economic and political interests, and the inherent difficulty of achieving fully integrated economic, social, and environmental planning. These forces would only have remained even more dominant, and these planning processes more fragmented, without the conceptual innovation of the EIA.

How, then, might the EIA be made more effective in the future? One incremental step would be for organizations using the EIA to insist on using it as a concise decision document among alternatives rather than an exhaustive inventory to justify a single preconceived action. Another could be to link it more explicitly to monitoring and mitigation measures. Still another would be to tie it more closely to related concepts and practices, such as adaptive management and environmental management systems. A more fundamental step would be to reaffirm environmentally sustainable development as an overarching policy criterion for approving development activities, and thus link the EIA directly with the concept—foreshadowed in the NEPA's policy statement—of sustainable development. A possible model for this might be the Dutch National Environmental Policy Plan (cf. de Jongh 1996; Netherlands 1998). The fulfillment of any of these steps, however, requires the policy commitment and political will to introduce and implement them.

In short, the EIA remains an important and influential conceptual innovation in environmental policy, even though it is too often implemented merely as a procedural requirement at the level of individual project proposals rather than integrated fully into strategic policy, program, and planning decisions.

Notes

1. Vermont Yankee Nuclear Power Corp. v. Natural Resources Defense Council Inc., 435 U.S. 519 (1978).

2. Strycker's Bay Neighborhood Council, Inc. v. Karlen, 444 U.S. 223 (1980).

3. Robertson v. Methow Valley Citizens Council, 490 U.S. 332 (1989) at 351.

4. There is now a large literature on the EIA, and an even larger one on environmental assessment practices worldwide. The most comprehensive review of the effectiveness of the EIA worldwide was conducted in 1996, cosponsored by the International Association for Impact Assessment and the Canadian Environmental Assessment Agency (Sadler 1996); a few other authors also have studied it (e.g., Wood 1997; Jay et al. 2007). For an unusually detailed comparison of the EIA in seven major countries, see Wood (2002). Much of what follows is based on these sources.

5. Law of the People's Republic of China on Evaluation of Environmental Effects. Accessed December 3, 2016, http://www.npc.gov.cn/englishnpc/Law/2007-12/06/content_1382122.htm.

References

Anderson, Frederick R. 1973. *NEPA in the Courts: A Legal Analysis of the National Environmental Policy Act*. Washington, DC: Resources for the Future.

Andrews, Richard N. L. 2006. *Managing the Environment, Managing Ourselves: A History of American Environmental Policy*. 2nd ed. New Haven, CT: Yale University Press.

Caldwell, Lynton K. 1963. "Environment: A New Focus for Public Policy?" *Public Administration Review* 23:132–139. doi:10.2307/973837.

Caldwell, Lynton K. 1976. "The National Environmental Policy Act in Retrospect and Prospect." *Workshop on the National Environmental Policy Act.* Report to the Subcommittee on Fisheries and Wildlife Conservation and the Environment of the House Committee on Merchant Marine and Fisheries, US House of Representatives, Serial No. 94-E, 69–86.

Caldwell, Lynton K. 1997. "Implementing NEPA: A Non-Technical Political Task." In *Environmental Policy and NEPA: Past, Present, and Future,* ed. Ray Clark and Larry Canter, 25–50. Boca Raton, FL: St. Lucie Press.

Cha, Yoon Jung, Myung-Pil Shim, and Seung Kyum Kim. 2011. "The Four Major Rivers Restoration Project." *UN Water International Conference: Water in the Green Economy in Practice: Towards Rio+20.* Zaragoza, Spain: UN Water. Accessed October 2, 2015, http://www.un.org/waterforlifedecade/green_economy_2011/pdf/session_8_water_planning_cases_korea.pdf.

Clark, Ray. 1997. "NEPA: The Rational Approach to Change." In *Environmental Policy and NEPA: Past, Present, and Future,* ed. Ray Clark and Larry Canter, 15–23. Boca Raton, FL: St. Lucie Press.

Clark, Ray. 2014. "Introduction: The History, the Hope, and the Reality." *Environmental Practice* 16:261–269. doi:10.1017/s1466046614000428.

de Jongh, Paul E. 1996. *The Netherlands' Approach to Environmental Policy Integration: Integrated Environmental Policy Planning as a Step towards Sustainable Development.* Washington, DC: Center for Strategic and International Studies. Accessed October 2, 2015, http://citeseerx.ist.psu.edu/viewdoc/download?doi=10.1.1.197.802&rep=rep1&type=pdf.

Equator Principles Association. 2013. *Equator Principles III.* Accessed October 2, 2015, http://www.equator-principles.com/index.php/ep3.

European Commission. 2013a. *Guidance on the Application of the Environmental Impact Assessment Procedure for Large-Scale Transboundary Projects.* Accessed October 2, 2015, http://ec.europa.eu/environment/eia/pdf/Transboundry%20EIA%20Guide.pdf.

European Commission. 2013b. *Guidance on Integrating Climate Change and Biodiversity into Environmental Impact Assessment.* Accessed October 2, 2015, http://ec.europa.eu/environment/eia/pdf/EIA%20Guidance.pdf.

Gibson, Robert B. 2012. "In Full Retreat: The Canadian Government's New Environmental Assessment Law Undoes Decades of Progress." *Impact Assessment and Project Appraisal* 30:179–188. doi:10.1080/14615517.2012.720417.

Houck, Oliver. 2009. "How'd We Get Divorced? The Curious Case of NEPA and Planning." *Environmental Law Reporter* 39 (7): 10645–10650.

International Association for Impact Assessment. 1999. *Principles of Environmental Impact Assessment Best Practice*. Accessed October 2, 2015, http://www.iaia.org/uploads/pdf/principlesEA_1.pdf.

International Association for Impact Assessment. 2009. *What Is Impact Assessment?* Accessed October 2, 2015, http://www.iaia.org/uploads/pdf/What_is_IA_web.pdf.

Jay, Stephen, Carys Jones, Paul Slinn, and Christopher Wood. 2007. "Environmental Impact Assessment: Retrospect and Prospect." *Environmental Impact Assessment Review* 27:287–300. doi:10.1016/j.eiar.2006.12.001.

Lindstrom, Matthew J., and Zachary A. Smith. 2008. *The National Environmental Policy Act: Judicial Misconstruction, Legislative Indifference, and Executive Neglect*. College Station: Texas A&M University Press.

Morgan, Richard K. 2012. "Environmental Impact Assessment: The State of the Art." *Impact Assessment and Project Appraisal* 30:5–14. doi:10.1080/14615517.2012.661557.

Nash, Harriet L. 2014. "Defining Appropriate Spatial and Temporal Scales for Ecological Impact Analysis." *Environmental Practice* 16:281–286. doi:10.1017/s1466046614000271.

National Environmental Policy Act (NEPA). 1970. "The NEPA Statute." Accessed April 11, 2017, https://energy.gov/sites/prod/files/nepapub/nepa_documents/RedDont/Req-NEPA.pdf.

Netherlands. 1998. *Third National Environmental Policy Plan*. Accessed October 2, 2015, http://www.un.org/esa/agenda21/natlinfo/action/netherla.htm.

Noble, Bram F. 2009. "Promise and Dismay: The State of Strategic Environmental Assessment Systems and Practices in Canada." *Environmental Impact Assessment Review* 29:66–75. doi:10.1016/j.eiar.2008.05.004.

Partidario, Maria do Rosario. 1996. "Strategic Environmental Assessment: Key Issues Emerging from Current Practice." *Environmental Impact Assessment Review* 16: 31–55. doi: 10.1016/0195-9255(95)00106-9.

Partidario, Maria do Rosario. 2012. *Strategic Environmental Assessment Better Practice Guide—Methodological Guidance for Strategic Thinking in SEA*. Lisbon: Portuguese Environmental Agency. Accessed October 2, 2015, http://www.iaia.org/pdf/special-publications/SEA%20Guidance%20Portugal.pdf.

Sadler, Barry. 1996. *Environmental Assessment in a Changing World: Evaluating Practice to Improve Performance*. International Association for Impact Assessment and Canadian Environmental Assessment Agency. Accessed October 2, 2015, http://www.iaia.org/pdf/EIA/EAE/EAE_10E.PDF.

Sax, Joseph L. 1971. *Defending the Environment: A Strategy for Citizen Action*. New York: Borzoi.

Song, Young-Il, and John Glasson. 2010. "A New Paradigm for Environmental Assessment (EA) in Korea." *Environmental Impact Assessment Review* 30:90–99. doi:10.1016/j.eiar.2009.05.008.

Stewart, Shannon C. 2014. "NEPA as a Tool for Strategic Analysis and Decision Making." *Environmental Practice* 16:316–322. doi:10.1017/s1466046614000337.

Trnka, Joseph, and Elizabeth Ellis. 2014. "Streamlining the National Environmental Policy Act Process." *Environmental Practice* 16:302–308. doi:10.1017/s1466046614000313.

United States. 1970. *National Environmental Policy Act of 1969*. Public Law 91–190, January 1.

US Congress. Senate. Committee on Interior and Insular Affairs and House Committee on Science and Astronautics. 1968. *Congressional White Paper on a National Policy for the Environment*. Report (Serial T), 90th Cong., 2nd sess.

US Congress. Senate. Committee on Interior and Insular Affairs. 1969a. *National Environmental Policy*. Hearing, April 16, 91st Cong., 1st sess.

US Congress. Senate. 1969b. *National Environmental Policy Act of 1969*. Report on S. 1075, no. 91–296, July 9, 91st Cong., 1st sess.

US Council on Environmental Quality (CEQ). 1970. "Statements on Proposed Federal Actions Affecting the Environment: Interim Guidelines." *Federal Register* 35 (April 30): 7391.

US Council on Environmental Quality (CEQ). 1978. "NEPA Regulations." *Federal Register* 43 (November 28): 55990.

US Council on Environmental Quality (CEQ). 1981. "Forty Most Asked Questions concerning CEQ's National Environmental Policy Act Regulations." *Federal Register* 46: 18026, March 23. Accessed October 2, 2015, http://energy.gov/sites/prod/files/G-CEQ-40Questions.pdf.

US Council on Environmental Quality (CEQ). 2014. *Effective Use of Programmatic NEPA Reviews*. Accessed October 2, 2015, https://ceq.doe.gov/current_developments/docs/Effective_Use_of_Programmatic_NEPA_Reviews_Final_Dec2014_searchable.pdf.

US Government Accountability Office. 2014. *National Environmental Policy Act: Little Information Exists on NEPA Analyses*. Report No. GAO-14–370. Accessed October 2, 2015, http://www.gao.gov/products/GAO-14-370.

Wood, Christopher. 1997. "What Has NEPA Wrought Abroad?" In *Environmental Policy and NEPA: Past, Present, and Future*, ed. Ray Clark and Larry Canter, 99–111. Boca Raton, FL: St. Lucie Press.

Wood, Christopher. 2002. *Environmental Impact Assessment: A Comparative Review*. 2nd ed. New York: Routledge.

Wood, Christopher, and Carys Jones. 1997. "The Effect of Environmental Assessment on UK Local Planning Authority Decisions." *Urban Studies* 34:1237–57.

World Bank. 1991. *Environmental Assessment Sourcebook*. Washington, DC: World Bank.

World Bank. 2011. *Strategic Environmental Assessment in Policy and Sector Reform*. Washington, DC: World Bank.

Wright, Andrew J., Sarah J. Dolman, Michael Jasny, E.C.M. Parsons, Doris Schiedek, and Sharon B. Young. 2013. "Myth and Momentum: A Critique of Environmental Impact Assessments." *Journal of Environmental Protection* 4:72–77. doi:10.4236/jep.2013.48a2009.

5 Environmental Risk: New Approaches Needed to Address Twenty-First Century Challenges

Michael E. Kraft

The concept of environmental risk is pervasive in contemporary environmental policy. There are legions of experts, within and outside government, who conduct, communicate, interpret, and debate studies designed to influence decisions on the myriad environmental risks faced today, from the health effects of air pollution to the impacts of climate change. This chapter focuses on the key conceptual innovations that lie behind these ideas about environmental risk, the practical implications of using environmental risk assessment and risk evaluation in governmental decision making, and the role that environmental risk concepts may play in the future.

The most striking innovation in environmental risk concerns our greatly improved capacity to identify and measure risks, and apply the ideas and methods in new ways to *the environment*—that is, human exposure to toxic chemicals in the air, water, or land, or the food we eat, and also the ecological impacts of human activity. Equally important are the decisions by policy makers, generally beginning in the 1970s, to incorporate established risk concepts into environmental legislation and then the administrative processes that agencies use to set acceptable levels of risk. These twin advancements in the ideas and methodology of environmental risk analysis and public policy processes have sparked considerable debate over the level of safety that society seeks, the costs of achieving it, and how best to balance often-conflicting values in policy decisions. The concept of environmental risk was appealing to key policy makers by the late 1970s and early 1980s, in part because they sought a professional and scientific way to respond to emerging political demands as well as legitimize both the new environmental policies and their implementation in the US Environmental Protection Agency (EPA) along with other federal and state agencies.

Environmental risk is likely to continue to be one of the most important concepts in policy for decades to come. At the same time, it is imperative to find other concepts, methods, and decision-making processes that can help to improve critical choices for the next generation of environmental challenges. These include decision making at the local level to foster sustainable communities as well as national and international responses to global threats such as climate change, loss of biological diversity, and the impacts on environmental systems of feeding and otherwise providing for the needs of over 9.5 billion people destined to inhabit the earth by the mid-twenty-first century. All these decisions require judgments about environmental risks and how to reduce them, and also involve choices about the relative roles of technical experts, the public, and policy makers in decision making that is both highly complex and often deeply contentious.

Environmental Risk: Origins and Evolution

Many analysts trace the origins of risk concepts to developments in the twentieth century. These include the use of probabilistic calculations in engineering and business, concern after World War II over the rapid growth of civilian nuclear power and exposure to radiation, and various aspects of defense and national security, including threats of nuclear war.

More directly related to the concept of environmental risk, concern also rose rapidly in the 1960s over the presence of environmental toxins, such as pesticides, mercury, lead, and other chemicals. These newly discovered risks were a central focus of public policy by the 1960s thanks to the efforts of policy entrepreneurs such as Rachel Carson, Barry Commoner, and other scientists who sought to alert the public and spur governmental action. Eventually the concept of environmental risk became a core component of environmental protection policy in the United States and other developed nations, especially during the 1970s and 1980s (Davies 1996; Fischhoff et al. 1981; Vogel 2012; Wiener et al. 2010).

The chapter returns later to the practical use of environmental risk concepts in governmental decision making. First it tries to clarify the nature of the conceptual innovation through a brief historical review of how the concept of risk evolved, was then linked with the environment, and eventually was incorporated into environmental decision making. The origin of the concepts of risk in general and environmental risk in particular is much older and varied than most students of environmental policy typically acknowledge.

The History of Risk and Risk Analysis

As historians of risk analysis remind us, humans have long dealt with problems of risk and uncertainty, and engaged in different forms of risk analysis to deal with natural disasters such as floods and famine, among other challenges. By the seventeenth and eighteenth centuries, early writing about probability theory appeared, as did life expectancy tables. Both set the stage for later conceptual and practical developments about risk. As Vincent T. Covello and Jeryl Mumpower (1986, 37) put it, modern risk analysis "has its twin roots in mathematical theories of probability, and in scientific methods for identifying causal links between adverse health effects and different types of hazardous activities."

Early analyses of risk were primitive by contemporary standards, and relied largely on personal observation rather than systematic data collection and assessment of possible causes of the phenomenon of concern. At the time, there were few scientific models of how biological, chemical, and physical processes might produce adverse effects, and no instruments capable of measuring them.

Nonetheless, governments sought to limit the impact of epidemic diseases, control pollution (especially in crowded cities), promote food safety, ensure the safety of buildings, enhance transportation safety, and prevent occupational injuries, among other actions, all forerunners of modern risk analysis and risk management. The more elaborate of such activities are relatively recent, dating from the eighteenth and nineteenth centuries, and becoming more common after the beginning of the Industrial Revolution during the nineteenth century, raising the level of public health and safety risks to new highs (Covello and Mumpower 1986).

Modern Risks

Despite these many early efforts to identify and respond to societal risks, the twentieth century sparked a quite-different reaction to risk and brought new pressures on governments to regulate technological risks—that is, risks that are derived from the use of modern technologies, such as electric power generation, chemical manufacturing, and intensive agricultural processes. This change can be linked to the nature of modern risks and new public attitudes about both the acceptability of such risks and legitimacy of governmental intervention in a free market economy to reduce or mitigate them.

These new risks were, as Covello and Mumpower (1986, 49) observed, "fundamentally different in both character and magnitude from those encountered in the past." Where the old risks were communicable diseases, infections, and food scarcity, for example, the new ones were unseen but highly feared exposure to radiation and toxic chemicals that affected people through the environment, such as through the air, water, and land (Dunlap et al. 1993; Fischhoff et al. 1981; Slovic 1987).

These risks included many of the conditions at which environmental policy has been directed, such as the accumulation of radioactive wastes from civilian nuclear power, widespread use of pesticides and industrial chemicals that enter the air and water on which we depend, unsafe drinking water, exposure to hazardous wastes, depletion of the stratospheric ozone layer, nuclear power plant accidents, and more recently, climate change. Innovations in the concept of environmental risk stimulated a great deal of new study and commentary about such risks, and improvements in methods for analyzing and comparing their impacts on public health as well as the environment. In these ways, the use of environmental risk helped to provide the public and policy makers with essential, if necessarily incomplete, information for developing public policy responses.

Conceptual Innovations in Environmental Risk and Their Effects

As this brief historical review suggests, the central innovations related to the concept of environmental risk lie in two interrelated developments. One is the joining of the concept of risk with the new idea of the environment, in combination with the greatly improved capacity to identify, measure, and manage the new risks through the use of scientific methods of data collection and analysis. These technical and managerial capacities eventually sparked widespread societal debate over what constitutes an acceptable level of environmental risk, and what forms of government involvement, such as standard setting and regulation, are justifiable in terms of the costs or burdens placed on businesses or the economy in the aggregate.

The second conceptual development lies in the way public officials came to rely on risk analysis as a means to engage that debate and try to depoliticize what had become by the 1980s an increasingly contentious environmental policy, particularly within the United States. Risk assessment and cost-benefit analysis became central tools in the effort to introduce more balance into environmental decision making than was evident in the 1960s and 1970s.

One good indicator of the second development can be seen in the posture of the EPA. By 1984, William Ruckelshaus, serving his second term as EPA administrator, officially endorsed "risk assessment and risk management" as the primary framework for the agency's decision making (Andrews 2006). Three years later, a major US EPA (1987) report on risk management in the agency noted that the "fundamental mission of the Environmental Protection Agency is to reduce risks."[1] That is, the EPA would not try to eliminate risks, in effect a zero-risk condition, but rather reduce them to a level that was acceptable in terms of the costs imposed on society. Moreover, these kinds of decisions were defended as the professional work of scientists and other experts in the agency as a way of responding to critics who often asserted that the agency's staff was merely advancing its own political agenda.

The view of risk management that Ruckelshaus and others offered differs from the more absolutist way of thinking about environmental policy goals that prevailed during the 1960s and 1970s. The US Clean Air Act Amendments of 1970, for example, called on the EPA to protect the public from "any known or anticipated adverse effects" associated with the major air pollutants, and the agency was not to consider costs of attainment in setting the standards. The US Clean Water Act, approved in 1972, set deadlines for the *elimination* of discharges of pollutants into navigable waters by 1985, and mandated that all the nation's waters were to be "fishable and swimmable" by 1983.

Much the same pattern developed in other areas of public health and safety. During the 1960s and 1970s, the US Congress and states approved a wide array of policies to protect occupational safety, consumer product safety, food and drug safety, vehicle safety, and public health (Harris and Milkis 1996). These actions were possible because at the time, trust in business was at a low point and public confidence in the capacity of government to take effective policy action was high. Other developed nations recognized much the same kinds of risks and took similar actions to regulate them in areas as diverse as genetically modified foods, nuclear power, automobile emissions, smoking, and new drug approval (Harrison and Hoberg 1994; Munton 1996; Vogel 2012; Wiener 2003; Wiener et al. 2010).[2]

By the 1980s, the political climate had changed, especially in the United States. Business interests and their supporters became far more active and effective in limiting the scope of such laws, in part through calls for balancing the costs of governmental actions against the popular goals of environmental protection, consumer protection, and public health and safety (Kraft and Kamieniecki 2007; Layzer 2012; Vig and Kraft 1984, 2016). The use of risk analysis and its associated methodology of cost-benefit analysis increasingly gained support, both inside government agencies and among critics of environmental policy (Stone 2012; Wildavsky 1988).

The impacts of these conceptual and public policy developments, as we will see later, were both positive and negative. Among the positive effects were many new efforts to measure and compare diverse risks, and set priorities for action in recognition of both limited agency resources and relentless criticism of the economic costs of regulation (Davies 1996; Fiorino 1990; Lave 1982). Even here, however, the long-term effects of such actions, as Daniel J. Fiorino (2014, 743) has noted, "did not match the high expectations that emerged around the national, regional, state, and local projects."

Somewhat more positively, public dialogue over societal risks was aided strongly by the development of risk assessment and communication methods. Yet here too the effects are mixed. Consensus grew in the scientific community over how to measure, compare, and discuss modern risks, such as those posed by nuclear power, radioactive wastes, and toxic chemicals (Davies 1996). Communicating about risks nevertheless became fraught with practical challenges even when risk professionals believed their assessments were well grounded in science (Dunlap et al. 1993; Stern and Fineberg 1996; Slovic 1987, 1993).

Part of the problem lies in the public understanding of risk. The public frequently interprets risks in a way that can differ greatly from scientific assessments. This in turn has created many conflicts over facility siting, environmental regulatory actions, and disclosure of information about chemicals released to the environment by industrial manufacturing, oil and gas drilling, and similar activities (Kraft 2014; Kraft et al. 2011). These conflicts are evident today in both Europe and the United States in disputes over genetically modified food and the use of nuclear power, where the public tends to be more worried about such risks than are scientists.

Among the negative effects of relying on the concept of environmental risk was an increased polarization over environmental policy, particularly over what public policies should seek to achieve, and what kinds of evidence ought to be considered in balancing the protection of health and the environment against the advancement of economic development. Environmentalists argued that the use of risk analysis raised the barrier for demonstrating the harmful effects of pollution, and made it easier for industry to challenge agency decision making. Indeed, by the 1990s, opponents of environmental regulation in the United States frequently sought to enact "regulatory reform" legislation that would make it even more difficult for the EPA and other agencies to approve new regulations by facilitating legal challenges to the science underlying those rules (Kraft 2016).

For these and other reasons, Ruckelshaus's goal of putting risk analysis and risk management at the center of EPA decision making ultimately fared poorly in depoliticizing the process. It also failed to convince EPA critics that the agency was merely exercising its professional and expert judgment as it set environmental standards. If anything, criticism of the agency, especially by business, conservative ideological groups, and Republican lawmakers, escalated after the 1980s (Layzer 2012; Vig and Kraft 2016).

Political Reactions to Risk Regulation Policies

As noted, the new environmental policies adopted in the 1960s and 1970s, particularly within the United States, were not universally praised, and some of their costs soon became the object of political rallying cries by conservatives and the business community against what they saw as expanding, unjustified governmental regulation. By the late 1970s, some initial reactions set in, and a full countermovement took root in the 1980s during the Reagan administration (Kraft and Kamieniecki 2007; Vig and Kraft 1984). Such critics also called into question the accuracy of many risk assessments conducted or promoted by environmental agencies (Layzer 2012; Vig and Kraft 2016). Conservative groups raised such objections even though in general they supported the use of risk assessment as one way to reduce regulation of what they characterized as minor risks through careful measurement and balancing of risk, costs, and benefits (Lave 1982; Wildavsky 1988).

In contrast, environmentalists often have been skeptical about both risk assessment and cost-benefit analysis, believing that they were more likely to be used to weaken environmental protection than to strengthen it, particularly in Republican administrations. They feared that many benefits of regulation, such as improved public health or reduced risks of climate change, might not be measured accurately or given much weight in agency decision making. They also have complained that formal requirements for conducting comprehensive risk assessments and cost-benefit analysis before taking any major regulatory action have hampered regulators' ability to deal with issues, and especially when agency budgets and staffing have been constrained (Vig and Kraft 2016). Thus, they say, regulators have not been able to keep pace with the multiplicity of environmental challenges they face.

Despite such general doubts about both the concept of risk assessment and methods, environmentalists have strongly supported increased regulation based on such risk assessment. The EPA's Clean Power Plan of 2015 is a case in point. Environmentalists and public health groups praised the agency's analysis of the risk of climate change and economic benefits of regulating existing coal-fired power plants (responsible for about 40 percent of US carbon emissions), whereas industry and political conservatives strongly criticized the agency's analysis and new regulatory rules that the Obama administration put forth (Davenport and Harris 2015).[3]

The antiregulatory movement of the past four decades has succeeded to some extent in changing the political landscape as proponents of strong environmental policy increasingly have struggled to defend existing regulatory policies. While public opinion surveys from the 1980s to the 2010s show continuing public concern about environmental problems and a desire to maintain protective regulatory action, they also indicate considerable uncertainty about the severity of those problems and rising apprehension over the economic costs of action (Daniels et al. 2013).

These subtle changes in public attitudes may help to explain an intriguing shift in emphasis in the United States in comparison to Europe. The United States has moved from a cautious approach to risk acceptance and emphasis on regulatory stringency in the initial wave of legislation adopted during the 1970s and 1980s to a position today where it is more tolerant of environmental risks than are most European countries, whose policy actions are more firmly rooted in the precautionary principle. Yet the level of risk that is sought varies from one problem area to another, too. The United States, for example, has been more cautious about exposure to tobacco smoke than has Europe, but less cautious about exposure to toxic chemicals (Vogel 2012; Wiener et al. 2010).

The Innovative Nature of Environmental Risk

Environmental risk refers to threats posed to human health and the natural environment, the causes of which are human activities, such as industrial manufacturing, energy generation, and energy and chemically intensive agriculture. The environment is the vehicle through which humans are exposed to the by-products of such activities.

As noted earlier, the conceptual innovations in environmental risk lie in the linking of the idea of risk and the environment, and also to our improved capacity to identify, measure, and manage these environmental risks (National Research Council 1983). The innovations include the way in which public officials sought to employ risk analysis to try to depoliticize and legitimate environmental policy that had become a lightning rod for conservative critics and the business community. Some explanation of how risk is defined and measured helps to judge the nature of these innovations, some of their effects beyond those already discussed, and how suitable the concepts and methods might be in the future.

Risk and Risk Assessment Explained

Environmental risk, much like other core concepts in environmental policy, is both simple and complex. At its simplest, risk refers to a fairly intuitive concept that most people can understand. Risk is about threats or dangers people face, from automobile accidents and air pollution to climate change. Risk professionals tend to define and measure risk as a function of the probability of occurrence of certain events along with the consequences or harms created by them. This involves two related actions: risk identification and risk assessment. Risk identification entails the recognition that a given set of conditions may pose a risk to society and therefore requires some kind of characterization. Risk assessment is more complex, and seeks to measure or estimate the extent or magnitude of the risk. The harm might be injury, disease, death, degraded environmental quality, or adverse economic impacts.

A long-standing assumption is that, as Ruckelshaus (1983) put it, "there is an objective way to assess risk" even if there is "no purely objective way to manage it." Hence, one characteristic of the conceptual innovation is a belief that there is a technical process for measuring or assessing risk that merits serious consideration as well as acceptance by others precisely because it is "objective" or scientific. As noted, Ruckelshaus sought to deflect political criticism of the agency by portraying its work as professional and scientific, and risk analysis was one set of methods that he thought might have that effect.

Despite this belief, careful reviews of how such risk assessments are conducted in practice highlight the complexities of the technical tasks, and the many choices that experts must make that involve some mix of science and policy considerations under conditions of uncertainty. Indeed, the National Academy of Sciences found about fifty areas in a typical risk assessment that involve some judgments of this kind (National Research Council 1983). As Lester Lave (1987, 294) remarked in the 1980s, "Current risk estimates are fraught with uncertainty."

Thus, despite the effort to characterize risk assessment as purely technical, scientific, or objective, the reality is that we have an inherently limited understanding of complex systems and the risks they present to society (Perrow 1984). Moreover, as Andy Stirling (2010, 1029) observes, "An overly narrow focus on risk is an inadequate response to incomplete knowledge," in part because a wide range of uncertainty often surrounds many risk assessments, and that uncertainty may not be conveyed well to the public and policy makers.

The problems can be even greater when applying risk concepts to environmental systems since frequently they are not as well defined or bounded as are buildings, bridges, highways, and nuclear power plants. Knowledge of how such environmental or ecological systems work remains quite limited, as does our capacity to measure the impact of human activities on them. This is evident, for example, when we try to calculate the impact of human actions, such as deforestation or the burning of fossil fuels, on global climate systems. In addition to a reliance on abstract modeling of such systems, allowance must be made for varying future conditions; this compounds the uncertainty, because they cannot be known precisely.

Significant disagreement also exists among risk experts, and often those who are presumed to have expert knowledge of a risky technology have been proven wrong. Sometimes they were excessively confident in the reliability of their methods, such as in the safety of the Fukushima Daiichi nuclear power plant in Japan that suffered a catastrophic accident because backup systems were disabled by a tsunami. In some cases, criticism from scientists external to the decision-making body was not available or was ignored as risk analysts produced faulty estimates (Freudenburg 1988).

These general obstacles to conducting reliable risk analyses apply just as much, if not more so, to environmental risks. Recall that risk analysis as a set of methods is derived from the study of physical systems and likelihood of failure, such as building collapse. But can the same highly formalistic engineering methods be applied to complex ecological systems or even the seemingly simple matter of determining the effect of, say, fine air pollution particles on human health?

Frances Lynn (1986, 14) reminds us of another constraint. There is a link, she says, "between political values, place of employment, and scientific beliefs." She found that scientists employed by industry "tended to be politically and socially more conservative than government and university scientists." They actually chose scientific assumptions that "decreased the likelihood that a substance would be deemed a risk to human health and increased the likelihood that a higher level of exposure would be accepted as safe." Even though risk experts have organized themselves into professional organizations such as the Society for Risk Analysis and founded scientific journals such as *Risk Analysis*, such differing values and perspectives are likely to continue, and they open the process of risk assessment to dispute.

Risk Evaluation and Management

Another part of the conceptual innovation is what happens following such an assessment. The second stage of the risk assessment process is one of judging the acceptability of risks, with or without governmental intervention. This stage often is called risk evaluation, although it may not be an entirely separate analytic process. It may be embedded in the way governments choose to act on a specific risk, which in turn may reflect prevailing social values and tolerances for risks, or simply the way in which risk professionals are accustomed to operating within a specific institutional or bureaucratic culture, such as the Nuclear Regulatory Commission or EPA (Kraft 1986; Lynn 1986; National Research Council 1983).

Who decides whether a given level of risk is acceptable? In daily life, all of us do. For environmental risks that make it onto societal or governmental agendas, society debates the severity of the problems and what ought to be done about them. Policy makers introduce and debate legislation, and may also commission studies and hold hearings to gain knowledge of the latest scientific estimates as well as the consequences of taking action on the problem. Organized interest groups and think tanks might conduct their own studies, and advance their arguments in any number of venues to try to influence the decisions.

Eventually a formal statute may be approved that renders a verdict on what government will do to address the risk, such as acting on climate change or releases of mercury to the air from coal-fired power plants. Those decisions are then subject to revision at any point should the evidence or societal preferences change. The term *risk management* is used to refer to the decisions that governments or other organizations ultimately make about taking action along with the particular actions that they do take, such as adopting and implementing environmental, health, and safety policies and regulations.

The assessment and management of risk as conceived in this way have long been used in industry and governments when dealing with estimates of accidents or fault, from seismic safety to building collapse, automobile accidents, airline accidents, industrial plant accidents, and damage from storms (Perrow 1984). What was new in the concept of environmental risk was the extension of these concepts and practices to the environment. Yet as noted just above, the concepts and methods of risk analysis may be far more problematic when applied to various kinds of environmental risk.

Other Limitations of the Concept of Environmental Risk

Much of the technical literature and practice on risk assessment is devoted to the identification and measurement of risk. For example, is a specific chemical linked in some way to health effects, and if so, what concentration or magnitude of the chemical to which humans are exposed (mercury, lead, or a certain pesticide) leads to identifiable adverse consequences?

Even the best risk assessments can be faulted on methodological grounds, particularly when such studies are combined with economic estimates of costs and benefits of risk reduction. The costs and benefits of improving health are always difficult to estimate, and many critics would further object on ethical grounds to representing human health, lives, and ecosystem functioning by somewhat-arbitrary economic values or market considerations.

In 2005, for example, the US Government Accountability Office (2005) identified four significant shortcomings in the EPA's economic analysis of its proposed options for controlling the release of mercury, including a failure to estimate "the value of the health benefits directly related to decreased mercury emissions." The agency concluded that those weaknesses limited the usefulness of the analysis for policy makers. Studies of regulatory decision making have documented these kinds of methodological limitations, and they have long been acknowledged by the regulatory officials, who have been quite open about weaknesses inherent in quantitative risk analysis (Hadden 1984; Kraft 1986; Lave 1982).

Public versus Expert Views of Risk

In addition to these kinds of concerns over limited or unreliable data and the methodology of risk assessment, society faces the long-standing recognition that risk professionals understand and measure risk in a way that differs significantly from the way ordinary people do. Briefly stated, risk professionals tend to define risk in a way that reflects technical and often-quantitative judgments based on the evidence for a given situation, whereas most nontechnical people give much greater weight to what are usually called social, cultural, or community factors (Hadden 1991). These factors can change over time, and affect judgments about whether governments should take action on environmental risks or not, or the extent of governmental involvement.

Since governmental policies and practices are expected to be responsive to the values and preferences of the broad public, potentially there is a conflict. At heart, this is the essence of two well-known EPA studies: *Unfinished Business* (US EPA 1987) and *Reducing Risk* (US EPA 1990). What the agency found, not surprisingly, was that the US public worries a great deal about some environmental problems, such as hazardous waste sites and groundwater contamination, which the EPA accords a relatively low rating of severity. The public exhibits far less concern or sense of urgency about other problems that the agency rates high in severity, including the loss of biodiversity, ozone depletion, climate change, and indoor air pollution.

The same differences in assessments of these varied risks seem to be evident in the 2010s. For instance, the recent use of shale fracturing has produced enormous quantities of gas and oil, lowered their costs (especially for natural gas), and strengthened the economies of many states. The general public has been supportive of the process, yet in the last several years, opposition to fracturing has risen sharply, and surveys now show a public desire to regulate the industry and at a minimum require disclosure of the chemicals used in fracturing because of their potential risk to public health (Brown et al. 2013). The industry, not surprisingly, continues to insist that any risk from the chemicals is immeasurably low and regulation is unnecessary.

Practical Implications of Environmental Risk

As made clear in the section above, the practical implications of the concept of environmental risk are found in the decision-making processes of governmental agencies. In the United States, environmental policy statutes and executive orders in effect mandate the use of risk analysis as well as cost-benefit analysis as part of the process of developing environmental quality or emissions standards. These standards reflect the maximum level of exposure to environmental risks such as toxic chemicals that is deemed to be acceptable as specified by the statutory language and other applicable decision criteria.

Balancing Risks and Costs in Decision Making

Statutory language is frequently explicit about risk regulation, but more often than not the laws require that administrative agencies seek to balance a safe level of risk with the costs that accompany the actions they propose to take. The US Safe Drinking Act of 1974, for example, states that it is intended to "assure that the public is provided with safe drinking water." The basis for EPA decision making is put as the protection of the public's health to the extent feasible, taking costs into consideration.

Even if the statute does not explicitly call for such balancing of risk reduction with the costs of action, the political reality, as discussed earlier, is that the EPA and other government agencies, including state agencies, need to take the economic and societal impacts of new standards and rules into account. Otherwise they would face intense objection from both industry and elected officials. It is a rare EPA rule proposal that is not accompanied by elaborate statistics on the number of lives saved, illnesses prevented, hospital admissions averted, and lost workdays avoided. All this is measured in the dollar value of health benefits in relation to the dollar value of compliance costs, typically for industry. Operating under executive orders that have been in effect in every presidential administration since Ronald Reagan's, the Office of Information and Regulatory Affairs, located in the White House Office of Management and Budget, reviews all major agency actions, including those from the EPA, and issues an annual report documenting these kinds of economic analyses.

Administrative Procedures, Legal Constraints, and Methodological Limitations

In addition to the requisite economic and risk analyses that accompany new rules, the administrative processes that agencies in the United States must follow provide extensive opportunities for stakeholders, and particularly business groups, to submit evidence and bring pressure to bear on agencies as they make decisions of this kind (Andrews 2016; Kraft and Kamieniecki 2007). If displeased with the final decision, one party or another may take the agency to court to try to have its decisions altered. This possibility stimulates the agency staff to exercise great caution in how it measures and acts on risks as well as to consider how stakeholders are likely to respond (Rinfret and Furlong 2013). One result is that an agency often combines a new standard and rules with considerable flexibility and accommodation for industry or the states to meet their obligations. The EPA did just that in its controversial 2014 proposal to regulate greenhouse gas emissions from coal-fired power plants (Davenport and Baker 2014).

It should be said that the EPA and other agencies are limited in how they can use risk analysis as a basis for decision making by some peculiarities of US legal history and administrative procedures. The Administrative Procedure Act of 1946 governs the rule-making process for the EPA and other independent or executive agencies. In effect, the act requires agencies to keep the public informed about new rules, provide for public involvement in the rule-making process (usually via mandatory public notice and comment periods), and establish standards for formal rule making and adjudication.

As a result, any major proposal is accompanied by the requisite scientific, economic, and other data; published notices in the *Federal Register* inviting public comment; review by the Office of Information and Regulatory Affairs; and publication of a final rule, accompanied by the agency's response to the major issues that are raised during the public participation stage. A roughly analogous process is found at the state level. In addition, generally the burden of proof in US regulation lies with the agencies, not the parties to be regulated. That is, US environmental laws typically force the government to prove that a chemical or practice is risky rather than to put that burden on industry. These procedural requirements and legal constraints pose significant barriers to the use of risk analysis in regulation.

European approaches to regulation differ from those in the United States, and are more likely to emphasize reliance on the precautionary principle, where protection of human health and the environment are considered to be too important to wait for scientific certainty before taking action. Furthermore, the burden of proof—for example, about a chemical's public health risk—may be placed on industry instead of government. The precautionary principle is particularly crucial when risk assessments involve a high degree of uncertainty or where little confidence exists in industry's own assessments. As is often the case, the US political system tends to err on the side of protecting industry from excessive regulation and the costs that may come with it, especially, although not exclusively, in Republican administrations.

There are other problems as well. For instance, historically scientists have examined risks on a chemical-by-chemical basis, such as the effects of arsenic in drinking water or lead in the air on public health. Yet the health effects of exposure to pollutants can be interactive and synergistic, and different methods are needed to capture such effects, if even possible. Similarly, ecological risk analysis, a much newer set of methods, is needed to provide reliable information about potential damage to ecosystems. Thus, regulatory decisions depend critically on improving environmental science and bringing it to bear more effectively on policy decisions.

The Role of the States in a Federal System

Compared to other nations, the US political system has a robust federal structure that gives the fifty states great autonomy in environmental policy.

As a result, some states go well beyond the federal government in addressing environmental problems and using risk analysis. The issue of comparative risk analysis has been favored by the federal government, although little headway in comparing and prioritizing risks is evident despite two EPA reports encouraging such an action (Davies 1996; Fiorino 2014). Yet scientists praised a 1994 report prepared for the California EPA for its careful review of evidence on dozens of environmental hazards and the risks posed by each. The two-year study by a hundred scientists was hailed especially for its careful explanation of methodologies, data sources, and assumptions behind the risk assessments (Stone 1994).

Additionally, many states have sought to do more than the federal government in addressing risks. As Barry Rabe (2016, 38) reports, thirty-four states "have adopted laws that move beyond federal standards in preventing risks from chemical exposure, such as bans of specific chemicals thought to pose health risks or comprehensive chemical management systems." Minnesota in particular has an impressive program that requires firms to submit toxic pollution prevention plans every year and emphasize treatment of "chemicals of concern." Along with other states, Minnesota established a multidisciplinary team to analyze environmental threats posed by "emerging contaminants" and respond quickly to them, including risks from nanotechnology. California, as is well known, has been the nation's leader in addressing the risks of climate change. Often these kinds of state initiatives lead eventually to federal action that builds on the states' activities.

Finally, it is worth noting that both the states and federal government increasingly rely on information disclosure rather than, or in addition to, regulation as a means of reducing risk (Hadden 1989; Kraft et al. 2011). Even before a 1986 federal law, the Emergency Planning and Community Right to Know Act, which created the Toxics Release Inventory program, similar right-to-know laws began appearing at the state and local levels. By the 2010s, states across the nation started enacting right-to-know laws about natural gas fracturing, sometimes in combination with regulation, as a response to rising public concern about the chemicals used in the fracturing process and their possible contamination of drinking water sources, among other risks (Kraft 2014).

New Approaches Needed for Third-Generation Environmental Challenges

One common way of distinguishing environmental policy changes in the United States over the past four decades—a way also applicable to other nations—is to posit three overlapping eras or epochs. These epochs involve regulating for environmental protection in the 1970s, seeking efficiency-based regulatory reform and flexibility in the 1980s and after, and moving toward sustainability or sustainable development beginning in the 1990s. Each is characterized by a distinctive way of defining environmental problems, the policy goals being sought, the use of certain concepts, implementation strategies, and policy tools, information and data management needs, and a prevailing political and institutional context. The new epochs bring new ideas and practices, but they also overlay and never entirely replace the previous epochs (Mazmanian and Kraft 2009).

The role of risk along with how it is conceived, measured, and incorporated into decision making is evident in all three epochs. The third or sustainability epoch, however, calls for more integrated and comprehensive assessment of environmental problems, and that suggests a more limited role for the concept of risk as well as the methods of risk analysis than was the case during the 1970s and 1980s. Using this epoch's framework allows some modest anticipation about how environmental risk and risk analysis might be used in the future, and what other analytic needs will arise to deal with evolving environmental problems.

Risk Management in the 1970s and 1980s

As this scheme suggests, environmental risk as a concept and risk analysis as a set of methods to study risk fit well within the first epoch of the 1970s when solving environmental problems, such as air and water pollution as well as the control of toxic chemicals, was assumed to be a relatively simple matter of identifying specific chemicals, measuring the magnitude of the threat to the environment and public health, and choosing an appropriate method of regulating the risk. This approach continued into the 1980s and later years, and also fit well within the prevailing ideas of the second epoch. As discussed earlier, this is when another dimension of the conceptual innovation became evident as the EPA and other agencies sought to depoliticize environmental management, and characterize it as more professional and scientific than political in nature. Such efforts fit well into the portrayal of decision making at this time as one where policy makers sought to balance risk reduction with the costs of action—essentially an efficiency-based policy reform. What was advanced here was the integration of risk analysis and cost-benefit analysis, and the idea of comparing and ranking risks according to their severity and priority for action (Davies 1996; Fiorino 2014).

Some critics argued, of course, that the enterprise of risk analysis was inherently limited by the data that were available, and the imperfect methods that existed for measuring risks to human health and the environment. Hence, from the 1980s to the 2010s, and particularly within the United States, there have been concerted efforts to pull back from what critics saw as overly intrusive or costly environmental protection regulation based on insufficient evidence (Vig and Kraft 2016; Rabe 2016). European nations have followed a somewhat different path, choosing to be more cautious in setting risk protection levels (Wiener et al. 2010; Vogel 2012).

Rethinking Environmental Risk for the Twenty-First Century

Looking ahead to the rest of the twenty-first century, it is likely there will be a continuation of present practices in dealing with many of the same kinds of environmental risks (polluted air and water, and exposure to toxic chemicals), but with one important change. As suggested by thinking from the third epoch, newer environmental challenges will require a different kind of response. Is the concept of environmental risk broad enough to permit evaluation of and action on third-generation environmental problems such as climate change, biodiversity loss, population and economic growth, or the many inequities in the distribution of risks, costs, and benefits across society, generations, and the world?

After all, risk analysis was designed to deal with specific risks such as exposure to particular chemicals. Can the same concepts and methods be used for estimating and evaluating risks for complex environmental problems of the future? Or are new concepts and methods needed to do so effectively? The language of risk is used today to discuss climate change and other global problems, but the context within which policy options are considered is quite different from the way risk analysis has been used to deal with chemical and other risks over the past forty years.

What about expectations for public dialogue about risks and participation in decision making under these circumstances? Can the public understand risk information well enough to foster a productive exchange about environmental policy choices? How will these societal conversations about risks and policy options be affected by the increasingly politicized issue-framing and disinformation campaigns that have become so common today? That polarization and political debate is evident in conflicts over climate change, its causes, and policy actions that might be taken (Dunlap et al. 2016; Guber and Bosso 2013).

Addressing third-generation issues is already difficult given their global scope, inherent uncertainties, low visibility, low salience, and reduced trust in science, scientists, and other experts—all evident in debates over climate change (DiMento and Doughman 2014; Dunlap and McCright 2015; Oreskes and Conway 2010). But might the problems be addressed more successfully through a different set of concepts rooted in sustainability or sustainable development? Would doing so work better at the local, regional, or state levels, at least initially, than at the national and global levels?

At the local level, the problems tend to be more concrete and understandable, and both the public and policy makers frequently appear to have great interest in solving problems that are demonstrably related to quality of life in the community. Many cities and regions, large and small, both in North America and Europe, have made remarkable progress in recent years in fostering energy efficiency and conservation, water conservation, restoration of water bodies, improved building efficiency, transportation planning, improved air quality, environmentally sensitive land use planning, and more (Mazmanian and Kraft 2009).

Experience of this kind suggests the continuing utility of the intriguing concept of environmental risk, especially when it is firmly linked to the ideas and practices of sustainable development. Actions at the local level as well as the national and global levels also speak to the value of governing risk through the private sector, epistemic communities of experts, and nonprofit organizations in addition to formal processes within governmental agencies (O'Neill 2013). The future should bring a rich diversity of risk-related ideas and practices that are reimagined and reformulated to match twenty-first-century environmental problems. Such a development will give students of environmental policy abundant opportunities to study how the concept of environmental risk is used in public discourse, policy analysis, and policy-making processes in government.

Notes

1. Much the same history can be found in other nations in the late twentieth and early twenty-first centuries, whether in the form of the precautionary principle or other guidelines for risk decision making (Vogel 2012).

2. Even when they shared many of the same purposes and assumptions about how to address such risks, nations established different kinds of regulatory styles and frameworks that reflected distinctive legal systems and cultures. The US style, for example, has been characterized as highly formal, adversarial, and legalistic, with considerable opportunity for public commentary on proposed risk regulation as well as participation by organized interest groups in both the regulatory agencies and court system. It also often involves significant conflict between the executive, legislative, and judicial branches of government.

3. For the full regulatory proposal along with the technical and economic analyses supporting it, see US EPA 2015. Industry and a coalition of some seventeen states have sued the agency over the plan, throwing the final decision into the federal courts.

References

Andrews, Richard N. L. 2006. "Risk-Based Decision Making: Policy, Science, and Politics." In *Environmental Policy*, ed. Norman J. Vig and Michael E. Kraft, 215–238. 6th ed. Washington, DC: CQ Press.

Andrews, Richard N. L. 2016. "The Environmental Protection Agency." In *Environmental Policy*, ed. Norman J. Vig and Michael E. Kraft, 151–170. 9th ed. Washington, DC: CQ Press.

Brown, Erica, Kristine Hartman, Christopher Borick, Barry G. Rabe, and Thomas Ivacko. 2013. *Public Opinion on Fracking: Perspectives from Michigan and Pennsylvania*. Ann Arbor, MI: Center for Local, State, and Urban Policy.

Covello, Vincent T., and Jeryl Mumpower. 1986. "Risk Analysis and Risk Management: An Historical Perspective." In *Risk Evaluation and Management*, ed. Vincent T. Covello, Joshua Menkes, and Jeryl Mumpower, 519–540. New York: Plenum. Originally published in *Risk Analysis* 5, no. 2 (1985): 103–120.

Daniels, David P., Jon A. Krosnick, Michael P. Tichy, and Trevor Tompson. 2013. "Public Opinion on Environmental Policy in the United States." In *The Oxford Handbook of U.S. Environmental Policy*, ed. Sheldon Kamieniecki and Michael E. Kraft, 461–486. New York: Oxford University Press.

Davenport, Coral, and Peter Baker. 2014. "Taking Page from Health Care Act, Obama Climate Plan Relies on States." *New York Times*, June 2. Accessed December 6, 2016, http://www.nytimes.com/2014/06/03/us/politics/obama-epa-rule-coal-carbon-pollution-power-plants.html.

Davenport, Coral, and Gardiner Harris. 2015. "Obama to Unveil Tougher Environmental Plan with His Legacy in Mind." *New York Times*, August 2, A1.

Davies, J. Clarence, ed. 1996. *Comparing Environmental Risks: Tools for Setting Government Priorities*. Washington, DC: Resources for the Future.

DiMento, Joseph F. C., and Pamela Doughman. 2014. *Climate Change: What It Means for Us, Our Children, and Our Grandchildren*. 2nd ed. Cambridge, MA: MIT Press.

Dunlap, Riley E., Michael E. Kraft, and Eugene A. Rosa, eds. 1993. *Public Reactions to Nuclear Waste: Citizens' Views of Repository Siting*. Durham, NC: Duke University Press.

Dunlap, Riley E., and Aaron M. McCright. 2015. "Challenging Climate Change: The Denial Countermovement." In *Climate Change and Society: Sociological Perspectives*, ed. Riley E. Dunlap and R. J. Brule. New York: Oxford University Press.

Dunlap, Riley E., Aaron M. McCright, and Jerrod H. Yarosh. 2016. "The Political Divide on Climate Change: Partisan Polarization Widens in the U.S." *Environment* 58 (5) (September–October): 4–22.

Fiorino, Daniel J. 1990. "Can Problems Shape Priorities? The Case of Risk-Based Environmental Planning." *Public Administration Review* 50 (January–February): 82–90. doi:10.2307/977298.

Fiorino, Daniel J. 2014. "Streams of Environmental Innovation: Four Decades of EPA Policy Reform." *Environmental Law* 44 (3): 723–760.

Fischhoff, Baruch, Sarah Lichtenstein, Paul Slovic, Stephen L. Derby, and Ralph L. Keeney. 1981. *Acceptable Risk*. Cambridge: Cambridge University Press.

Freudenburg, William R. 1988. "Perceived Risk, Real Risk: Social Science and the Art of Probabilistic Risk Assessment." *Science* 242 (October 7): 44–49. doi: 10.1126/science.3175635.

Guber, Deborah Lynn, and Christopher J. Bosso. 2013. "Issue Framing, Agenda Setting, and Environmental Discourse." In *The Oxford Handbook of U.S. Environmental Policy*, ed. Sheldon Kamieniecki and Michael E. Kraft, 437–460. New York: Oxford University Press.

Hadden, Susan G., ed. 1984. *Risk Analysis, Institutions, and Public Policy*. Port Washington, NY: Associated Faculty Press.

Hadden, Susan G. 1989. *A Citizen's Right to Know: Risk Communication and Public Policy*. Boulder, CO: Westview Press.

Hadden, Susan G. 1991. "Public Perception of Hazardous Waste." *Risk Analysis* 11 (1): 47–57. doi:10.1111/j.1539-6924.1991.tb00568.x.

Harris, Richard A., and Sidney M. Milkis. 1996. *The Politics of Regulatory Change: A Tale of Two Agencies*. 2nd ed. New York: Oxford University Press.

Harrison, Kathryn, and George Hoberg. 1994. *Risk, Science, and Politics: Regulating Toxic Substances in Canada and the United States*. Montreal: McGill-Queen's University Press.

Kraft, Michael E. 1986. "The Political and Institutional Setting for Risk Analysis." In *Risk Evaluation and Management*, ed. Vincent T. Covello, Joshua Menkes, and Jeryl Mumpower, 413–434. New York: Plenum.

Kraft, Michael E. 2014. "Using Information Disclosure to Achieve Policy Goals: How Experience with the Toxics Release Inventory Can Inform Action on Natural Gas Fracturing." *Issues in Energy and Environmental Policy* 6 (March): 1–18.

Kraft, Michael E. 2016. "Environmental Policy in Congress." In *Environmental Policy: New Directions for the Twenty-First Century*, ed. Norman J. Vig and Michael E. Kraft, 103–127. 9th ed. Washington, DC: CQ Press.

Kraft, Michael E., and Sheldon Kamieniecki. 2007. *Business and Environmental Policy: Corporate Interests in the American Political System*. Cambridge, MA: MIT Press.

Kraft, Michael E., Mark Stephan, and Troy D. Abel. 2011. *Coming Clean: Information Disclosure and Environmental Performance*. Cambridge, MA: MIT Press.

Lave, Lester, ed. 1982. *Quantitative Risk Assessment in Regulation*. Washington, DC: Brookings Institution.

Lave, Lester, 1987. "Health and Safety Risk Analyses: Information for Better Decisions." *Science* 236 (April 17): 291–295. doi: 10.1126/science.3563509.

Layzer, Judith A. 2012. *Open for Business: Conservatives' Opposition to Environmental Regulation*. Cambridge, MA: MIT Press.

Lynn, Frances M. 1986. "The Interplay of Science and Values in Assessing and Regulating Environmental Risks." *Science, Technology, and Human Values* 11 (2): 40–50.

Mazmanian, Daniel A., and Michael E. Kraft, eds. 2009. *Toward Sustainable Communities: Transition and Transformations in Environmental Policy*. 2nd ed. Cambridge, MA: MIT Press.

Munton, Don, ed. 1996. *Hazardous Waste Siting and Democratic Choice*. Washington, DC: Georgetown University Press.

National Research Council. 1983. *Risk Assessment in the Federal Government: Managing the Process*. Washington, DC: National Academy Press.

O'Neill, Kate. 2013. "Global Environmental Policy Making." In *The Oxford Handbook of U.S. Environmental Policy*, ed. Sheldon Kamieniecki and Michael E. Kraft, 230–256. New York: Oxford University Press.

Oreskes, Naomi, and Erik M. Conway. 2010. *Merchants of Doubt: How a Handful of Scientists Obscured the Truth on Issues from Tobacco Smoke to Global Warming*. New York: Bloomsbury Press.

Perrow, Charles. 1984. *Normal Accidents: Living with High-Risk Technologies*. New York: Basic Books.

Rabe, Barry G. 2016. "Racing to the Top, the Bottom, or the Middle of the Pack? The Evolving State Government Role in Environmental Protection." In *Environmental Policy*, ed. Norman J. Vig and Michael E. Kraft, 33–57. 9th ed. Washington, DC: CQ Press.

Rinfret, Sara R., and Scott R. Furlong. 2013. "Defining Environmental Rule Making." In *The Oxford Handbook of U.S. Environmental Policy*, ed. Sheldon Kamieniecki and Michael E. Kraft, 437–460. New York: Oxford University Press.

Ruckelshaus, William. 1983. "Science, Risk, and Public Policy." *Science* 221 (September 9): 1026–1028. doi: 10.1126/science.6879200.

Slovic, Paul. 1987. "Perception of Risk." *Science* 236:280–285. doi:10.1126/science.3563507.

Slovic, Paul. 1993. "Perceived Risk, Trust, and Democracy." *Risk Analysis* 13:675–682. doi:10.1111/j.1539-6924.1993.tb01329.x.

Stern, Paul C., and Harvey V. Fineberg, eds. 1996. *Understanding Risk: Informing Decisions in a Democratic Society*. Washington, DC: National Academy Press.

Stirling, Andy. 2010. "Keep It Complex." *Nature* 468 (December 23–30): 1029–1031. doi: 10.1038/4681029a.

Stone, Deborah. 2012. *Policy Paradox: The Art of Political Decision Making*. 3rd ed. New York: W. W. Norton.

Stone, Richard. 1994. "California Report Sets Standard for Comparing Risks." *Science* 266 (October 14): 214. doi: 10.1126/science.7939654.

US Environmental Protection Agency (US EPA). 1987. *Unfinished Business: A Comparative Assessment of Environmental Problems*. Washington, DC: US EPA.

US Environmental Protection Agency (US EPA). 1990. *Reducing Risk: Setting Priorities and Strategies for Environmental Protection*. Washington, DC: US EPA.

US Environmental Protection Agency (US EPA). 2015. *Clean Power Plan for Existing Power Plants*. Accessed December 5, 2016, https://www2.epa.gov/cleanpowerplan/clean-power-plan-existing-power-plants.

US Government Accountability Office. 2005. *Observations on EPA's Cost-Benefit Analysis of Its Mercury Control Options*. Washington, DC: Government Accountability Office.

Vig, Norman J., and Michael E. Kraft, eds. 1984. *Environmental Policy in the 1980s: Reagan's New Agenda*. Washington, DC: CQ Press.

Vig, Norman J., and Michael E. Kraft, eds. 2016. *Environmental Policy: New Directions for the Twenty-First Century*. 9th ed. Washington, DC: CQ Press.

Vogel, David. 2012. *The Politics of Precaution: Regulating Health, Safety, and Environmental Risks in Europe and the United States*. Princeton, NJ: Princeton University Press.

Wiener, Jonathan B. 2003. "Whose Precaution After All? A Comment on the Comparison and Evolution of Risk Regulatory Systems." *Duke Journal of Comparative and International Law* 13:207–262. doi:10.2139/ssrn.460262.

Wiener, Jonathan B., Michael D. Rogers, James K. Hammitt, and Peter H. Sand, eds. 2010. *The Reality of Precaution: Comparing Risk Regulation in the United States and Europe*. Washington, DC: RFF Press.

Wildavsky, Aaron. 1988. *Searching for Safety*. New Brunswick, NJ: Transaction Books.

6 Critical Loads: Negotiating What Nature Can Withstand

Karin Bäckstrand

This chapter analyzes the concept of critical loads, which represents a science-based approach to grounding policy on notions of nature's *carrying capacity* or *ecosystem tolerance limits*. The idea of nature's toleration limits has found concrete expression in the concept of *critical loads*, which can be formulated as the maximum amount of pollutants that an ecosystem can tolerate without being changed or damaged in the long term. The critical loads concept reflects efforts to base environmental policy on scientifically determined critical thresholds for ecosystems, encapsulating the idea that pollution should be kept within the bounds of what nature can withstand. This chapter focuses on European transboundary air pollution, which has been the primary site for pioneering efforts putting the idea of critical loads concept into practice. It is therefore a prime example of usable or policy-relevant science in the context of UN and EU transboundary air pollution regulation (Lidskog and Sundqvist 2004).

The critical loads idea represents a conceptual innovation in environmental policy; it has operationalized the notion of *limits* central to environmental debates into both domestic and international policy. Peter Haas and David McCabe (2001, 327) argue that "the concept was virtually revolutionary in diplomacy because it assigned differential national obligations based on the carrying capacity of vulnerable ecosystems rather than a politically equitable (and arbitrary) emission cut that is universally applied." The 1999 Gothenburg Protocol targeting the interlinked problems of acidification, eutrophication, and ground-level ozone has been framed as a "smart protocol," and one of the most scientifically advanced ones ever signed (Lidskog and Sundqvist 2011, 8). The critical loads notion is a conceptual innovation that matters in terms of its successful inroad into international and national air pollution agreements, and is relevant for several reasons.

First, it invokes powerful ideas and metaphors of ecological limits, such as that we should regulate with an eye to "what nature can withstand." Second, it has been operationalized into policy and serves as a foundation for international air pollution agreements. Third, the career of the critical loads concept is closely tied to new modes of interfacing science with policy through integrated assessment models.

The first section of the chapter argues that critical loads qualifies as a conceptual innovation situated between broad ideas of carrying capacity and more specific policy tools, such as *risk assessment*. The second section traces the origin and emergence of the concept in international air pollution agreements over the past thirty years. Building on previous work, a discursive understanding of the rise of the concept is advanced (Bäckstrand 2001) by analyzing the different discourses on critical loads in the public, international negotiations, and scientific arenas. The fourth section explores how the concept has been put into practice in UN air pollution diplomacy, while the concluding section looks at the significance and impact of the concept.

Conceptual Innovation

The critical loads concept can be seen as a middle range idea, between the grand concepts of *sustainable development*, *planetary boundaries*, and *precautionary principle*, and the more focused policy tools of *environmental risk* and *environmental assessment*. While the concept of critical loads is confined to transboundary air pollution treaties, the underpinning idea of toleration limits also surfaces in domestic environmental legislation. For example, the US Clean Water Act includes the related concept of Total Maximum Daily Loads: the calculation of the maximum amount of pollutants that a water body can receive and still meet the water quality standards (US Environmental Protection Agency 2014). Furthermore, the US 1990 Clean Air Act mandates that the Environmental Protection Agency set National Ambient Air Quality Standards for pollutants such as carbon monoxide, lead, nitrogen dioxide, and ozone. Similarly, the 2-degree temperature target evokes the idea of nature's toleration limits: it is defined as the maximum allowable warming to avoid dangerous anthropogenic interference in the climate (Randalls 2010). This target has been developed to guide the understanding of and decision making about climate change. The 2-degree and aspirational

1.5-degree temperature targets were formulated by the European Union in its climate policies, and are included in the new global climate treaty—the Paris Agreement that was adopted in December 2015.

The critical loads notion also resembles the concept of planetary boundaries introduced in 2009 (Rockström et al. 2009). By identifying nine such boundaries for sub-ecosystems, a "safe operating space for humanity" was mapped out; the analysis quantified six and argued that three had been transgressed (climate change, changes in global nitrogen cycle, and biodiversity loss). In the end, the reference to the planetary boundaries concept was omitted from the Rio+20 agreement—*The Future We Want*—signed at the UN Conference on Sustainable Development in Brazil in 2012. While the broad public is more familiar with the concept of planetary boundaries, the critical loads concept addresses a more specialist audience of experts and diplomats in EU and UN air pollution policy. Yet, in contrast to planetary boundaries, it has made a difference to policy beyond the world of ideas and narratives.

It has been argued that critical loads represented a revolution in science-policy interaction in regulating transboundary environmental problems (Levy 1993, 100; Wettestad 2000, 3). For the first time in international environmental diplomacy, the basis for negotiating emission reduction targets was informed by scientifically derived assessments of ecosystem sensitivities as opposed to arbitrarily chosen political proposals. The innovativeness lies in the ambition to operationalize the overarching notions of limits or carrying capacity of ecosystems—a central idea in the discourse on sustainable development. The "innovativeness" in the critical loads concept is linked to how science is harnessed to policy, and how the idea of limits was developed as a scientific concept and subsequently translated into policy. The career of the critical loads concept is tied to new modes of interfacing science with policy through integrated assessment models. During the 1990s, environmental diplomacy in the context of the Convention on Long-Range Transboundary Air Pollution (CLRTAP) increasingly relied on computerized modeling in negotiating critical loads–based protocols. In the negotiation of the 1988 Nitrogen Protocol, the critical loads approach was mentioned as a conceptual framework for future protocols after a successful submission by Canada and Sweden that abatement policies should take into account the critical loads of ecosystems (Siebenhüner 2011, 100).

The scientific uncertainties associated with the critical loads approach, however, make "the concept more appropriately described as an instrument of environmental risk assessment rather than a carrier of information of a definite threshold response to a certain stressor" (Barkman 1998, 17). The development of critical loads illustrates the powers at work when a commonsense idea is legitimated by scientific "facts." The framing of a specific scientific interpretation of critical loads into popular narratives loaded with moral perceptions and identities ascribed to nature is key to understanding the success of the concept. As the concept has been put into practice as a decision framework for the negotiation of international air pollution agreements, it has turned into a discourse on risk assessment and cost-effectiveness, as will be discussed in the following sections.

Definitions of Critical Loads

The underlying idea of critical loads is that policies should be based on *ecosystem tolerance levels*—that is, the level of pollutant deposition should be set so as to avoid long-term significant change in the structure and function of ecosystems. This is in line with an emphasis on effect-oriented environmental policies "in which the emission reduction is guided by the actual impacts that emissions from a particular source have on sensitive ecosystem, human health and material" (Amann 1995, 41). The aim of the effect-oriented strategies incorporating the concept of critical loads is to reduce the emissions of air pollutants to a level where, ultimately, the critical load will no longer be exceeded (Swedish NGO Secretariat on Acid Rain 1998, 4).

As illustrated in figure 6.1, critical loads or levels are the quantitative upper limits on atmospheric deposition or ambient air quality.[1] If the threshold is exceeded, certain harmful effects, such as acidification, eutrophication, or ozone exposure, may occur in the long term. There is a time lag; in the short term, it is not likely that a link between critical loads exceedance and adverse effects can be observed. If the critical loads are exceeded during a longer time, though, the risks of ecosystem damage increase. The concept of *target loads* is based on critical loads that have been defined on scientific grounds. Yet target loads take political, economic, and technological feasibility into consideration (Swedish NGO Secretariat on Acid Rain 1995, 2). Keeping the target loads below the critical load creates a margin of safety. If they are higher than the critical load, a certain risk of damage is accepted. Hence, while science is used as a tool to establish

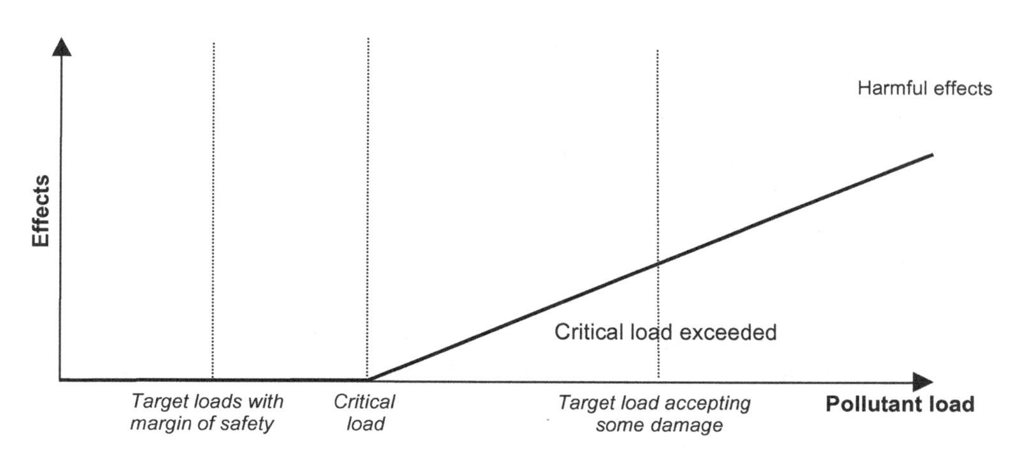

Figure 6.1
The critical loads concept
Source: Nilsson 1989.

toleration limits, the negotiations determine what risk is politically acceptable and at what cost.

The Origin and Emergence of the Critical Loads Concept

This section traces the scientific discovery of acidification and political context for adopting the critical load concept as a core principle in international air pollution agreements. The 1979 CLRTAP provided an institutional framework for dealing with the threats of regional air pollution. In the early 1990s, the critical load formed a basis for developing ecosystem-oriented abatement strategies in the CLRTAP regime. Today, attention is increasingly directed toward the role of the European Commission in transboundary air pollution negotiations, and close institutional linkages between the CLRTAP and EU processes (Selin and VanDeveer 2011).

The History of Acidification and Transboundary Air Pollution Negotiations

At its inception, the CLRTAP regime was framed around the problem of acidification caused by emissions of sulfur dioxide from the combustion of fossil fuels. The acidification of lakes and soils was a widespread problem in northern Europe and Scandinavia, where the bulk of the sulfur emissions stemmed from foreign sources. In the late 1980s, the interconnected dimensions of air pollution were gradually recognized, as the focus turned to nonsulfuric compounds. The problem of eutrophication caused by excess nitrogen deposition was tied to the loss of biodiversity as well as the corrosion of cultural heritage and buildings.

In the 1990s, attention was increasingly directed toward the human health impacts of the photochemical smog exposure that plagued southern Europe. Scientific evidence indicated that ozone exposure was a threat to human health as it impaired lung function and triggered respiratory diseases. Ground-level ozone also has impacts on ecosystems by causing damage to agricultural crops and trees. When global environmental problems such as climate change came to dominate the international agenda, less priority was given to the transboundary air pollution agenda. In 2012, fine particulate matter and short-lived climate forcers such as black carbon were added in the Revised Gothenburg Protocol (UNECE 2010). Sweden became the first country to ratify the Gothenburg Protocol in late 2015 (Ågren 2015).

In the early twentieth century, a decline in fish populations was observed in Norwegian lakes. Scientists linked this problem to the increased acidification of freshwater. But the issue failed to generate any significant political attention. In the late 1960s, scientists brought air pollution to the public's attention in Sweden (Lundgren 1998). A Swedish chemist, Svante Odén (1968), published a scientific report demonstrating that acidification in Scandinavia was attributable to the long-range transport of sulfur dioxide emissions from the United Kingdom and Central Europe. Nevertheless, the debate on acidification was triggered in Sweden as early as October 1967, when Odén published his scientific findings in *Dagens Nyheter*, a major Swedish daily newspaper. The long-range transport of air pollutants was portrayed as an "insidious chemical war." Other countries, however, remained skeptical of the scientific findings that the acidification of Scandinavian ecosystems was precipitated by the long-range transport of sulfur compounds. The Swedish and Norwegian governments raised the issue at the UN Conference on Human Environment in Stockholm in 1972 through a scientific report on transboundary air pollution (Bolin 1972). Still, the issue did not receive as much notice as the Swedish and Norwegian governments had hoped.

In 1977, though, the Organisation for Economic Co-operation and Development (1977) published an alarming report confirming earlier findings that air pollution did travel long distances of hundreds of kilometers. In addition, the study showed that several countries received more pollution from abroad than they generated domestically. The Cold War fed into the politics of transboundary air pollution as the establishment of the convention in 1979 was largely a product of the atmosphere of détente in East–West relations in the mid-1970s (Chossudovsky 1989).

The negotiation of a convention for airborne pollutants started in 1978, and in November 1979 the CLRTAP was signed by thirty-three parties—thirty European states along with the United States, Canada, and the European Commission. The framework convention was the first treaty to tackle atmospheric pollutants and was largely a result of lobbying by Nordic countries, which together with Canada demanded legally binding commitments to reduce sulfur emissions. Fifty-one parties now have ratified the convention. The Working Group on Strategies, which includes representatives of all parties to the convention, has negotiated all protocols. The Executive Body took the final decisions on the basis of prepared documents. A third body formed within the CLRTAP was the Working Group on Effects. Its task was to conduct research on the impacts of sulfur emissions and other airborne pollutants on natural ecosystems, human health, materials, forestry, and agriculture. Eight binding protocols negotiated under the auspices of the UN Economic Commission for Europe have subsequently been added to the CLRTAP regime, of which six have relevance for addressing environmental problems related to sulfur, nitrogen, and volatile organic compounds.[2]

From the National to the International Context

In 1974, Canadian scientists introduced the concept of critical loads when they identified critical thresholds for eutrophication in Canadian lakes. At the same time, Swedish researchers at the Swedish Environmental Protection Agency (SEPA) developed the critical load concept and engaged in research collaboration in the Nordic countries (Bäckstrand and Selin 2000; Nilsson 1989; Swedish NGO Secretariat on Acid Rain 1995). Yet it was only in the late 1980s that the concept was first introduced in the CLRTAP negotiations. As the issue of the acidification of soil and surface water in Europe became politicized, Swedish scientists and government officials promoted the critical load concept as a guiding principle for international air pollution control The first attempt to define tolerance limits for ecosystems was initiated in Sweden and Canada at the end of the 1970s, when the maximum deposition levels of the acidifying pollutants of lakes were determined. By comparing lakes exposed to various levels of sulfuric compounds, scientists tried to establish a critical threshold below which no acidification of lakes would occur.

The first official recognition of these efforts took place at the 1982 Stockholm Conference on Acidification, where the critical loads concept was accepted by policy makers. The SEPA, in collaboration with the Nordic Council of Ministers, arranged a number of scientific workshops in order to facilitate knowledge exchange as well as mobilize scientific and political support for the concept of critical loads. In 1986, a workshop was arranged in Oslo to test the idea of critical loads in the international research community (Nilsson 1986). An agreement on the utility of critical loads as a viable policy approach was reached after initial skepticism and even outright resistance among some scientists (Bäckstrand 2001, 130). The workshop produced the following working definition of critical loads: "The highest load that will not cause chemical changes leading to long-term harmful effects on the most sensitive ecosystems" (Nilsson 1986).

In 1988, a workshop was held in Sweden to develop a new sulfur protocol for critical loads. It was offered in the context of criticism of the 1985 sulfur protocol, where uniform flat-rate emission reductions of 30 percent were perceived as lacking a scientific basis (Nilsson and Grennfelt 1988). Representatives from the CLRTAP member states and government-designated scientific experts participated, and they agreed on a definition of critical loads that has been widely accepted: "A quantitative estimate of an exposure to one or more pollutants, below which significant harmful effects on specified sensitive elements of the environment do not occur according to present knowledge" (ibid., 9). The shift between the previous definition of "most sensitive ecosystems" to "significant harmful effects" with the addition "according to our present knowledge" can be interpreted as a shift to a more "politically qualified concept" that increased the political acceptability of critical loads (Ishii 2011, 178).

Integrated assessment models constituted a cornerstone for linking science and policy. One of the first and most successful is the Regional Acidification Information and Simulation (RAINS) model developed in the 1980s by scientists at the International Institute for Applied System Analysis (Hordijk and Amann 1997). The RAINS model combines data on societal activities causing transboundary air pollution with data on the availability of abatement options and their costs. The model calculates the transport of air pollutants across Europe, the influence on air quality, and impacts on human health, the built environment, and ecosystems. Its unique feature lies in its ability to derive an optimal and cost-effective distribution of emission reductions across Europe to meet air quality objectives as well as reduce the adverse impacts of acidification, eutrophication, and human health. RAINS was a useful tool to interface science with policy, according to the leading environmental modelers Leen Hordijk and Markus Amann (ibid., 337), who noted,

> In essence, RAINS is a scenario-generating device that helps users understand the impacts of future actions—or inaction—and to design strategies to achieve long-term environmental goals at the lowest possible cost. With a few hours of training, scientists, bureaucrats, politicians and other non-technical users can pose any number of "What if ... ?" RAINS gives answers to such questions, usually within minutes.

Discursive Contestation

The critical loads concept is linked to powerful discourses on ecosystem tolerance that have a wide appeal among various actors, such as industry, scientists, environmental nongovernmental organizations (NGOs), officials at environmental protection agencies, and policy makers. In the public arena, the discourse on critical loads as *critical thresholds* is predominant, and figures in official statements, parliamentary debates, the media, and NGO information campaigns. Swedish scientists were leading advocates of the critical load concept, arguing that it was "politically neutral and relied solely on natural science to justify targets for emission reductions" (Patt 1998, 7). NGOs embraced the concept of critical loads at the 1982 Stockholm Conference, which led to the NGOs Stop Acid Rain Campaign a few years later. In public discourse, the concept focuses on the most vulnerable ecological systems (thus reflecting the concept of *nil damage*) and was heralded as a useful policy instrument building on the precautionary principle (Lidskog and Sundqvist 2004). As the critical load concept has been tailored into a decision framework for multilateral air pollution diplomacy, however, it has been transformed from a discourse on critical thresholds to one about risk management and cost-effectiveness.

The Public Arena: Critical Loads as Critical Thresholds

One critical loads discourse is embedded in a normative framework about the prospects for using science to define nature's toleration limits as a base for environmental policy. Knowledge brokers, who translate complex scientific information into convincing policy narratives that can be comprehended by laypersons, have skillfully articulated this discourse. Research administrators at the SEPA, NGOs, and regional authorities produced simplified and accessible accounts of the critical loads concept. The public discourse on critical loads articulates the notion of *critical threshold,* signaling an absolute protection of ecosystems from damage. The threshold concept implies that ecosystems can only absorb perturbations induced by, for example, acidifying compounds within certain limits without significantly changing their structure and function. Critical loads for various air pollutants will be determined so as to avoid perturbations that damage ecosystems or human health, or in the worst case, cause irreversible changes. The threshold concept of critical loads that circulates in SEPA documents (Bertills and Hannenberg 1995, 80; SEPA 1996, 6) and governmental proposals thus conveys the idea that if scientifically determined critical loads are implemented, there will be an absolute protection of ecosystems. The concept has evolved as a demand from societal actors to motivate further emissions reductions. It serves as a legitimate framework based on scientific assessments of what level of stress different ecosystems can withstand without suffering long-term damage.

The Regulatory Arena: Critical Loads as Risk Assessment

Among the scientific community of modelers and negotiators in the CLRTAP, there is a second discourse on critical loads: *risk assessment.* This discourse has emerged in the process of employing the concept as a tool to negotiate politically accepted levels of ecosystem tolerance—so-called target loads. Maps depicting critical loads, and the importation of these maps into integrated assessment models such as RAINS, were a crucial foundation for international negotiations on abatement scenarios in the CLRTAP. In this discourse, the critical loads approach can be seen as a device for assessing and comparing the risks associated with different levels of ecosystem damage. Decision makers can manage risks by choosing among abatement strategies with different levels of ecosystem protection and costs that are politically acceptable. Interim targets are adopted, since emission reduction strategies leading to a complete attainment of critical loads are regarded as too costly in the short term. Consequently, the discourse on critical loads as risk assessment is in tension with the notion of an *absolute* protection of ecosystems as found in the threshold idea. The recognition that critical loads may entail a relative protection of ecosystems is clearly articulated by a scientist involved in the mapping of the concept: "Critical loads are not really about what nature can withstand. It is the level of acceptable damage or acceptable change to the ecosystems that we can agree upon" (Bäckstrand 2001, 139).

In a negotiation process, environmental protection is balanced against abatement costs, technical feasibility, and political acceptance. When critical loads were adopted for the CLRTAP, the use of percentiles implied that there was a relative versus absolute protection of ecosystems. The parties accepted the risk that the critical loads might be exceeded in sensitive areas and that there would be a risk for ecosystem damage in the long term. The process of identifying and calculating critical thresholds for terrestrial and water ecosystems is hampered by the difficulty of selecting an indicator organism that can adequately represent the ecosystem. Since ecosystem properties and boundaries are vague, the representativeness of the indicator will entail uncertainties for the whole ecosystem. Parties were initially presented with ten scenarios derived from the RAINS model, which estimated the levels of ecosystem protection associated with different emissions cuts and the corresponding costs of such emission reductions (Haas and McCabe 2001). In choosing scenarios, policy makers were faced with the trade-off between ensuring the protection of a relatively high percentage of the ecosystem (90 percent when based on the tenth percentile) and preventing abatement costs on the European level from skyrocketing.

The Economic Arena: Critical Loads as Cost-Effectiveness

In reality, costs, not what nature can withstand, are central, as stated by Lars Björkbom (cited in Bäckstrand 2001, 139), the former Swedish chair of the CLRTAP's Working Group on Strategies of LRTAP. The discourse on critical loads as cost-effectiveness was pivotal in negotiating the air pollution agreements. This discourse was constituted by use of the RAINS model as decision support tool for reaching agreements. The advantages of flexible, cost-effective, and differentiated emission cuts are contrasted with previous uniform, centralized, and flat-rate emission strategies. Moreover, cost-benefit analysis in the CLRTAP regime was revitalized in the 1990s and emerged in parallel with the critical loads approach (Patt 1998). The RAINS model, with its optimization features, could identify the cost-minimal allocation of abatement measures. The area in which the exceedance of critical loads is the highest can be targeted, which also makes the overall marginal costs lower. The marginal cost for cutting sulfur dioxide in Sweden, beyond the already-achieved 80 percent reduction, was considerably higher than for countries in central Europe.

Contesting the Critical Loads Concept

Despite the widespread institutionalization of critical loads in the CLRTAP regime, scientists have voiced skepticism about it. The scientific basis of the concept has been criticized (Bäckstrand 2001, 141). The discourse as *weak science* has been articulated by soil and water scientists outside the CLRTAP research networks who warned against oversimplification in the representation of ecosystems as well as the use of steady-state instead of dynamic modeling (Ishii 2011, 179). The argument is that modeling and mapping are fraught with uncertainties, partly because of the use of steady-state modeling. Scientific evidence indicates that critical load numbers would be considerably lower if dynamic modeling was applied. This line of criticism came from ecologists who targeted the basic conceptual and mathematical models as well as criteria underlying the mapping. The critical loads approach was regarded as policy-relevant environmental science that failed to live up to standards of what constitutes good science.

One line of criticism articulated by some ecologists concerns the modeling—that is, the use of biochemical models to calculate critical loads. Since the understanding of natural systems is incomplete, the models can never accurately describe natural processes, and therefore are poor and simplistic representations of complex ecosystems. Furthermore, the chemical criterion underlying the calculation and modeling of critical loads has been criticized (ibid., 179). This is a chemical measurement selected by modelers to create a quantitative link between acid pollutants loads and ecosystem damage. Some ecologists, though, regard this as a pure laboratory product created by modelers with little experience of fieldwork. The scientific credibility issue has been de-emphasized by the imperative of providing policy makers with accessible and rapid knowledge that can form a basis for negotiating air pollution agreements.

In contrast to a risk acceptance approach, some critics argue that policies should be based on the precautionary principle, which prescribes preventive action despite scientific uncertainties. Zero levels of pollution or ecosystem stress should be the explicit and ultimate goal, since even tiny levels of pollution can produce adverse physical effects on ecosystems. Critics also are concerned that in a situation where critical loads are fully attained, there will be no incentive to make additional reductions.

The critical loads concept was contested in the CLRTAP negotiations. Yet its main rationale was that a critical load for acidification allowed decision makers to identify cost-efficient levels of environmental protection. Sweden, Canada, and the Netherlands were in favor of critical loads as a science-based approach that replaced the previous, biased, politicized, and arbitrary reliance on uniform emission cuts. Moreover, as "victim" countries of transboundary pollution, Sweden and Canada contended that critical loads would provide a stronger base for policy, because a science-based approach would strengthen their claims (Haas and Stevens 2011, 327). While initially a skeptic, the United Kingdom came to support the concept owing to its strong scientific basis and predictions that uniform, negotiated emission cuts might be more costly than emission ceilings derived from critical loads. The Soviet Union also endorsed the concept since it was formally couched in a natural science paradigm rather than a market-based idiom such as cost-benefit analysis. Canada supported the concept for the same reasons as Sweden. Germany accepted it in theory, but ultimately wanted to base policies on the principle of the best available technology. Switzerland and Austria were in favor of the concept, too, but only if there could be full attainment of critical loads. Without safety margins, the concept would conflict with the precautionary principle. In contrast, the Mediterranean countries and France were passive in negotiations of the first protocol based on critical loads (Patt 1998).

The 1999 Gothenburg Protocol highlights both environmental and human health problems associated with transboundary air pollution. The argument for a multipollutant, multieffect strategy, skillfully advanced by scientists, is that it would be an incentive for more states to participate, since the types of air pollution problems differ among geographic regions in Europe (Grennfelt et al. 1994). Southern Europe's main concern is ground-level ozone, while northern Europe is exposed to acidification. An emphasis on a single pollutant or effect would not bring all the parties to the negotiating table.

Practical Implications

This section analyzes how the critical loads concept has been incorporated into governance practice as an instrument of international air pollution regulation. For the first time ever in environmental diplomacy, integrated assessment models have been used directly to aid decision making. The concept was viewed as a step forward compared to earlier flat-rate, politically negotiated reduction targets. The latter were regarded as nonscientific, arbitrary, unfair, and economically ineffective (Siebenhüner 2011, 99). Uniform emission cuts fail to take into account that ecosystems vary in their sensitivity to pollution, and that the costs of reducing emissions vary across Europe. In contrast, the critical loads concept favors equity in the distribution of environmental effects over equity of reduction commitments among countries. Distributing reduction commitments according to principles of ecosystem effects and nature's toleration limits represented an innovation in international environmental diplomacy.

Critical loads has been described as "a procedure for developing optimized abatement strategies by which differentiated emissions reductions are obtained on the basis of scientifically derived critical values" (UNECE 1991, 3). By importing maps as one "module" in the RAINS model, strategies for emission reductions in acidification, eutrophication, and tropospheric ozone may be obtained. With its optimization features, it can identify the cost-minimal allocation of abatement measures. The marginal cost for cutting sulfur dioxide in Sweden is considerably higher compared to countries in central Europe. The ideal application of the principle of cost-effectiveness in relation to critical loads involves two steps: determining the ecosystem sensitivity and reaching agreement on emissions levels below the critical load; and applying the principle of cost-effectiveness by optimizing emission cuts and transforming these into differential national commitments. In negotiating scenarios with different levels of ecosystem protection, however, these steps were in practice blended.

At its inception, the CLRTAP regime was framed around acidification and sulfuric compounds. The scientific assessment process along with the institutional framework were largely organized around the acidification problem and driven by actors, such as the Scandinavian countries, that were harmed by the problem. In the 1990s, the transboundary pollution agenda broadened as scientific assessment and policies emerged on nonsulfuric compounds, such as volatile organic compounds, persistent organic pollutants, and heavy metals. The "first-generation" protocols set uniform reduction targets and timetables for all states. The 1985 sulfur protocol was signed in Helsinki and obliged states to reduce sulfur emissions by 30 percent from the 1980 levels by 1993. Under the nitrogen oxide protocol signed in Sophia in 1988, states committed to freezing nitrogen oxide emissions at the 1987 levels by 1994 and negotiating to further reduce emissions based on the critical loads. Finally, the volatile organic compounds protocol was signed in Geneva in 1991; states agreed to reduce these compounds by 30 percent (Bäckstrand and Selin 2000; Selin and VanDeveer 2011).

The "second-generation," effect-oriented protocols contained geographically differentiated abatement strategies based on critical loads that reflected the sensitivity of ecosystems to airborne pollutions. European-wide critical loads were compiled into maps consisting of grid cells of 150 by 150 kilometers each. These maps were a critical building block in air pollution negotiations. They enabled a comparison of data on ecosystem sensitivity with maps depicting the actual deposition or atmospheric concentrations of pollutants (Levy 1995, 62). Applying the critical loads concept to air pollution agreements was a complex endeavor that required data on emission inventories and maps from more than thirty countries.

The first protocol containing a critical loads approach was the Second Sulfur Protocol on Further Reductions of Sulfur Emissions. It was signed by twenty-eight countries in Oslo in 1994 and replaced the expired 1985 sulfur protocol. The aim of the second sulfur protocol is to "ensure that sulfur deposition do not, in the long term, exceed critical loads" (UNECE 1994). The political objective of the new protocol was to reduce the gap between the 1990 sulfur deposition and critical loads by 60 percent in every square of the European Monitoring and Evaluation Program grids throughout Europe in the most cost-effective way between 1990 and 2000. A key element in the protocol was that the different emission reduction targets were based on both the extent of ecosystem damage caused by emissions and the estimated cost for reducing those emissions. The national commitments adopted in the protocol closed 60 percent of the gap between existing measures and those needed to avoid exceeding the critical loads (Haas and McCabe 2001, 329; Levy 1995, 62). Already in 1992, states had decided that it would be too costly to reach the critical loads targets, and lowered their ambition to reduce the difference between current emissions and critical loads by 50 percent, thereby protecting 90 percent of the ecosystems across Europe. In the following years' negotiation round in the Working Group on Strategies, it was agreed that a 60 percent gap closure could be achieved at the same cost, meaning that critical loads for acidification would be exceeded in "only" 7 percent of the ecosystems. This was a considerable improvement compared to the 1985 sulfur protocol with uniform emission cuts where 30 percent of the ecosystems remained above the critical level. The parties found that setting target loads was a complex and arbitrary process. The interim target in the 1994 sulfur protocol (i.e., a 60 percent gap closure) can be seen as a target load, since full attainment of the critical load was perceived as being too costly.

In fall 1994, negotiations started on a second protocol regulating nitrogen and other airborne pollutants. The future protocol was modeled on the second sulfur protocol, where critical loads served as a foundation. In Gothenburg in 1999, twenty-four European countries along with the European Union, the United States, and Canada signed the Protocol to Abate Acidification, Eutrophication, and Ground-level Ozone.[3] The protocol encompasses several pollutants (nitrous oxides, volatile organic compounds, and ammonia) and regulates several effects (acidification, eutrophication, and formation of ground-level —ozone) simultaneously. The scientific rationale behind this is that on its own, or in interaction with other types of pollutants, a pollutant is the source of more than one effect. Nitrous oxides, for example, contribute to both acidification and eutrophication as well as indirectly to the formation of ground-level ozone. The protocol represents a more integrated approach as both environmental and health impacts are addressed by setting emission ceilings for four pollutants to be reached by 2010.

The negotiations for the Revised Gothenburg Protocol for the reduction of air pollutants by 2020 started in 2007, but proceeded more slowly than anticipated. Initially the protocol was to be finalized at the end of 2011. In May 2012, after five years of negotiations, the Revised Gothenburg Protocol was adopted (Ågren 2012). It can be conceived as a "third-generation" agreement since it takes into account the interlinked effects of acidification, eutrophication, ground-level ozone accumulation, particulate matter, and black carbon (Siebenhüner 2011, 102). The protocol broke new ground by including targets for fine particulates (particulate matter 2.5), specifically addressing the short-lived climate forcer black carbon (soot).[4] The inclusion of particulate matter is expected to create synergies between air pollution and climate change mitigation. Overall, the EU member states should reduce emissions of sulfur dioxide by 59 percent, nitrogen oxides by 6 percent, volatile organic compounds by 22 percent, and particulate matter by 22 percent, from 2005 to 2020. The Gothenburg Protocol is expected to bring significant reductions in air pollution, leading to health improvements, such as decreased mortality from exposure to particulate matter 2.5 and ground-level ozone by 27 and 11 percent, respectively. Forest and freshwater ecosystem exposures to acid deposition above critical loads are expected to decrease by more than 55 percent. Less improvement is expected for eutrophication, though, where critical loads excess in ecosystems will fall by only 20 percent. Environmental NGOs have voiced criticism of the low ambition level of the new protocol and regard it as a missed opportunity (Ågren 2012, 4).

In the European Union, there have been parallel efforts to incorporate an integrated approach to transboundary air pollution control. Yet EU policies on air pollution remain highly fragmented (Selin and VanDeveer 2011). In March 1997, after a Swedish proposal in the Council of Ministers, the EU strategy for acidification was adopted, containing an interim goal of a 50 percent gap closure by 2010. The full attainment of the critical load for acidification was regarded as too expensive. Nevertheless, demands from countries in southern Europe to address the formation of low-level ozone accelerated the process toward an integrated approach to air pollution management. The CLRTAP and EU processes were linked together, and in 1999, the European Commission proposed the National Emission Ceilings (NEC) Directive, based on a multipollutant/multieffect strategy that relied on integrated assessment modeling. The International Institute for Applied System Analysis provided the modeling input for negotiating the Gothenburg Protocol and preparing the NEC directive. The GAINS model, launched in the early 2000s, extended the multipollutant/multieffect approach of the earlier RAINS model.

The twin processes of negotiating the Gothenburg Protocol and drafting the NEC Directive highlight the more active role of the European Union as an actor in CLRTAP diplomacy. The institutional links between the European Union and the CLRTAP have gradually been strengthened. In 2001, the NEC Directive was adopted; this set 2010 upper emission limits for each member state for all four pollutants related to acidification, eutrophication, and ground-level ozone. After some delay the new NEC Directive for air pollution emission targets beyond 2010 was adopted by the European Parliament at the end of 2016, and entered into force on the 31 December of that year. As part of the 2005 Thematic Air Pollution Strategy, at the end of the European Union's "Year of Air" 2013, the European Commission presented the new Clean Air Policy package, which included proposals for emission reductions for six main air pollutants by 2030. These targets were essentially identical to those in the 2012 Gothenburg Protocol, which environmental and health NGOs criticized for being weak and not legally binding after 2025 (Ågren 2016).

The Future of Critical Loads

What lies in the future for the concept? In the adopted long-term strategy for CLRTAP 2010–2020 (UNECE 2010), the science-based decision-making and critical loads–based approaches are framed as essential components, and the links between science and policy should be further strengthened. After thirty years of air pollution diplomacy, it has been argued that the critical loads concept faces new challenges with public attention shifting from air pollution to global issues such as climate change. Some challenges are outlined in the long-term strategy, which includes developing health targets, creating synergies between air pollution and climate mitigation, expanding critical loads to other regions of the world, and engaging the public in air pollution policies. What are the challenges for the future of the concept?

First, there is the challenge of environmental effectiveness and performance. Despite the success in consolidating the concept as a foundation for EU and UN air pollutants agreements, the targets for emissions reductions to be achieved by 2020 in the 2012 Gothenburg Protocol and forthcoming EU air pollution directive are inadequate to achieve the long-term objective of not exceeding critical loads to protect human health and ecosystems.

Apart from significant damage to ecosystems, agricultural crops, and materials as well as corrosion, air pollution causes nearly half a million premature deaths in the European Union annually, in addition to health impacts such as hospitalization and millions of lost working days (Ågren 2012). More than 95 percent of citizens in urban areas in Europe are exposed to harmful levels of particulate matter and ozone. According to a recent report, air pollution globally takes 3.3 million lives a year (Ågren 2015). The costs of health-related air pollution damage for the European Union are estimated to be between €330 and €940 billion per year, with more than a quarter of a million premature deaths annually anticipated by 2030 with the revised targets (Ågren 2014, 2016). The disappointment with progress in the Revised Gothenburg Protocol and EU air policies are manifested in a letter to the EU commissioners signed by major environmental NGOs and citizens groups voicing concern that the targets are insufficient (European Environmental Bureau 2013; Ågren 2012, 2).

Second, there is a need to address the global governance of air pollution and troposphere as well as the interaction between air pollution and climate change, including developing strategies to reduce short-lived climate forcers, such as black carbon (Grennfelt et al. 2013, 3). The CLRTAP has in its long-term strategies emphasized the synergies between the control of air pollution and global warming. The inclusion of critical loads for particulate matter in the Gothenburg Protocol represents a win-win situation for improved air quality and reduced climate change. The regulation of transboundary air pollutants most detrimental to health, such as particulate matter and tropospheric ozone, are central. In the CLRTAP, there is a stepwise improvement of multipollutant/multieffect strategies and the development of health-oriented targets based on critical levels for intercontinental flows of particulate matter. In revising the Gothenburg Protocol, there have been calls to develop the critical loads concept. This entails developing new concepts and ideas, such as including recovery processes and their relations to critical loads, and incorporating these in the RAINS model (Lidskog and Sundqvist 2004, 218).

A third challenge is public legitimacy. In the 2013 Eurobarometer, nearly 80 percent of Europeans thought that the European Union should propose additional measures to combat air quality–related problems in Europe (European Environmental Bureau 2013). The critical loads concept has been hammered out in a rather-technocratic manner during three decades without engaging the public. In this vein, critical loads represents a scientization of policy making, in which experts and negotiators predominate at the expense of citizens, stakeholders, and NGOs. In contrast to the concepts of sustainable development and the precautionary principle, which are more salient in public discourse and citizen debates, the debate on critical loads has been limited to specialized and expert-dominated forums. For the CLRTAP to maintain its future credibility and legitimacy, it needs to rely more on what Ulrich Beck (1992) calls reflexive scientization, which entails involving civil society and other stakeholders in expert decision making (Bäckstrand 2001; Lidskog and Sundqvist 2011, 17). The need for more accountable air pollution policies is reflected in the CLRTAP long-term strategy that stresses the role of public participation along with access to information, stakeholder inclusion, and public communication (UNECE 2010; Grennfelt et al. 2013).

Conclusion

The concept of critical loads represents a conceptual innovation that plays a significant role in environmental discourse and policy practice. Ideas of nature's toleration and ecological limits are manifest in the 2-degree temperature target for climate change and planetary boundaries concept. While the 2- and 1.5-degree targets made their way into the 2015 Paris Climate Agreement, the planetary boundaries idea was omitted from the Rio+20 outcome, despite its widespread appeal as a heuristic for biophysical limits. In comparison, the critical loads concept has had a significant policy impact through its incorporation in domestic legislation and regulation as well as international air pollution agreements in the United Nations and European Union since the early 1990s.

The critical loads concept has played a pioneering role in international environmental diplomacy with regard to how science is harnessed for policy. The idea of nature's toleration limits was transformed into a scientific concept of ecosystem sensitivities, which in turn served as a foundation for a new generation of science-driven EU and UN air pollution agreements. Critical loads have been framed as a successful example of the interaction of science and policy, and integrated assessment models such as RAINS have been tools in environmental diplomacy. Environmental scientists and modelers play a key role in presenting scenarios for ecosystem sensitivities and deriving abatement strategies to use in negotiating politically and economically acçeptable risks of ecosystem degradation.

That the concept has been applied in the full policy cycle from agenda setting to implementation, domestically and internationally, demonstrates that ideas and arguments matter. Discursively, it has gained popularity and legitimacy as it invokes powerful and commonsense metaphors that we should ground policy and regulation on scientific assessment about "what nature can withstand." The success of the critical loads concept is related to it being "constructively ambiguous." It appeals to multiple audiences—scientists, NGOs, and environmental activists—in evoking the idea of basing policy on nature's toleration limits. It also appeals to policy makers, bureaucrats, and diplomats as a prime illustration of usable and policy-relevant science.

Finally, the critical load concept has proved to be a fruitful and highly adaptive innovation that has gradually been applied to new problem areas (such as health or eutrophication) beyond its initial framing around acidification as well as new pollutants (ground-level ozone, ammonia, and particular matter). Yet despite the robust scientific foundation of air pollution diplomacy, critical loads for acidification and eutrophication are still exceeded throughout Europe. The emission reduction targets stipulated in the Gothenburg Protocol and the EU NEC Directive are not sufficient to protect ecosystems across Europe, or to avoid adverse health impacts and mortality from air pollution. The lingering gap between the politically negotiated targets and scientifically derived critical loads are indicative of the limits of science "speaking truth to politics."

Notes

1. The critical load is a measure of the maximum amount of deposition (acidification or eutrophication) that an ecosystem can endure; critical levels refer to the maximum atmospheric concentration of pollutants.

2. The 1979 CLRTAP, which came into force in 1983, now has fifty-one parties out of the fifty-five members in the UN Economic Commission for Europe.

3. The protocol has been referred to as the multipollutant/multieffect protocol since it addresses the environmental effects and health impacts of air pollution as well as several pollutants. It was also described as the "Super Protocol" because it aimed to address the interconnected dimensions of air pollution. I will refer to it as the Gothenburg Protocol.

4. Particulate matter 2.5 stands for particulate matter in the air of less than 2.5 micrometers in diameter.

References

Ågren, Christer. 2012. "New Gothenburg Protocol Adopted." *Acid News*, no. 2, June. Gothenburg: Air Pollution and Climate Secretariat.

Ågren, Christer. 2014. "A New EU Clean Air Strategy up to 2030." *Acid News*, no. 1, March. Gothenburg: Air Pollution and Climate Secretariat.

Ågren, Christer. 2015. "Air Pollution Takes 3.3 Million Lives a Year." *Acid News*, no. 4, December. Gothenburg: Air Pollution and Climate Secretariat.

Ågren, Christer. 2016. "New Watered-Down EU Air Pollution Targets." *Acid News*, no. 3, October. Gothenburg: Air Pollution and Climate Secretariat.

Amann, Markus. 1995. "Strategies for Reducing Emissions of Air Pollutants in Europe: The Role of Integrated Assessment." In *Proceedings of the 10th World Clean Air Congress*, 4:39–48.

Bäckstrand, Karin. 2001. "What Can Nature Withstand? Science, Politics, and Discourses in Transboundary Air Pollution Diplomacy." PhD diss., Lund University.

Bäckstrand, Karin, and Henrik Selin. 2000. "Sweden—A Pioneer of Acidification Abatement." In *International Environmental Agreements and Domestic Politics: The Case of Acid Rain*, ed. Arild Underdal and Kenneth Hanf, 87–108. Aldershot, UK: Ashgate.

Barkman, Andreas. 1998. *Critical Loads—Assessments of Uncertainty*. Lund, Sweden: Department of Chemical Engineering II, Lund University.

Beck, Ulrich. 1992. *Risk Society: Towards a New Modernity*. London: Sage Publications.

Bertills, Ulla, and Peter Hannenberg, eds. 1995. *Acidification in Sweden: What Do We Know Today?* Report 4422. Stockholm: SEPA.

Bolin, Bert. 1972. *Sweden's Case Study for the United Nations Conference on Human Environment: Air Pollution across National Boundaries*. Stockholm: Norstedt.

Chossudovsky, Evgeny. 1989. *East–West Diplomacy for Environment in the United Nations*. New York: UN Institute for Training and Research.

European Environmental Bureau. 2013. *Revision of the Thematic Strategy of Air Pollution and Accompanying Legislative Proposals for Cleaner Air in Europe: Letter to All EU Commissioners*. Brussels: European Environmental Bureau.

Grennfelt, Peringe, Öystein Hov, and Dick Dwerwent. 1994. "Second Generation Abatement Strategies for NOx, NH3, SO2, and VOCs." *Ambio* 22 (7): 425–433.

Grennfelt, Peringe, Anna Engleryd, John Munthe, and Ulrika Hååard, eds. 2013. *Saltsjöbaden V: Taking International Air Pollution Policies in the Future*. TemaNord 571. Copenhagen: Nordic Council of Ministers.

Haas, Peter M., and David McCabe. 2001. "Amplifiers or Dampeners: International Institutions and Social Learning in the Management of Global Environmental Risks." In *Learning to Manage Global Environmental Global Risks*, ed. Social Learning Group, 323–348. Vol. 1. Cambridge, MA: MIT Press.

Haas, Peter M., and Casey Stevens. 2011. Organized Science, Usable Knowledge, and Multilateral Environmental Governance. In *Governing the Air: The Dynamics of Science, Policy, and Citizen Interaction*, ed. Rolf Lidskog and Göran Sundqvist, 125–162. Cambridge, MA: MIT Press.

Hordijk, Leen, and Markus Amann. 1997. "How Science and Policy Combined to Combat Air Pollution." *Environmental Law and Policy* 37 (4): 336–340.

Ishii, Atsushi. 2011. "Scientists Learn Not Only Science but Also Diplomacy: Learning Processes in the European Transboundary Air Pollution Regime." In *Governing the Air: The Dynamics of Science, Policy, and Citizen Interaction*, ed. Rolf Lidskog and Göran Sundqvist, 163–194. Cambridge, MA: MIT Press.

Levy, Marc. 1993. "European Acid Rain: The Power of Tote-Board Policy." In *Institutions for the Earth: Sources of Effective International Environmental Protection*, ed. Peter Haas, Robert Keohane, and Marc Levy, 75–132. Cambridge, MA: MIT Press.

Levy, Marc. 1995. "International Co-operation to Combat Acid Rain." In *Green Globe Yearbook on International Cooperation on Environment and Development*, ed. Helge Ole Bergesen, Georg Parrman, and Øystein B. Thomesson, 59–68. Oxford: Oxford University Press.

Lidskog, Rolf, and Göran Sundqvist. 2004. "From Consensus to Credibility: New Challenges for Policy Relevant Science." *Innovation* 17 (3): 205–226. doi:10.1080/1351161042000241144.

Lidskog, Rolf, and Göran Sundqvist. 2011. "Transboundary Air Pollution Policy in Transition." In *Governing the Air: The Dynamics of Science, Policy, and Citizen Interaction*, ed. Rolf Lidskog and Göran Sundqvist, 1–38. Cambridge, MA: MIT Press.

Lundgren, Lars. 1998. *Acid Rain on the Agenda: A Picture of Chain of Events in Sweden, 1966–68*. Lund, Sweden: Lund University Press.

Nilsson, Jan, ed. 1986. *Critical Loads for Nitrogen and Sulphur*. Report from a Nordic Working Group, Miljörapport, 11. Copenhagen: Nordic Council of Ministers.

Nilsson, Jan. 1989. "Kritisk belastning eller vad naturen tål." In *Vad tål naturen?*, ed. Lars J. Lundgren. Solna: Naturvårdsverket.

Nilsson, Jan, and Peringe Grennfelt, eds. 1988. *Critical Loads for Sulphur and Nitrogen*. Report from a workshop held in Skokloster, Sweden, Nord, March 19–24, 15. Copenhagen: Nordic Council of Ministers.

Odén, Svante. 1968. The Acidification of Air and Precipitation and Its Consequences in the Natural Environment. Ecology Committee Bulletin, No. 1. Stockholm: Swedish National Research Council.

Organisation for Economic Co-operation and Development. 1977. *The OECD Program on Long-Range Transport of Air Pollutants: Summary Report*. Paris: Organisation for Economic Co-operation and Development.

Patt, Anthony. 1998. "Analytical Frameworks and Politics: The Case of Acid Rain in Europe." ENRP Discussion Paper E-98–20. Cambridge, MA: Harvard University.

Randalls, Samuel. 2010. "History of 2 Degree Target." *Wiley Interdisciplinary Reviews: Climate Change* 1 (4): 598–605. doi:10.1002/wcc.62.

Rockström, Johan, Will Steffen, Kevin Noone, Åsa Persson, F. Stuart Chapin III, Eric Lambin, Timothy M. Lenton, et al. 2009. "Planetary Boundaries. Exploring a Safe Operating Space for Humanity." *Ecology and Society* 14 (2): 32. doi:10.1111/j.1540-5842.2010.01142.x.

Selin, Henrik, and Stacy VanDeveer. 2011. "Institutional Linkages and European Air Pollution Politics." In *Governing the Air: The Dynamics of Science, Policy, and Citizen Interaction*, ed. Rolf Lidskog and Göran Sundqvist, 61–92. Cambridge, MA: MIT Press.

Siebenhüner, Berndt. 2011. "Transboundary Science for Transboundary Air Pollution Policies in Europe." In *Governing the Air: The Dynamics of Science, Policy, and Citizen Interaction*, ed. Rolf Lidskog and Göran Sundqvist, 93–124. Cambridge, MA: MIT Press.

Swedish Environmental Protection Agency (SEPA). 1996. *Facts about Swedish Policy on Acid Rain*. Stockholm: SEPA.

Swedish NGO Secretariat on Acid Rain. 1995. "Critical Loads: So Much and No More." *Environmental Factsheet* (6): 1–7.

Swedish NGO Secretariat on Acid Rain. 1998. "LRTAP Convention. Air Pollution Treaty." *Environmental Factsheet* (8): 1–4.

UN Economic Commission for Europe (UNECE). 1991. *The Critical Load Concept and the Role of the Best Available Technology and Approaches*. EB. AIR/WG.5/R.24/Rev.1. Geneva: United Nations.

UN Economic Commission for Europe (UNECE). 1994. *United Nations Economic Commission for Europe: Protocol to the Convention of Long-Range Transboundary Air Pollution on Further Reduction of Sulphur Emissions*. UN ECE/EB AIR 40. Geneva: United Nations.

UN Economic Commission for Europe (UNECE). 2010. *Long-Term Strategy for the Convention on Long-Range Transboundary Air Pollution and Action Plan for Its Implementation.* EC/EB.AIR/106/Add.1. Decision 2010/18. Geneva: United Nations.

US Environmental Protection Agency. 2014. "Implementing Clean Water Act Section 303(d): Impaired Waters and Total Maximum Daily Loads (TMDLs)." Accessed December 6, 2016, https://water.epa.gov/lawsregs/lawsguidance/cwa/tmdl/overviewoftmdl.cfm.

Wettestad, Jørgen. 2000. "From Common Cuts to Critical Loads: The ECE Convention on Long-Range Transboundary Air Pollution (CLRTAP)." In *Science and Politics in International Environmental Regimes*, ed. Steinar Andresen, Tora Skodvin, Arild Underdal, and Jørgen Wettestad, 95–122. New York: Manchester University Press.

7 Adaptive Management: Popular but Difficult to Implement

Judith A. Layzer and Alexis Schulman

Popularized by scientists in the 1970s, *adaptive management* is an integrative, multidisciplinary approach to managing landscapes and natural resources. It has been recommended for use in managing water resources, national parks, endangered species and their habitats, wildlife, and even agriculture (Klerkx et al. 2010; Parkes et al. 2006; Rogers 2003). The central goal of adaptive management is to improve natural resource management by learning how ecosystems respond to human intervention, and then adjusting actions in response to that learning.

Adaptive management stands in contrast to *conventional management*, whose proponents placed great faith in the ability of traditional scientific inquiry to produce generalizable and predictive knowledge that could be used to effectively control natural processes. Conversely, a key premise of adaptive management is that management always takes place under conditions of uncertainty, which derives from multiple sources—not least of which is the inherent and unique complexity of ecological systems. Adaptive management's crafters hoped to enable managers to act in the face of this uncertainty, particularly in situations where rigid adherence to the status quo, despite disastrous results, was becoming the norm. To this end, adaptive management blends the advantages of learning by doing and scientific empiricism in order to produce knowledge that is both salient and valid. Adaptive management involves a structured, cyclic process of developing hypotheses about how natural systems respond to management interventions, designing management projects to test those hypotheses, monitoring the results, and adjusting management on the basis of what is learned. More recently, proponents have touted adaptive management as a way to enhance the resilience of social ecological systems.

Adaptive management has been widely embraced by government agencies in the United States, Canada, Europe, Africa, the Caribbean, and Australia as well as international institutions like the United Nations.[1] A variety of nonprofit conservation organizations, including the the Nature Conservancy, Wildlife Conservation Society, and World Wildlife Fund, have produced guidelines and toolkits for implementing the idea. In the early 2000s, Foundations of Success created a guide to adaptive management that aimed to bring some conceptual clarity to the term, drawing on examples from projects in Zambia, Papua New Guinea, and British Columbia.

Despite adaptive management's broad appeal, many critics complain that it rarely works in practice as prescribed in theory. As J. B. Ruhl and Robert L. Fischman (2005, 426) lament, most implemented adaptive management programs resemble "'adaptive management-lite,' a watered-down version of the theory that resembles ad hoc contingency planning more than it does 'learning-by-doing.'" One reason adaptive management has not always delivered on its promise to make natural resource management more "rational" is that in the real world of policy making, scientists and natural resource managers must contend with advocates who have conflicting values and goals. Scientists and managers also operate in the context of institutions—both organizations and laws—that create particular constraints and opportunities, and are generally inflexible and resistant to change. In recognition of these sociopolitical realities, the focus of much adaptive management practice and scholarship has shifted to governance, especially collaboration with stakeholders, transformation of the institutions responsible for management, and the process of social learning.

The Origins and Evolution of Adaptive Management

The concept of adaptive management was born out of dissatisfaction with the conventional approach to land and natural resource management, which typically focused on a single target at a single scale, and therefore was unresponsive to emerging social demands and almost inevitably led to ecological collapse.[2] According to ecologist C. S. "Buzz" Holling (1995, 8), "Success in controlling an ecological variable that normally fluctuated led to more spatially homogenized ecosystems over landscape scales. It led to systems more likely to flip into a persistent, degraded state, triggered by disturbances that previously could be absorbed. This is the definition for loss of resilience." Initial success in managing individual resources also triggered institutional changes, as management agencies shifted their objectives to improving their efficiency—at spraying pesticides, fighting fires, or releasing hatchery fish—rather than restoring ecological health. As Holling adds, "The very success in managing a target variable for sustained production of food or fiber apparently leads inevitably to an ultimate pathology of less resilient and more vulnerable ecosystems, more rigid and unresponsive management agencies, and more dependent societies" (8).

Even as they were asked to provide static targets for management, scientists recognized that the optimal way to manage any complex biological system is almost always unclear because of four main sources of uncertainty (Allen et al. 2011). The first source of uncertainty is structural: scientists never have a complete understanding of the dynamics of a given ecosystem. The second source is semirandom environmental variation. The third is the partial controllability of management actions, which are implemented in unpredictable ways and with poorly understood consequences. And the fourth is partial observability: you cannot directly observe the state of nature but instead must rely on monitoring data that are inherently partial. Despite these uncertainties, managers and scientists were routinely asked to prepare comprehensive and detailed environmental impact assessments before taking any significant action. Furthermore, the predictions made in environmental assessments were rarely tested against real-world outcomes because of a paucity of monitoring, so scientists did not have the opportunity to revise their models in light of new information (Karkkainen 2005).

Growing dissatisfaction with the conventional management approach dovetailed with revisions in ecology that emerged in the 1950s, and were well established by the 1970s and 1980s. In the "balance of nature" view, rooted in the early twentieth-century work of Frederic Clements, nature is "tightly organized, interdependent, and highly coevolved" (Barbour 1996, 233). Species assemblages move in an orderly fashion toward a stable, climax state; if disturbed—by fire, floods, logging, or other events—natural associations will recover their original composition through succession. Over time, however, field studies led ecologists to challenge this static perspective, and replace it with a view of nature as "complex, fuzzy edged, and probabilistic" (ibid.). According to the "flux of nature" (or heterogeneity) paradigm, species that invade after a disturbance can be highly variable and determined by chance; both early and late successional species can be present at all times; large and small disturbances are inherent parts of the internal dynamics of successional cycles; and some disturbances can transform an ecosystem into an alternative, stable state (Holling 1995). From this vantage point, ecosystems are unpredictable moving targets; they coevolve with management, and surprises (that is, unexpected ecosystem responses) are inevitable.[3] Thus, management must be flexible and adaptive, monitoring must lead to corrective responses, and experiments should probe the continually changing reality of the world (Gunderson et al. 1995; Holling 1995).

Adaptive management purports to address uncertainty while taking into account the possibility of surprise through the following elements common to most definitions of the concept: explicitly stated management objectives (which according to the National Research Council [2004] are regularly revisited and redefined); a range of management options; a model of the system being managed that provides a basis for predictions about its response to possible management actions; monitoring and evaluation of outcomes of management actions; mechanisms for incorporating learning into subsequent decisions; and a collaborative structure for stakeholder participation and learning (Holling 1995).

This structured decision-making process has its roots in several other fields, including business, experimental science, systems theory, and industrial ecology (Williams and Brown 2012). It emerged in natural resource management in the 1950s, when Raymond J. H. Beverton and Sidney J. Holt articulated an adaptive philosophy with respect to fisheries management (Williams 2011). But it was Holling (1978) and his colleagues who provided a more precise explication of adaptive management for land and natural resources in 1978, with the publication of an edited volume titled *Adaptive Environmental Assessment and Management.* Directed at experts responsible for conducting environmental impact assessments, that anthology included a set of case studies purporting to demonstrate the efficacy of an adaptive approach. Through these cases, the authors aimed to demolish beliefs in the possibility of generating environmental management policies that yield stable economic, social, and political results, and do not require extensive modification over time. They argued that natural resource managers needed to both generate better information about natural systems' response to management actions and improve their ability to act in the face of the uncertain, unexpected, and unknown. They recommended the following adjustments to practice: introducing environmental dimensions at the beginning of the policy design process, devising experiments aimed at producing information about the system, and including monitoring and remediation in the planning process.

Also in 1978, Carl Walters and Ray Hilborn, both fisheries biologists, published "Ecological Optimization and Adaptive Management" in the *Annual Review of Ecology and Systematics*. In this article, Walters and Hilborn noted that it is rarely possible to predict even the short-term impacts of management interventions in fisheries. Historically, they observed, fishery managers developed deterministic prediction models, then hedged against uncertainty by adopting slightly more conservative behavior than the models predicted to be optimal. Mistakes and failures were seldom treated as opportunities for learning. Like Holling and his colleagues, Walters and Hilborn (1978) urged the adoption of an adaptive approach that would enable managers to avoid four common counteradaptive responses to new information: preoccupation with a stable equilibrium, adoption of policies based on the most optimistic assumptions, infrequent review and revision of parameter estimates, and unwillingness to discard initial analyses and parameter estimates. Walters (1986) also published his now-classic *Adaptive Management of Renewable Resources*. In it, he characterized four features basic to adaptive management: bounding management problems and recognizing constraints, representing existing knowledge in models that identify assumptions and generate predictions, representing uncertainty and identifying alternative hypotheses, and designing policies to provide opportunities for learning while continuing to supply resources for use.

While Holling, Walters, and Hilborn softened up the community of natural resource managers and scientists to adaptive management, it was physicist and environmental studies professor Kai Lee who popularized the concept with his book *Compass and Gyroscope*. Drawing on Holling's 1978 volume, Lee (1993, 53) argued that "because human understanding of nature is imperfect, human interactions with nature should be experimental." Adaptive management, he continued, "applies the concept of experimentation in the design and implementation of natural-resource and environmental policies." Unlike his predecessors, however, Lee emphasized the social and political dimensions of adaptive management. Early conceptual models of adaptive management assumed scientists and managers were the key process actors, and other stakeholders would participate through formal channels, such as public hearings and comment periods. Echoing proponents of "collaborative management," Lee observed that minimal stakeholder participation reproduced the same adversarial approach to natural resource management that had reigned since the 1970s. Furthermore, it generally failed to produce social learning; instead, sidelined resource users and advocates who disagreed with management objectives would fault the adaptive management process while availing themselves of other bargaining venues (i.e., by appealing to congressional representatives) or challenging regulations in court. By contrast, Lee contended, adaptive management in collaboration with stakeholders might produce learning more reliably. At the same time, Lee raised doubts about the practicality of such an approach based on his personal experience during the 1980s with the Northwest Power Planning Council.

Since the publication of these seminal works, there has been considerable debate over what, exactly, constitutes adaptive management, and the definition has evolved over time as scientists and managers gain experience with the concept. Most scholars and practitioners agree that at its core, adaptive management involves a process of learning about ecosystem responses to management through experience (i.e., "learning by doing") and adjusting management in response to that learning. There is also general agreement that the process of adaptive management involves the following steps: identifying management goals, specifying management options, creating a rigorous statistical process for interpreting the system's response to management interventions (usually involving the creation of quantitative conceptual models and/or a rigorous experimental design), implementing management actions, monitoring system response to management interventions, and adjusting management in response to monitoring (Westgate et al. 2013). And most distinguish adaptive management from what it is not: trial-and-error management, which involves trying something, and if it doesn't work as hoped, trying something else—an ad hoc revision of strategy over time. Rather, adaptive management entails a structured decision-making process in which scientists work with managers and stakeholders to purposefully design management in order to learn specific things about how a particular system functions.

Craig R. Allen and Lance H. Gunderson (2011) offer a simple example of how adaptive management might be implemented to conserve an endangered species like the least tern. Suppose there is broad agreement that reduced habitat is the culprit behind the tern's decline. But uncertainty remains as to what habitat element exactly has been reduced and how managers can recover the species through habitat improvements. Perhaps woody or herbaceous vegetation is encroaching on tern habitat; alternatively, some change in habitat conditions may have increased the population of nesting predators; or foraging habitat may have declined, resulting in reduced survival of the young. An adaptive management approach might involve designing concurrent replicated manipulations of vegetation and predator populations to see which, if any, produces an increase in tern populations. Even if one manipulation produces a positive response, managers must then figure out how best to reproduce that effect—using herbicide, physically manipulating the environment with bulldozers, or simply restoring the spring pulse flows of rivers.

As this example suggests, adaptive management requires a change in how scientists approach their charge. Specifically, scientists must be open to integrative modes of scientific inquiry that draw on multiple sources of evidence—as opposed to reductionist and narrowly disciplinary approaches that achieve consensus over time by examining only small pieces of larger systems (Holling 1995). Some scholars emphasize the centrality of modeling to adaptive management. According to Walters (1997), adaptive management should begin with a concerted effort to integrate existing knowledge into dynamic models capable of generating predictions about the impacts of alternative policies. Models, he says, can clarify the problem as well as enhance communication among scientists, managers, and stakeholders, screen out policy options unlikely to do much good, and identify key knowledge gaps that make model predictions suspect.

Also essential for many is the inclusion of stakeholders. The structure of stakeholder involvement varies from *participation limited* processes to *integrated adaptive management* where the public engages as "peers and partners with their manager and scientists colleagues" (Stankey et al. 2003, 13). Theoretically, integrated adaptive management ensures the realistic bounding of management problems as well as social learning that enables adjustment in response to new information (Allen and Gunderson 2011; Gunderson et al. 1995; Schreiber et al. 2004). And finally, scholars agree that monitoring in adaptive management is not for the purpose of checking compliance with regulations or making sure a management action has been carried out.

Instead, monitoring should aim to facilitate learning by collecting information on the parameters identified in the initial model(s) of the system and in relation to management goals (Schreiber et al. 2004).

In recognition that management opportunities are heterogeneous, scholars and practitioners distinguish between *active* and *passive* adaptive management. The former involves treating management interventions as deliberate experiments designed to test hypotheses and reduce uncertainty. If properly designed, such experiments can simultaneously produce short-term yields and better information for long-term management (Walters and Hilborn 1978). The latter involves using previous experience with similar systems to construct the best possible model of a system, making predictions based on that model, and then revising the model as new information becomes available as a result of management and monitoring (ibid.). Active adaptive management is closer to the original conception; passive adaptive management is a pragmatic adjustment to the concept in recognition that active adaptive management is impractical or infeasible in a given situation. Regardless of the type of adaptive management adopted, the learning pursued is in service of management, not simply for the sake of advancing science.

Adaptive Management in Practice

Fishery managers were among the first to try experimental management, with practice preceding the development of adaptive management theory. Walters and Hilborn (1978) explain that in the late 1970s, yellowfin tuna in the eastern Pacific Ocean were being subjected to a deliberate policy of overexploitation as a means of determining the maximum sustainable yield. Similar experiments were conducted in the 1950s with Pacific halibut and salmon. These experiments highlighted what the authors call the "dual control problem": large-scale management interventions produce short-term payoffs that damage the system but can generate long-term learning that can help reduce damages in the long run.

In the 1980s, Keith Sainsbury and his colleagues applied adaptive management to groundfish management in northwestern Australia—using an approach that Lee (1999) characterizes as the high-water mark in adaptive management practice. Sparked by a decline in two highly valued species, this project involved ascertaining whether it was changes in catch methods that caused the decline and if the situation could be reversed. Scientists developed four hypotheses to explain the decline, but existing data were inadequate to distinguish among them. So modelers simulated five different management strategies that differentiated among the four hypotheses. They found that one strategy, if applied for five years, would yield the greatest returns. Managers adopted this strategy and monitored the results. According to participants, despite changes in institutions and management jurisdiction, long-term experimentation yielded outcomes that led to improvements in the management of this fishery (Sainsbury 1991; Schreiber et al. 2004).

Notwithstanding these promising beginnings, critics—many of them the original proponents of the concept—contend that adaptive management has been far more influential as an idea than as a way of doing conservation, and that many of the examples cited as successful actually depart from the theoretical ideal. For example, Walters (1997) reports, "I have participated in 25 planning exercises for adaptive management of riparian and coastal ecosystems over the past 20 years; only seven ... have resulted in relatively large-scale managed experiments, and only two of these experiments would be considered well planned in terms of statistical design." Lee (1999) concludes that the sophisticated requirements for monitoring and analysis implicit in adaptive management are hard to maintain in the face of conflict, which is the norm in natural resource management. Robin Gregory, Dan Ohlson, and Joseph Arvai (2006, 2411) bemoan the fact that although adaptive management "has been elevated to a position at the forefront of ecological science and environmental management," its track record vis-à-vis implementation is weak, and many initiatives that bear its name exhibit few, if any, of adaptive management's essential characteristics. And a review by Martin J. Westgate, Gene E. Likens, and David B. Li (2013, 134) confirms that "despite the extensive literature on adaptive management, there are surprisingly few practical, on-ground examples of adaptive management."

Three Prominent Illustrations of Adaptive Management

The most often-cited example of successful adaptive management in the United States involves the management of waterfowl in North America. Two scientists at the US Fish and Wildlife Service (FWS), Byron K. Williams and Fred A. Johnson (1995, 430), explain that although North American waterfowl are "among the world's most thoroughly investigated biota," the impacts of waterfowl harvest regulations are highly uncertain. This is partly because the regulations themselves are complex; it is also because regulations "chase" population (that is, limits are relaxed when populations appear abundant and are tightened when they appear low), so harvest and environmental impacts are confounded. In any case, scientists can only partially observe the impacts of regulations, and there are biological mechanisms they do not yet understand. Given these uncertainties, managers struggle to make decisions that are consistent with long-term conservation goals. And because waterfowl hunting is lucrative, there are substantial material benefits associated with reducing that uncertainty.

In the early 1990s, dissatisfied with the existing management regime, in which scientists simply furnished information and managers made ostensibly wise decisions, Williams and Johnson prescribed adaptive management for regulations on sport hunting for North American waterfowl. They argued that there were two essential preconditions for adaptive management to be effective: a clearly specified objective for harvest management, and a large-scale monitoring program that would deliver information about the biological attributes of interest. They pointed out that modeling programs already existed, as did the capacity to analyze the data those programs yielded. But what was occurring was essentially passive adaptive management. What they wanted to do was use regulations to pursue understanding of regulatory effects. As long as they had a clear objective, they could design regulations to generate information that would allow them to discriminate among four hypotheses (models) about how the system responds to different harvest regulations. (They did not have complete freedom, of course; regulations had to be acceptable to hunters and also enforceable.)

In 1995, the FWS adopted an adaptive harvest management program for North American waterfowl. Under the new approach, scientists focus data collection on the information most useful for management, and science is centered on hypotheses about how the managed system responds to management actions. The regulatory alternatives consist of three packages, corresponding to liberal, moderate, and conservative hunting regulations. Each package specifies daily bag limits and season lengths for each of the four waterfowl flyways. The FWS has developed four models that capture key sources of the uncertainty in mallard population responses to harvest.

The FWS and Canadian Wildlife Service, along with various state and provincial partners, carry out relevant monitoring programs (i.e., monitoring with a clear management focus). The adaptive harvest management process incorporates these elements into an annual cycle of adopting regulation and then assessing its effects using all four models. Models that predict well are given more weight than those that predict poorly. The annual assessment feeds into next year's regulations.

In the first decade of adaptive harvest management, mallard breeding population estimates fluctuated between six and twelve million. Ostensible benefits of the adaptive harvest management program include the following: debates have shifted from disputes over science to a clear focus on objectives and management alternatives; it has become apparent that many management strategies are biologically sustainable, and so optimality is determined by social values and stakeholder demands; and finally, efforts to fine-tune management to exploit additional small sources of variation are likely to have large costs while yielding only marginal benefits (Nichols et al. 2007).

Although the North American Waterfowl Management Plan is generally described as a strong example of adaptive management, conservation biologist Gary K. Meffe and his coauthors (2002) contend that it is merely an instance of passive adaptive management. They point out that, first, biologists collect data about harvest levels and population characteristics based on regulations that are enacted, rather than designing regulations specifically to generate the information they need. Second, biologists cannot set up true reference (control) areas; instead, they rely on previous years' baseline data, which furnish less powerful comparisons for deducing cause and effect. And third, future policy decisions are not tied to the outcomes of learning; decision makers are not obligated to modify harvest levels in accordance with new knowledge. Despite these criticisms, the adaptive harvest management program remains among the most uncontroversial illustrations of adaptive management—perhaps because it benefited from a clear and uncontested objective: to maximize long-term, cumulative harvest utility.

Another widely cited US example of adaptive management involves changing the operation of the Glen Canyon Dam—although evaluations of this project are even more ambiguous than those of the adaptive harvest management program. The Glen Canyon Dam was completed in the 1960s.

It was intended to alleviate potential constraints on the use of Colorado River water by states in the Upper Colorado River Basin. Authorized by the Colorado River Storage Project in 1956 and completed by the Bureau of Reclamation in 1963, the dam is located in Arizona, about fifteen miles upstream of Lee's Ferry. The dam created a reservoir, Lake Powell, which has become a valued recreational amenity as well as a water storage facility; it also produces hydropower. At the same time, the dam has wrought major ecological changes. Prior to the dam's construction, the Colorado River in the Grand Canyon experienced highly variable flows and sediment levels, and water temperatures varied dramatically over the course of the year. Construction of the dam eliminated most of the seasonal and annual variation in the river's flow, with any remaining variation in the service of hydropower generation. Moreover, the dam traps about 84 percent of the sediment that formerly entered the Grand Canyon, so sandbars and beaches in the canyon are shrinking. And discharges from Lake Powell, drawn from its lower depths, are cold around the year. One consequence is that of the eight species of native fish that once inhabited the canyon, four are no longer found there. One of the species that remains, the humpback chub, is listed as endangered under the federal Endangered Species Act because its largest remaining population is in the Grand Canyon (Feller 2008; Lovich and Melis 2007).

In 1992, Congress passed the Grand Canyon Protection Act, which mandated an environmental impact study on changes to the operation of the Glen Canyon Dam. The final environmental impact study, completed in 1995, prescribed operating the dam based on modified, low-fluctuating flows using the Adaptive Management Program. To facilitate the program, the interior secretary established a twenty-five-member Adaptive Management Work Group comprising stakeholders from federal and state agencies, hydropower interests, Native American tribes, and environmental groups.

According to Jeffrey Lovich and Theodore S. Melis (2007), ecologists with the US Geological Survey, review of the Adaptive Management Program after ten years found that operating the dam according to the modified, low-fluctuating flows alternative had not mitigated the influence of dam regulation with respect to either thermal or hydrologic changes, or the sand supply limitation downstream. The program did conduct two experimental releases that yielded useful information. For instance, the ecosystem's response to those releases made it clear that the scientists' model of the source of sediment for sandbar replenishment was inaccurate. The experimental releases also may have boosted the population of the humpback chub between 2001 and 2008.

Nevertheless, the Adaptive Management Working Group has been unable to agree on a permanent change in the flow regime that might benefit the chub in the long run. Legal scholar Joseph M. Feller (2008) attributes the Adaptive Management Program's failure to its insistence on collaborative decision making. Specifically, Feller argues that collaborative decision making has stifled adaptive management by making agreement among stakeholders a prerequisite to changes in dam operations. The kinds of changes in dam operations that would benefit the chub—particularly eliminating daily fluctuations in flow and increasing water temperature—hamper nonnative sport fisheries and hydropower production, both of which are important to some stakeholders.

Journalist April Reese (2009) reaches the same conclusion, reporting that "even as the science on the river's health points in the direction of flow changes that would mimic natural conditions, members of the twenty-five-member Adaptive Management Work Group remain entrenched along old battle lines that pit producers and Colorado River Basin states against environmental groups and wildlife agencies concerned about the river's ecological decline." Gridlock among stakeholders is exacerbated by the fact that Congress failed to articulate clear goals and priorities for adaptive management of Glen Canyon Dam operations; as a result, federal agencies have been reluctant to disturb the expectations of current user groups (Susskind et al. 2010; Zellmer and Gunderson 2009). The result, as Reese (2009) points out, is a policy stalemate that keeps the dam operating under the same conditions as in 1992, when Congress passed the Grand Canyon Protection Act.

A final prominent case is the application of adaptive management to the Great Barrier Reef Marine Park. The Great Barrier Reef is the world's largest coral reef system, encompassing 2,900 reefs and 900 islands in a 348,000-square-kilometer zone off the coast of Queensland, Australia. The reef is an important cultural and aesthetic resource, and also contributes substantially to the local economy, primarily through tourism. In 1975, concern over worsening ecological conditions in the Great Barrier Reef led the Australian government to establish the multiple use Great Barrier Reef

Marine Park and Great Barrier Reef Marine Park Authority (GBRMPA), a managing body that regulates park activities; drilling, mining, and exploration in the park were prohibited. After initially challenging the central government's authority, Queensland signed a comanagement agreement in 1979 with the commonwealth enabling the joint management of the reef. In 1981, the UN Educational, Scientific, and Cultural Organization declared the Great Barrier Reef a World Heritage Site.

The GBRMPA's main management tool is a zoning and permitting system that allows or prohibits particular uses in designated areas of the reef. These uses include fishing, the primary extractive industry in the Great Barrier Reef. As fishing pressure in the reef increased in the early 1990s, state and federal regulators found themselves with a shortage of information on which to base management decisions. Of particular concern was the impact of line and spearfishing, which were believed to have the greatest potential impact on reef ecology (Mapstone et al. 1996). In 1993, the newly formed national Cooperative Research Center for the reef initiated and financed the Effects of Line Fishing (ELF) Program. The primary function of ELF was the design and implementation of a seascape-wide controlled experiment—an active adaptive management approach that had been championed for several years by regulators and scientists, including Walters and Sainsbury, as the "most effective mechanism[s] for assessing empirically the responses of targeted stocks ... to changes in fishing pressure" (ibid., i). For two years, ELF researchers conducted preliminary studies, including a computer simulation of the impacts of a large-scale experiment on the red coral fish population, before settling on an adaptive management design.

Between 1995 and 1996, ELF researchers canvased the approach with relevant stakeholders. The experiment—actually a series of experiments to be conducted over nearly a decade using replicated reefs that were open, closed, or reopened to fishing—was immediately opposed by some environmental advocates and fishers (Mapstone et al. 2004; Walters 1997). Environmental groups cited the risks of opening previously protected reefs, while fishers rejected specific closures. The program faced an additional institutional challenge: the federal Parliament would need to amend the GBRMPA zoning provisions to allow for the unusual rezoning that the ELF experiment necessitated (Mapstone et al. 2004). Although facing dissent, ELF had the backing of the key decision makers—the GRBMPA and Fisheries Queensland—and also benefited from the support of a handful of important conservation and fisheries organizations. In November 1996, Parliament passed the necessary amendments, and the GBRMPA approved them in January 1997, allowing ELF to commence. Owing to controversy over the project and its implementation in a World Heritage Site, an international review panel was formed in 1998. In another victory, the panel recommended that the program continue to its set end date of 2006 (ibid.).

In a 2004 summary report, ELF researchers argued that results from the experimental manipulations provided "the most convincing evidence" that the GBRMPA's use of closed areas (i.e., reserves) had been effective in protecting Great Barrier Reef fish populations; they further maintained that the continued use of reserves would likely sustain healthy population densities "in spite of an active GBR fishery" (ibid., iii). The link between these findings and subsequent park policy is more tenuous. Between 2003 and 2004, the GBRMPA undertook a massive rezoning of the marine park, which among other changes increased the proportion of the park closed to fishing from 5 to 33 percent. Although the rezoning process was initiated in 1999, according to Terence P. Hughes and his colleagues (2007), the results from ELF influenced these changes to park management. Researchers also note that the successful rezoning—which has produced some improvements in the ecological health of the Great Barrier Reef (McCook et al. 2010)—owes as much to favorable political conditions. These conditions include a multiple use mandate that nonetheless clearly prioritizes ecological conservation; a centralized governance system with strong national and state leadership along with a commitment to reef protection; and a powerful tourism industry that generally supports conservation efforts (Evans et al. 2014; Gershman et al. 2011; Hughes et al. 2007).

Why Is Adaptive Management So Difficult in Practice?

Scholars cite a variety of challenges to implementing adaptive management. Foremost among them is intractable disagreement among stakeholders. According to many proponents, it is essential to garner stakeholder agreement on management goals that serve as the basis against which progress can be measured; they observe that failure to do so can lead some parties to reject the results of an adaptive management process (Stankey et al. 2003). But critics regard combining adaptive management with consensus-based decision making as adding a potentially insurmountable hurdle to changing management in response to new information (Doremus et al. 2011; Zellmer and Gunderson 2009).

A second political obstacle to adaptive management is inflexibility or unwillingness to experiment among organizations with management responsibility (Ladson and Argent 2002). Agencies with precautionary cultures or statutory mandates are especially likely to resist experimentation. Gunderson, Holling, and Stephen S. Light (1995, 495) bemoan the "extreme nature of the recalcitrance or inertia of institutions and the almost pathological inability to renew or restructure." Walters (1997) attributes risk aversion—that is, reluctance to experiment—among organizations to the desire to protect self-interest by bureaucrats. Experimentation, he says, can be threatening to agency scientists who feel the need to appear certain to maintain public credibility, perceive adaptive management as imperiling individual research programs and funding streams, and wish to avoid imposing immediate costs on stakeholders by unsettling existing management regimes.

Agencies also operate within legal frameworks that may further discourage experimentation. In the United States, for example, administrative law reflects an earlier era's belief in linearity and finality when it came to resource management planning. Under the National Environmental Policy Act and Administrative Procedure Act, requirements for extensive front-end, predictive analyses, public participation periods, and judicial review impose significant costs on active adaptive management (Biber 2013).

Other, more prosaic reasons given for adaptive management failures include the lack of adequate time and/or money, the failure of scientists to understand the need for information that is directly relevant to management, failure by management agencies to clearly define their responsibilities for implementing adaptive management, a tendency among scientists to overstate their ability to measure complex functional relationships through experimentation, a lack of leadership or "institutional champion," hijacking of management goals for research interests, and difficulties translating learning into practice (Allen and Gunderson 2011; Gilmour et al. 1999; Gregory et al. 2006; Lee 1999; Walters 1986, 1997).

Rather than focusing on adaptive management failures, some scholars have tried to identify the conditions under which adaptive management is likely to be effective. Lee (1999) contends that adaptive management "is likely to be worthwhile when laboratory style precision seems infeasible but trial-and-error seems too risky. And that's much of the time in conservation." But others are more skeptical. Walters (1986) points out that adaptive management is costly and slow in many situations, so managers need to decide whether adaptive management is appropriate (how much will they learn, and is it worth the cost and time?). Gunderson (1999) suggests that adaptive management may be effective only if there is resilience in the ecological system and flexibility among stakeholders in the social system. Meffe and his colleagues (2002) contend that adaptive management works best when large differences in the performance of the system are predicted between management intervention and reference area, data collection is practical (relatively inexpensive), and all stakeholders accept the basic model on which the process is based. The latter is only likely to be true, they observe, if the model is developed in collaboration with stakeholders. They add that groups need to be willing to change management on the basis of new knowledge, and need stable leadership and funding.

Gregory, Ohlson, and Arvai (2006) conclude that whereas experimental adaptive management may be appropriate for narrow, well-defined problems at a relatively small scale, large-scale land use planning under climate change is not a hospitable context for adaptive management. Similarly, Walters (1997) asserts that effective adaptive management initiatives tend to involve relatively simple institutional settings, with a single lead management agency along with a few dedicated people who have organized and maintained the initiative (institutional champion[s]). By contrast, he notes, adaptive management has failed in complex settings like the Everglades and Columbia River Basin. Ecologist Westgate and his coauthors (2013) concur, based on their comprehensive review of adaptive management–related journal articles, that adaptive management may not be well suited to testing management options associated with some large-scale ecological phenomena, or factors important at multiple spatial scales or across multiple land tenures. This conclusion is ironic, given that adaptive management is most visibly associated with—and arguably most needed in—complex, landscape-scale initiatives.

Rather than focusing on the appropriate scale, some scholars emphasize the nature of the management challenge. Williams and E. D. Brown (2012) suggest that adaptive management makes sense when there is substantial uncertainty about the impacts of management, it is realistic to expect we can reduce uncertainty, and reducing uncertainty can actually improve management. Similarly, legal scholars Holly Doremus and her coauthors (2011) argue that adaptive management "is especially appropriate when uncertainties make management choices difficult, but the prospects for reducing uncertainty appear good." This is because adaptive management requires more resources than conventional management, and imposes unfamiliar demands on stakeholders and institutions. The authors propose that managers be required to undertake an explicit, formal analysis of the prospects for learning and its expected value for management before employing adaptive management.

Salvaging Adaptive Management

Adaptive management is now embedded in the directives of land and natural resources management agencies in the United States and around the world. Yet adaptive management is not universally embraced. Many environmentalists remain ambivalent or skeptical about adaptive management, which they regard as a way of proceeding with controversial management actions on the grounds that corrections can be made in the future. They also worry that adaptive management gives managers too much discretion, thereby reducing accountability and creating the risk of agency capture by development interests. At the same time, some resource users worry that adaptive management does not provide them with adequate certainty, making it more difficult for them to plan (Doremus et al. 2011). This kind of distrust among stakeholders, combined with sticky institutions, has undermined efforts to implement adaptive management. But, as Westgate, Likens, and Li (2013) conclude, despite the difficulties associated with implementing adaptive management, there is no alternative, viable, or clearly superior framework.

Therefore, most contemporary adaptive management scholarship reflects a growing recognition of the interrelatedness of social ecological systems and the central importance of governance. Scholars recognize that for adaptive management to work, governance systems must be reconfigured or transformed from prediction and control to management as learning (Pahl-Wostl 2007). Ideally, such transformed systems would exhibit resilience, defined as the ability of a desired social ecological system to reorganize and renew itself following disturbance. Resilience is a function of the degree to which a social ecological system is capable of self-organization, and extent to which the system can enhance its capacity to learn and adapt.

Recent work has been devoted to exploring how social transformations occur and prescribing ways of enabling them. Some scholars have argued that insights from transition management can help advance such transformations (Pahl-Wostl 2007; van der Brugge and van Raak 2007). Transition management emerged in response to the failures of Dutch environmental policy; it was rooted in recognition of the difficulty of changing sociotechnical systems (van der Brugge and van Raak 2007). Proponents offered transition management as a governance theory with associated tools. Like adaptive management, transition management stresses the need for continuous learning and adjustment. Unlike adaptive management, which originally emphasized ecology, transition management has from its inception focused on the social aspects of management. According to Rutger van der Brugge and Roel van Raak (2007), one element of transition management that might prove useful for adaptive management is transition arenas, which are informal networks of innovators/visionaries who can lay the groundwork for a transformation in management.

Other scholars have investigated the possibilities of adaptive comanagement, an approach that proponents believe builds resilience in social ecological systems. Adaptive comanagement systems are "flexible, community-based systems of resource management tailored to specific places and situations, supported by, and working with, various organizations at different levels" (Olsson et al. 2004). Key attributes of adaptive comanagement include an emphasis on the diverse needs and distribution of power among stakeholders; the effort to build on culturally embedded, formal and informal rules and norms; formation of horizontal and vertical linkages as well as networks to foster trust and social learning; the inclusion of many types and sources of knowledge, and the shared development of that knowledge among stakeholders; and the enhanced capacity among resource management organizations to respond proactively to uncertainty. As an example of adaptive comanagement, Per Olsson, Carl Folke, and Thomas Hahn (2004) describe how unconnected actors across a landscape in southern Sweden were mobilized and reconfigured into a coherent whole pursuing ecosystem management—a self-organizing process that occurred in response to a threat to the area's cultural and ecological values. Central to this transformation, which took place over about a decade, was a local policy entrepreneur who initiated dialogues, mobilized social networks across scales, and coordinated people, information flows, and activities.

Conclusions

Since it was first articulated in the 1970s, adaptive management has been widely embraced by scientists, managers, and decision makers as the most promising approach to natural resource management. Adaptive management appeals to the scientifically sophisticated who have soundly rejected the outmoded ecological assumptions of conventional management; recognize how little is actually known about managed natural systems; and "are drawn to the trustworthiness of experimentation as a way to establish reliable knowledge" (Lee 1999). By purportedly addressing uncertainty and (increasingly) embracing the collaboration of stakeholders, adaptive management also appears to offer a pathway out of the conflict, paralysis, and environmental degradation that have afflicted many management programs.

Notwithstanding adaptive management's conceptual popularity, practitioners have struggled to transfer it from theory to practice. Hundreds of adaptive management programs have been attempted, but few are deemed "successes" (Rist et al. 2013). Researchers have cataloged numerous challenges to adaptive management's application, largely stemming from the sociopolitical realities in which it must take place. Even so, with no clear viable alternative, interest in its use persists. On the one hand, some researchers argue that far from being a panacea, adaptive management should be viewed as just one approach in the suite of management tools; it is only appropriate for a limited number of cases and may not be amendable to large, complex management problems. Instead of shrinking the domain of adaptive management's applicability, however, others have turned their attention to the broader sociopolitical system. For adaptive management to work, they assert, processes of learning and adjustment must take place across the entire linked socioecological system. The emerging concept of adaptive comanagement thus represents one way adaptive management may retain its broad relevancy, particularly in those cases for which it is perhaps most needed: complex landscape-scale initiatives.

Notes

1. For the United States, see the Department of the Interior's strategic plan for fiscal year 2011–2016, which calls for adaptive management as part of the agency's mission to provide a scientific foundation for decision making (Williams and Brown 2012). For Canada, see the Canadian Environmental Assessment Agency (2015) and British Columbia's strategy for managing BC Crown forestland, which covers more than two-thirds of the province. For Europe, see the Integrated Coastal Zone Management and EU Water Directive.

2. The focus on a single species and scale as well as the assumption of equilibrium and fixed carrying capacity dominated not only resource extraction but also landscape conservation/preservation efforts (Rogers 2003).

3. As Lance H. Gunderson (1999) points out, local surprises can be created by broader-scale processes about which there is little knowledge; there can also be cross-scale surprises, in which larger-scale fluctuations intersect with slowly changing internal variables to create an alternative stable state, as has occurred in the Chesapeake Bay; and there can be genuine novelty, when new variables and processes transform a system into a new state, as with the invasion of exotic species or construction of a dam.

References

Allen, Craig R., Joseph J. Fontaine, Kevin L. Pope, Ahjond S. Garmestani. 2011. "Adaptive Management for a Turbulent Future." *Journal of Environmental Management* 92 (5): 1339–1345. doi:10.1016/j.jenvman.2010.11.019.

Allen, Craig R., and Lance H. Gunderson. 2011. "Pathology and Failure in the Design and Implementation of Adaptive Management." *Journal of Environmental Management* 92:1379–1384. doi:10.1016/j.jenvman.2010.10.063.

Barbour, Michael. 1996. "Ecological Fragmentation in the Fifties." In *Uncommon Ground: Rethinking the Human Place in Nature*, ed. William Cronon, 233–255. New York: W. W. Norton and Company.

Biber, Eric. 2013. "Adaptive Management and the Future of Environmental Law." *Akron Law Review* 46 (4): 933–962.

Canadian Environmental Assessment Agency. 2015. "Operational Policy Statement Addressing 'Purpose of' and "Alternative Means' under the Canadian Environmental Assessment Act, 2012." Accessed December 9, 2016, http://www.ceaa.gc.ca/default.asp?lang=En&n=1B095C22-1.

Doremus, Holly, William L. Andreen, Alejandro Camacho, Daniel A. Farber, Robert L. Glicksman, Dale Goble, Bradley C. Karkkainen, et al. 2011. "Making Good Use of

Adaptive Management." Center for Progressive Reform, White Paper No. 1104, April. Accessed December 12, 2016, http://www.progressivereform.org/articles/Adaptive_Management_1104.pdf.

Evans, Louisa S., Natalie C. Ban, Michael Schoon, and Mateja Nenadovic. 2014. "Keeping the 'Great' in the Great Barrier Reef: Large-Scale Governance of the Great Barrier Reef Marine Park." *International Journal of the Commons* 8 (2): 396–427. doi:10.18352/ijc.405.

Feller, Joseph M. 2008. "Collaboration and the Colorado River: Collaborative Management of Glen Canyon Dam: The Elevation of Social Engineering Over Law." *Nevada Law Journal* 8:896–941.

Gershman, Dave, Julia Wondolleck, and Steven Yaffee. 2011. "Marine Ecosystem-Based Management in Practice: Great Barrier Reef Marine Park." Work product of the Ecosystem Management Initiatives Program, University of Michigan. Accessed December 12, 2016, http://webservices.itcs.umich.edu/drupal/mebm/?q=node/56.

Gilmour, Alistair, Greg Walkerden, and James Scandol. 1999. "Adaptive Management of the Water Cycle on the Urban Fringe: Three Australian Case Studies." *Conservation Ecology* 3 (1): 11. Accessed December 12, 2016, http://www.ecologyandsociety.org/vol3/iss1/art11.

Gregory, Robin, Dan Ohlson, and Joseph Arvai. 2006. "Deconstructing Adaptive Management: Criteria for Applications to Environmental Management." *Ecological Applications* 16 (6): 2411–2425. doi:10.1890/1051-0761(2006)016%5B2411:damcfa%5D2.0.co;2.

Gunderson, Lance H. 1999. "Resilience, Flexibility, and Adaptive Management—Antidotes for Spurious Certitude?" *Conservation Ecology* 3 (1). Accessed December 9, 2016, http://www.ecologyandsociety.org/vol3/iss1/art7.

Gunderson, Lance H., C. S. Holling, and Stephen S. Light. 1995. "Barriers Broken and Bridges Built: A Synthesis." In *Barriers and Bridges to the Renewal of Ecosystems and Institutions*, ed. Lance H. Gunderson, C. S. Holling, and Stephen S. Light, 489–532. New York: Columbia University Press.

Holling, C. S., ed. 1978. *Adaptive Environmental Assessment and Management*. New York: John Wiley and Sons.

Holling, C. S. 1995. "What Barriers? What Bridges?" In *Barriers and Bridges to the Renewal of Ecosystems and Institutions* , ed. Lance H. Gunderson, C. S. Holling, and Stephen S. Light, 1–34. New York: Columbia University Press.

Hughes, Terence P., Lance H. Gunderson, Carl Folke, Andrew H. Baird, David Bellwood, Fikret Berkes, Beatrice Crona, et al. 2007. "Adaptive Management of the Great Barrier Reef and the Grand Canyon World Heritage Areas." *Ambio* 36 (7): 586–592. doi:10.1579/0044-7447(2007)36%5B586:amotgb%5D2.0.co;2.

Karkkainen, Bradley C. 2005. "Panarchy and Adaptive Change: Around the Loop and Back Again." *Minnesota Journal of Law, Science, and Technology* 7:59–77.

Klerkx, Laurens, Noelle Arts, and Cees Leeuwis. 2010. "Adaptive Management in Agricultural Innovation Systems." *Agricultural Systems* 103 (6): 390–400. doi:10.1016/j.agsy.2010.03.012.

Ladson, Anthony R., and Robert M. Argent. 2002. "Adaptive Management of Environmental Flows: Lessons for the Murray-Darling Basin from Three Large North American Rivers." *Australasian Journal of Water Resources*, 5 (1):89–102.

Lee, Kai. 1993. *Compass and Gyroscope: Integrating Science and Politics for the Environment*. Washington, DC: Island Press.

Lee, Kai. 1999. "Appraising Adaptive Management." *Conservation Ecology* 3 (2). Accessed December 10, 2016, http://www.ecologyandsociety.org/vol3/iss2/art3.

Lovich, Jeff, and Theodore S. Melis. 2007. "The State of the Colorado River Ecosystem." *International Journal of River Basin Management* 5 (3): 207–221. http://sbsc.wr.usgs.gov/products/pdfs/P0151_10sep.pdf.

Mapstone, Bruce D., Anthony D. M. Smith, and Robert A. Campbell. 1996. "Design of Experimental Investigations of the Effects of Line and Spear Fishing on the Great Barrier Reef." CRC Reef Research Technical Report No. 7. Townsville, Queensland: CRC Reef Research Centre. Accessed December 12, 2016, http://rrrc.org.au/wp-content/uploads/2014/03/Technical-Report-07.pdf.

Mapstone, Bruce D., Campbell R. Davies, Lorne R. Little, André E. Punt, Adam D. M. Smith, Francis Pantus, Dong C. Lou, et al. 2004. "The Effects of Line Fishing on the Great Barrier Reef and Evaluations of Alternative Potential Management Strategies." CRC Reef Research Centre Technical Report No 52. Townsville, Queensland: CRC Reef Research Centre. Accessed December 12, 2016, http://rrrc.org.au/wp-content/uploads/2014/04/Technical-Report-52.pdf.

McCook, Laurence J., Tony Ayling, Mike Cappo, J. Howard Choat, Richard D. Evans, Debora M. De Freitas, Michelle Heupel, et al. 2010. "Adaptive Management of the Great Barrier Reef: A Globally Significant Demonstration of the Benefits of Networks of Marine Reserves." *Proceedings of the National Academy of Sciences of the United States of America* 107 (43): 18278–18285. doi:10.1073/pnas.0909335107.

Meffe, Gary K., Larry A. Nielsen, Richard L. Knight, and Dennis A. Schenborn. 2002. "Adaptive Management." In *Ecosystem Management*, 95–111. Washington, DC: Island Press.

National Research Council. 2004. *Adaptive Management for Water Resources Project Planning*. Washington, DC: National Academies Press.

Nichols, James D., Michael C. Runge, Fred A. Johnson, and Byron K. Williams. 2007. "Adaptive Harvest Management of North American Waterfowl Populations: A Brief

History and Future Prospects." *Journal of Ornithology* 148 (Supp. 2): S343–S349. doi:10.1007/s10336-007-0256-8.

Olsson, Per, Carl Folke, and Thomas Hahn. 2004. "Social-Ecological Transformation for Ecosystem Management: The Development of Adaptive Co-Management of a Wetland Landscape in Southern Sweden." *Ecology and Society* 9 (4). Accessed December 12, 2016, http://www.ecologyandsociety.org/vol9/iss4/art2.

Pahl-Wostl, Claudia. 2007. "Transitions towards Adaptive Management of Water Facing Climate and Global Change." *Water Resources Management* 21:49–62. doi:10.1007/s11269-006-9040-4.

Parkes, John P., Alan Robley, David M. Forsyth, and David Choquenot. 2006. "Adaptive Management Experiments in Vertebrate Pest Control in New Zealand and Australia." *Wildlife Society Bulletin* 34 (1): 229–236. doi:10.2193/0091-7648(2006)34%5B229:ameivp%5D2.0.co;2.

Reese, April. 2009. "Colorado River Adaptive Management Program Needs Overhaul, Critics Say." *Greenwire*, May 7. Accessed December 12, 2016, http://www.eenews.net/stories/77694.

Rist, Lucy, Adam Felton, Lars Samuelsson, et al. 2013. "A New Paradigm for Adaptive Management." *Ecology and Society* 18 (4): 63. doi:10.5751/ES-06183-180463.

Rogers, Kevin H. 2003. "Adopting a Heterogeneity Paradigm: Implications for Management of Protected Savannas." In *The Kruger Experience: Ecology and Management of Savanna Heterogeneity*, ed. Johan T. du Toit, Kevin H. Rogers, and Harry C. Biggs, 41–58. Washington, DC: Island Press.

Ruhl, J. B., and Robert L. Fischman. 2005. "Adaptive Management in the Courts." *Minnesota Law Review* 95 (2): 424–484.

Sainsbury, Keith J. 1991. "Application of an Experimental Approach to Management of a Tropical Multispecies Fishery with Highly Uncertain Dynamics." *ICES Marine Science Symposia* 193:301–320.

Schreiber, E. Sabine G., Andrew R. Bearlin, Simon J. Nicol, and Charles R. Todd. 2004. "Adaptive Management: A Synthesis of Current Understanding and Effective Application." *Ecological Management and Restoration* 5 (3): 177–182. doi:10.1111/j.1442-8903.2004.00206.x.

Stankey, George H., Bernard T. Bormann, Clare Ryan, Bruce Shindler, Victoria Sturtevant, Roger N. Clark, and Charles Philpot. 2003. "Adaptive Management and the Northwest Forest Plan." *Journal of Forestry* 101 (1): 40–46.

Susskind, Lawrence, Alejandro E. Camacho, and Todd Schenk. 2010. Collaborative Planning and Adaptive Management in Glen Canyon: A Cautionary Tale." *Columbia Journal of Environmental Law* 35 (1): 1–56.

van der Brugge, Rutger, and Roel van Raak. 2007. "Facing the Adaptive Management Challenge: Insights from Transition Management." *Ecology and Society* 12 (2). Accessed December 12, 2016, http://www.ecologyandsociety.org/vol12/iss2/art33.

Walters, Carl. 1986. *Adaptive Management of Renewable Resources*. New York: Macmillan.

Walters, Carl. 1997. "Challenges in Adaptive Management of Riparian and Coastal Ecosystems." *Conservation Ecology* 1 (2). Accessed February 22, 2017, http://www.consecol.org/vol1/iss2/art1.

Walters, Carl, and Ray Hilborn. 1978. "Ecological Optimization and Adaptive Management." *Annual Review of Ecology and Systematics* 9:157–188. doi:10.1146/annurev.es.09.110178.001105.

Westgate, Martin J., Gene E. Likens, and David B. Li. 2013. "Adaptive Management of Biological Systems: A Review." *Biological Conservation* 158 (February): 128–139. doi:10.1016/j.biocon.2012.08.016.

Williams, Byron K. 2011. "Adaptive Management of Natural Resources: Framework and Issues." *Journal of Environmental Management* 92:1346–1353. doi:10.1016/j.jenvman.2010.10.041.

Williams, Byron K., and E. D. Brown. 2012. "Adaptive Management: The U.S. Department of the Interior Applications Guide." Adaptive Management Working Group, US Department of the Interior, Washington, DC. Accessed December 9, 2016, https://www.usgs.gov/sdc/doc/DOI-Adaptive-Management-Applications-Guide-27.pdf.

Williams, Byron K., and Fred A. Johnson. 1995. "Adaptive Management and the Regulation of Waterfowl Harvests." *Wildlife Society Bulletin* 23:430–436.

Zellmer, Sandra B., and Lance H. Gunderson. 2009. "Resilience and Environmental Law Reform Symposium: Why Resilience May Not Always Be a Good Thing: Lessons in Ecosystem Restoration from Glen Canyon and the Everglades." *Nebraska Law Review* 87:893–949.

8 Sustainable Development: Linking Environment and Development

Oluf Langhelle

Sustainable development is both a simple and notoriously complex concept that has become part of the standard vocabulary of governments, international organizations, business organizations, local and regional authorities, and nongovernmental organizations (NGOs). It has also permeated all academic disciplines—be it political science, sociology, economics, architecture, business studies, and more. It is a highly disputed and contested concept, too, detested by some and celebrated by others. The World Commission on Environment and Development (WCED) crafted the concept of sustainable development in the period 1983–1987. The WCED's conception of sustainable development was unique in its perspective on global challenges, diagnosis, and proposed policy directions.

This chapter starts with the concept's origin and establishes what the innovation of sustainable development represented. Thereafter, it traces the concept's antecedents and the context in which it evolved. Then it examines what has happened with this notion in its thirty years of existence. Sustainable development has not only diversified conceptually in the years following the release of *Our Common Future*; it also has been challenged by competing and alternative concepts and interpretations. Some of the neighboring ideas of sustainable development are therefore discussed, before turning to the key questions: What has been the impact of sustainable development? Has sustainable development made any difference at all?

These questions are discussed in light of two different but equally important issues that have permeated the whole complex of global environment and development linkages since the late 1960s: the issue of (environmental) limits and the question of economic growth. Following Sharachchandra M. Lélé (1991, 609), who argues that "development is *a process of directed change*," and definitions of development thus "embody" both the *objectives* of the process and *means* of achieving the objectives, the chapter first explores the objectives of sustainable development and then the means of achieving it.

Sustainable Development as a Conceptual Innovation

The description of sustainable development provided by the WCED (1987) in *Our Common Future* is by far the most widely cited definition of the concept. It is referred to as "the canonical definition" (Lipschutz 2012, 480), the "standard definition" (Kates et al. 2005, 10), "a watershed" (Redclift 1992, 33), and a "paradigm shift" (Bernstein 2002, 2), to mention a few. As the mandate of the WCED reveals, sustainable development existed in the vocabulary before the commission was established. The WCED (1987, ix) was explicitly asked by the UN General Assembly "to propose long-term environmental strategies for achieving sustainable development by the year 2000 and beyond." Moreover, the commission was asked to recommend ways that concern for the environment could be translated into greater and more efficient cooperation between developing and developed countries, and define "common and mutually supportive objectives," "shared perceptions of long-term environmental issues," and "aspirational goals for the world community" (ibid.).

The formulation of sustainable development as defined by the commission was an explicit attempt to bring together, merge, and reconcile the (global) environment and development discourses that had been around since the late 1960s. As Steven Bernstein (2002, 5) argues, the report "marked the first real synthesis of the environment and development agendas." This particular synthesis was the WCED's innovation—one that included, among others, "a compromise between competing values, including growth, conservation, and inter- and intra- generational equity" (ibid.) as well as a specific operationalization of sustainable development in political, economic, and institutional terms (Engfeldt 2009, 110). What the exact content of this compromise was, however, is highly contested, and there are numerous interpretations of what the WCED meant by sustainable development. It contained descriptions and elements of various and, to some extent, competing pathways.

When the commission was established, it was placed in the middle of discursive struggles over a number of environmental and developmental issues for which there was some, but not much, guidance in terms of what was meant by sustainable development. Defining sustainable development therefore became of utmost importance for the WCED—to find an overarching concept that could guide and structure the world's environmental and developmental challenges. Or as described by Nitin Desai (1986, 1) in his note on the concept of sustainable development to all the commissioners, "It is ... essential that development policy in the broadest sense and environmental policy be integrated in a common framework. The concept of 'sustainable development' can provide the basis for such integration." What the commission did was to give this term new meaning with far more political content (Brundtland 1997, 79).

When the commission's mandate was discussed, some clearly wanted environmental issues to constitute the core of the WCED's work. Norwegian prime minister and head of the commission Gro Harlem Brundtland's response here is important for both the meaning of sustainable development and an understanding of how the WCED (1987, xi) interpreted the mandate itself:

> This would have been a grave mistake. The environment does not exist as a sphere separate from human actions, ambitions, and needs, and attempts to defend it in isolation from human concerns have given the very word "environment" a connotation of naiveté in some political circles.

Brundtland later explained this aspect of the commission's work as a way of catching and taking seriously the growing skepticism in developing countries toward the environmental concerns of the West. WCED member Sonny Ramphal is portrayed as a major advocate of this view, claiming (according to Brundtland 1997, 78) that some environmentalists are "more concerned about panda bears than human beings, and more concerned about increasing the number of bicycles in the Third World rather than that we should acquire trucks." This orientation obviously contributed to a more development-oriented approach within the commission (Adams 1990; Lafferty and Langhelle 1999).

The Objectives of Sustainable Development

There are two root words in sustainable development. The term *sustainable* is derived from the Latin *sus tenere*, meaning "to uphold" (Redclift 1993,

4), and does not necessarily denote any normative judgment. *Development*, on the other hand, "adds a value judgment by implying a desired (i.e., by implication, a desirable) evolution of human society" (Borowy 2014, 2). One of the innovative components of sustainable development was precisely that it in a novel way defined a normative goal that should serve as an aspirational goal for the world community. The commission defined sustainable development as "development that meets the needs of the present without compromising the ability of future generations to meet their own needs." This definition, according to the WCED (1987, 43) report, contains within it two key concepts:

- the concept of "needs," in particular the essential needs of the world's poor to which overriding priority should be given; and
- the idea of limitations imposed by the state of technology and social organization on the environment's ability to meet present and future needs.

The centrality of needs placed sustainable development on the development side of the environment-development nexus, with human needs satisfaction as the major objective (ibid.). Needs included sustenance, basic health, work, energy, housing, water supply, and sanitation. As stated in *Our Common Future*, "Our message is [first and foremost] directed toward people, whose well-being is the ultimate goal of all environment and development policies" (xiv). In terms of development, social justice is equivalent to the satisfaction of human needs along with equal opportunities both within and between generations (44, 46). Sustainable development, then, requires meeting the basic needs of all, thereby extending to all the opportunity to fulfill aspirations for a better life (8), and the concern for social equity between generations must "logically be extended to equity within each generation" (43).

The qualification that development also must be sustainable is a constraint placed on this goal. In terms of the environment, this is formulated as "the minimum requirement" for sustainable development: "At a minimum, sustainable development must not endanger the natural systems that support life on Earth: the atmosphere, the waters, the soils, and the living beings" (44–45). Hence, sustainable development defines a development trajectory that reconciles ecological sustainability (the environment) with intra- and intergenerational justice (development) at the global level, and to which all nations can aspire (Langhelle 2000). In short, as the title *Our Common Future* indicates, there is only one boat, and we either make it seaworthy and livable for everyone, or we may all sink (Lafferty and Langhelle 1999).

The Context of Sustainable Development Objectives

This formulation of sustainable development did not appear in a vacuum but rather in the context of a number of earlier attempts to reconcile environment and development. Some of the earlier formulations of environment and development linkages are especially important for understanding the discursive struggles as well as the context in which sustainable development was born.

In the process leading to the Stockholm Conference in 1972, both developed and developing countries were "unenthusiastic" about an environmental conference (Borowy 2014, 31). For developing countries in particular, the stakes were high. As argued by Iris Borowy (ibid.), "To people in the South, debate about limited resources or pollution did not promise protection against the deterioration of their lives—like it did to many in the North—but the threat of regulations which would stand in the way of an improvement of their lives." The Founex meeting in Switzerland in 1971 was organized to address the concerns of developing countries and overcome North–South tensions. The report from the meeting, as described by Borowy (36), was "pivotal" in changing the direction of the international environment/development debate. Lars-Göran Engfeldt (2009, 58) labels it as "groundbreaking," and Maurice Strong (quoted in Johnson 2012, 13) characterized the meeting as "the most important single event in the run-up to Stockholm."

The aim of the Founex meeting was to produce "a viable synthesis" of the environment/development relationship (ibid.). In the "Founex Report" (1971, 1), it was argued that "the current concern with environmental issues had emerged out of the problems experienced by the industrially advanced countries," and that these problems were "largely the outcome of a high level of economic development." And while these problems "tend to accompany the process of development," the major environmental problems of developing countries were seen as "essentially of a different kind," notably those of poverty and the "very lack of development of their societies." Life itself was endangered by "poor water, housing, sanitation and nutrition, by sickness and natural disasters." Hence, the cure for these problems was "development." As the process of development accelerated, however, the environmental problems associated with development would become more significant in developing countries. Thus environmental issues should be "recognized," and perceived as a "widening of the development concept" to be integrated in planning and policy making (2). The report represented a first move toward the reconciliation of environment and development.

The Stockholm Declaration on Human Environment and the Action Plan for the Human Environment echoed many of the conclusions from the "Founex Report." The Stockholm Declaration also contained several principles that influenced and reverberated in *Our Common Future*. Most fundamentally, it was stated in the Stockholm Declaration that "man has acquired the power to transform his environment in countless ways and on an unprecedented scale." With growing evidence of human-manufactured harm to the environment, it was now time to "shape our action throughout the world with a more prudent care for their environmental consequences." Parts of principle 1 can be read as an early expression of the equation that was to become the definition of sustainable development: "Man has the fundamental right to freedom, equality and adequate conditions of life, in an environment of a quality that permits a life of dignity and well-being, and he bears a solemn responsibility to protect and improve the environment for present and future generations." The Stockholm Declaration, though, still placed environment and development as "two sets of interests side by side" (Bernstein 2002, 67).

The 1980 *World Conservation Strategy* is often seen as one of the first uses of the term sustainable development. The report was prepared by the International Union for the Conservation of Nature and Natural Resources (IUCN 1980), and published with support from the World Wildlife Fund and UN Environment Program. Sustainable development appears in a couple of places in the report. It is hard, however, to find a place where the concept of sustainable development is explicitly defined (Adams 1990, 49). Instead, conservation and development are defined separately, and the main message is that "conservation and sustainable development are mutually dependent" (IUCN 1980, 10, 11). The definition of conservation is actually closest to what was to become sustainable development in *Our Common Future*. Conservation was defined as "the management of human use of the biosphere so that it may yield the greatest sustainable benefit to present generations while maintaining its potential to meet the needs and aspirations of future generations" (3, 11). In William M. Adams's (1990, 49) interpretation, the *World Conservation Strategy* represented "a significant repackaging of conservation," which was not a viable solution for a North–South agreement on an environment and development synthesis.

Hence, the innovation of the concept of sustainable development as defined in *Our Common Future* was that it bridged environment and development concerns at a global scale. It did so by introducing pragmatism and vagueness. It put people first by asserting that what is to be sustained in the process of development is the ability to cover human needs and secure equal opportunities, now and in the future. As James Meadowcroft (1997, 430) argues, it is not a particular institution, nor a specific pattern of activity, nor a given environmental asset that is supposed to be sustained but rather a process—"a process of sustainable development." This has two important implications.

First, it implies that not every environmental problem is necessarily a sustainable development issue. This is crucial because it partly determines how environmental effects are to be judged from a sustainable development perspective. It is as a prerequisite for development that the injunction to conserve plants and animals in *Our Common Future* must initially be understood. It is because the environment is vulnerable to destruction through development itself that the constraint of sustainability is placed on the goal of development (WCED 1987, 46).

Second, it implies that an activity that is not itself sustainable could be a part of an ongoing process that *is* sustainable (Meadowcroft 1997). This applies not only to social behavior but also to activities like the consumption of renewable and nonrenewable resources. If this were not so, nonrenewable resources could not be consumed at all. Forestry provides a good illustration of this point, where it is asserted that it is possible to speak of a sustainable reduction of the physical stock (WCED 1987, 127).

The question of what constitutes a specific instance of sustainable development thus becomes a matter of degree. Clearing forests for farming can be part of sustainable development if certain conditions are present. Moreover, it is apparent that sustainable development may have quite-different implications for different countries, depending on the level of development, availability of resources, size of population, level of need satisfaction, and the possibilities of substitution between natural and human-created capital. The fact that sustainable development by necessity has to be contextualized is precisely what allowed a number of different actors to adhere to its development objectives. The UN General Assembly endorsed *Our Common Future* on December 11, 1987.

Conceptual Evolution, Critics, and Adjacent Concepts

That the above understanding of sustainable development has been heavily criticized is well known. It has been dubbed an oxymoron and criticized for being "shamefully anthropocentric"; too development oriented; too environment oriented; blatantly social democratic; fuzzy about needs, wants, and desires; hopelessly naive in implying that "you can have your cake and eat it too"; and in general, obscuring underlying complexities and contradictions.

The most notable adjustment to the mainstream understanding of the concept, however, happened in the follow-up to *Our Common Future* within the United Nations. Already in the action plan from the Earth Summit in Rio de Janeiro, Agenda 21, there were signs of a further development of the concept of sustainable development, most notably in presenting it as composed of different dimensions. *Our Common Future* does not explicitly make such a distinction (Meadowcroft 2000; Holden 2007). In Agenda 21, section I is labeled "Social and Economic Dimensions." Section II is called "Conservation and Management of Resources for Development," and addresses fourteen different environmental challenges (United Nations 1993). Hence, for all practical purposes, Agenda 21 operates with a concept of sustainable development in which there are three dimensions: the social, the economic, and the environmental. It also represented another innovation in terms of defining sustainable development.

In the document "Programme for the Further Implementation of Agenda 21," adopted by the UN General Assembly in 1997, this had become the standard understanding of sustainable development: "Economic development, social development and environmental protection are interdependent and mutually reinforcing components of sustainable development" (United Nations 1997, para. 23). The same "interdependent and mutually reinforcing pillars" are found in the "Plan of Implementation of the World Summit on Sustainable Development" (United Nations 2002, 2). In the outcome document of the UN Commission on Sustainable Development (UNCSD) Rio+20, "The Future We Want," sustainable development is understood to "ensure the promotion of economically, socially and environmentally sustainable future for our planet and for present and future generations" (United Nations 2012, 1).

The dividing of sustainable development into different but interrelated dimensions also had some resemblance to the ways sustainable development was operationalized in the debates between neoclassic and ecological economists. The economic sustainability debate frames sustainable development as a question of the composition of capital (between natural, human-made, human, and social capital) necessary in order to sustain welfare over time. According to Eric Neumayer (2013, 8), most economists would accept the following definition of an economic concept of sustainable development: "Development is defined ... to be sustainable if it does not decrease the capacity to provide non-declining per capita utility for infinity," bluntly leaving out the first objective of the sustainable development: intragenerational justice. Sustainability in economic terms comes in two main competing versions: weak and strong. Neumayer (ibid.) refers to weak and strong sustainability as two "paradigms" of sustainable development, with radically different policy implications. What is to be sustained in this formulation, however, is human welfare in time.

The concept of sustainability may also have other meanings. It may be used simply to denote things that can be carried forward in time indefinitely without any reference to sustainable development (Ekins 1993). Some argue that it fits better to talk about sustainability than sustainable development when referring to developed countries like the United States (Fiorino 2010, 586n1). Others have counterposed sustainability and sustainable development, and reserve sustainability to indicate "something more radical and stronger than the 'reformist' notion of sustainable development," which incorporates concern for the natural world "as more than just a means to human development" (Barry 2002, 438). Others see sustainable development as a threat to sustainability, primarily because of its close association with economic growth (Springett and Redclift 2015), and therefore prefer sustainability to sustainable development. Likewise, John Robinson (2004, 370) identifies sustainable development as associated more with economic growth, technical fixes, and a managerial and incremental approach, while sustainability as preferred by many NGOs and academic environmentalists focuses more on changes in individual attitudes toward nature along with "the ability of humans to continue to live within environmental constraints." These issues bring us directly to questions of limits and growth.

The Means of Sustainable Development

A common critique of *Our Common Future* is that it is vague about what is needed in order to achieve sustainable development (Dryzek 1997, 2013). From its frame of reference, however, the commission argued that a set of critical objectives—or strategic imperatives—follow from the concept. The means of achieving the objectives included: reviving growth; changing the quality of growth; meeting essential needs for jobs, food, energy, water, and sanitation; ensuring a sustainable level of population; conserving and enhancing the resource base; reorienting technology and managing risk; and merging environment and economics in decision making (WCED 1987, 49). Two issues that directly link to these strategic objectives have been at the center of the critique: the question of economic growth, and the issue of physical, ecological, or environmental limits.

Limits and Growth

With severe global poverty and economic stagnation in the developing world at the time, the commission bluntly stated that "what is needed now is a new era of economic growth." Yet it should be a growth that "is forceful and at the same time socially and environmentally sustainable" (ibid., xii). For many, this prescribes more of what causes environmental problems than what cures them. The positions taken on these issues were nonetheless connected to subtle but extremely important conceptual (and empirical) changes in *Our Common Future* expressed partly through the second of the "key concepts" entailed in the definition of sustainable development. This is "the idea of limitations imposed by the state of technology and social organization on the environment's ability to meet present and future needs" (43).

The awareness that meeting the needs of the present and expanding the opportunities for a better life to all might have environmental costs was fundamental to the conception of sustainable development. It was argued that with expected population growth, "a five- to tenfold increase in world industrial output can be anticipated by the time world population stabilizes sometime in the next century. Such growth has serious implications for the future of the world's ecosystems and its natural resource base" (213). This was, according to Jim MacNeill (1990, 109), OECD environment director and later general secretary for the WCED, precisely the "tacit question" that the commission had to struggle with. The growth needed to meet human needs and aspirations "translates into a colossal new burden on the eco-sphere" (MacNeill et al. 1991, 27).

Two critical issues therefore had to be solved. First, are there limits, and how should they be conceived? And second, is it possible to change the content of economic growth, and if so, how? The optimism in *Our Common Future* is fundamentally based on its view about the possibility of changing the content of growth to avoid limits. On the one hand, it was asserted that growth has no set limits, because the "limits" can be "manipulated" by technology and social organization. On the other hand, the report contends, "But ultimate limits there are, and sustainability requires that long before these are reached, the world must ensure equitable access to the constrained resource and reorient technological efforts to relieve the pressure" (WCED 1987, 45). So there are limits, but they are both flexible and absolute.

The commission maintained that different limits hold for the use of energy, materials, water, and land (ibid.). In fact, it argued that the limits likely to be exceeded first are the availability of energy and the biosphere's capacity to absorb the by-products of energy use. These limits may be approached far sooner than those imposed by other material resources, because of the depletion of oil reserves and carbon dioxide buildup leading to global warming (58–59). Indeed for climate change, they may already be at hand. MacNeill, Pieter Winsemius, and Taizo Yakushiji (1991, 27) thus contended that the "maxim of sustainable development was not 'limits to growth,' it was 'the growth of limits.'"

Limits could be avoided, and economic growth could become environmentally and socially sustainable, though, only if industrialized nations continued the recent shifts in the content of their growth toward less material- and energy-intensive activities, by improvements in efficiency in materials and energy use, and making the content of growth more equitable in its impact (WCED 1987, 51–52). These conditions were further elaborated in *Our Common Future*, and they should be seen as complementary aspects of a progrowth position (Langhelle 1999).

Arguably, economic growth was advocated as "a means rather than as an aim" in the report (Holden 2007, 29). Still, *Our Common Future* made "reviving growth the top strategic priority" (Bernstein 2002, 64), with room for different interpretations and possible pathways. It placed itself between the proponents and opponents of economic growth that predated the report, but far closer to the proponents of growth. *Limits to Growth* (Meadows et al. 1972) was "the unspoken—and sometimes spoken—negative model for the Brundtland Commission," and "any resemblance to this study had to be avoided at all costs." The expression "limits to growth" was "constantly avoided" (Borowy 2014, 155), and there are no references to *Limits to Growth* in *Our Common Future*.

More influential was the work within the OECD. In the late 1960s, senior officials within the OECD were already questioning what was seen as an "inadequate societal response to problems resulting from the unprecedented economic growth and environmental burdens" (ibid., 28). In 1970, the OECD became the first international organization to establish a commission in charge of environmental questions. Increasingly, the center of OECD work on the environment revolved around how to reconcile environmental protection with market policies and how to achieve this in practice. Hence, in the OECD *Observer*, MacNeill addressed what he perceived to be shifts in the scope and focus of environmental policy. MacNeill (1979, 40) expected a "profound debate to come" based on "the view that even anticipatory environmental policies cannot attack the fundamental forces underlying the continued degradation of the environment." This debate raised the questions of "alternative lifestyles and alternative growth patterns compatible with the maintenance of a healthy environment." MacNeill held that he personally believed the "relationship between economic development and environmental quality is positive, at least in the long-term," but that it was "difficult to demonstrate with available data and methodologies" (41).

In Bernstein's (2002, 52) interpretation, the OECD gradually moved toward the view that with the proper use of economic instruments, the reciprocal and positive linkages between environmental protection policies and economic growth could be secured, and "economic growth and environmental protection could be compatible." This also became the core message of *Our Common Future*, but hinges on the premise that the content of growth can actually be changed to avoid environmental limits.

Overall, the means that followed from the concept of sustainable development were mixed. It included a call for multilateralism as well as shared responsibility for global environmental and development problems due to economic and ecological interdependence. Developed countries should assist in the eradication of poverty by substantially increasing development aid, transferring environmentally sound technology, promoting freer and fairer trade, and inducing changes in consumption and production in both developed and developing countries. In addition, developed and developing countries together should triple the total expanse of protected areas in order to conserve a representative sample of the earth's ecosystems. This constituted a global partnership for environment and development (Langhelle 1999), a norm complex that Bernstein (2002, 68–69) describes as "managed sustainable growth" due to its proposed "mix of command-and-control regulation and economic/marked-based incentives for environmental management."

The direction the follow-up would take, however, was to be determined by the "realities of the international political economy" (69).

The Evolution and Influence of Sustainable Development

As the discussion above illustrates, sustainable development in the Brundtland version contained both reformist and radical elements along with a number of possible pathways for future action. Which direction did sustainable development take after *Our Common Future*? Which version(s) of it became part of politics and policies? What has been its impact? The following addresses these questions by looking at efforts to implement sustainable development policies at global, national, and local levels.

Sustainable Development at the Global Level

The resolution of the UN General Assembly that endorsed *Our Common Future* asked all governing bodies, organs, organizations, and programs of the UN system to review their policies, programs, budgets, and activities with the aim of contributing to sustainable development. Since then, the United Nations has actively pursued a sustainable development agenda institutionally as well as through summits and meetings addressing its different aspects.

The UN Conference on Environment and Development (UNCED) process in the United Nations started with the preparations for the Earth Summit in Rio de Janeiro in 1992. Even before Rio, a number of countries—Canada, Denmark, the Netherlands, Finland, Japan, Norway, France, Italy, the United Kingdom, and Sweden—had debated *Our Common Future* in parliament and initiated domestic action to follow up on the report (Borowy 2014, 169). These countries were also actively involved in the preparations for Rio. The Earth Summit was described as a "landmark event" that "launched a new partnership for sustainable development" (United Nations 1997, 1). It resulted in a number of agreements, most notably the Rio Declaration on Environment and Development, Agenda 21, the Forest Principles, the UN Framework Convention on Climate Change (UNFCCC), and the Convention on Biological Diversity (CBD). Of these, only the two conventions where legally binding. They were also negotiated independently of the main UNCED process, but on schedules set so that they could be signed at Rio. In the aftermath of the Rio Summit, the UN General Assembly established the Commission on Sustainable Development in 1992 as the first UN body on sustainable development.

Agenda 21 had several important impacts on the sustainable development agenda. First, it strengthened the focus on what Borowy sees as a serious omission in *Our Common Future*: northern affluent lifestyles. As argued by Borowy (2014, 206–7), this was an issue that "haunted" the commission's discussions: "Clearly, the commissioners were aware of this point. ... [T]he question of consumption lifestyles was related to the issue of global limits, which was clearly crucial and would not go away." As Neil Carter (2007, 219) points out, it was an issue containing "political dynamite." Agenda 21 sharpened the focus on unsustainable patterns of production and consumption (see chapter 4 in Agenda 21), but consumption remained deeply controversial.

Second, Agenda 21 strengthened and structured the follow-up in its call for national action plans for sustainable development and the inclusion of "major groups": women, children and youths, indigenous peoples, local authorities, workers and trade unions, business and industry, and farmers and the technological community (see chapters 23–32 in Agenda 21). It is hard to assess the importance of this, but a number of countries developed sustainable development plans under the heading of Agenda 21 or sustainable development (Lafferty and Meadowcroft 2000), and local authorities in a number of countries developed local Agenda 21 plans for their communities (Lafferty and Eckerberg 1998). Business, NGOs, and other major groups became engaged with sustainable development in the years that followed.

The Earth Summit was significant in another respect. In Bernstein's (2002, 89) interpretation, the UNCED outcomes were "more definite on the promotion of a liberal and growth oriented economic order and less so on ensuring ecological viability." According to Bernstein, the "form of governance that emerged emphasized one particular pathway from the concept of 'sustainable development,' a neoliberal pathway where environmental concerns were reformulated in the context of a liberal international economic order" (3). In contrast to the "Keynesian-like compromise" of *Our Common Future*, liberal environmentalism "promotes market and other economic mechanisms (such as tradable pollution permit schemes or the privatization of commons) over 'command-and-control' methods (standards, bans, quotas, and so on) as the preferred method of environmental management" (7).

Equally crucial, though, was the signing of the two legally binding conventions, the UNFCCC and CBD, which addressed the two most important environmental problems identified in *Our Common Future*: climate change and loss of biological diversity. The impact of these two issues on policy development within the European Union and many other countries should not be underestimated. How much of that can be attributed to the concept of sustainable development is another matter, but both conventions drew heavily on the framework of sustainable development. For example, the "ultimate objective" of the UNFCCC is more or less identical to what was formulated as the minimum requirement of sustainable development in *Our Common Future*: "stabilization of greenhouse gas concentrations in the atmosphere at a level that would prevent dangerous anthropogenic interference with the climate system" (United Nations 1992). Yet with the UNFCCC and CBD conventions, two of the most critical sustainable development issues were in effect lifted out of the UNCED process.

In 2000, at the Millennium Summit in New York, the Millennium Declaration was unanimously adopted. It set out a series of time-bound targets, with a deadline of 2015, known since the Millennium Summit as the Millennium Development Goals. Among the goals were to eradicate extreme poverty and hunger, and ensure environmental sustainability. Based on the relative success and progress on the Millennium Development Goals, the outcome document from UNCSD at Rio+20 called for Sustainable Development Goals that "should address and incorporate in a balanced way all three dimensions of sustainable development and their inter-linkages" (United Nations 2012, 43).

On September 25, 2015, the UN General Assembly adopted the 2030 Agenda for Sustainable Development containing 17 Sustainable Development Goals and 169 targets to be implemented by 2030. Among them are goals to "end poverty in all its forms everywhere" (goal 1); "ensure access to affordable, reliable, sustainable and modern energy for all" (goal 7); "ensure sustainable consumption and production patterns" (goal 12); and "take urgent action to combat climate change and its impacts" (goal 13) (United Nations 2015, 15–23).

These goals are described as "integrated and indivisible and balance the three dimensions of sustainable development"; they are meant to "stimulate action over the next 15 years in areas of critical importance for humanity and the planet" in an attempt to transform "our world" (ibid., 1). As with sustainable development, the Sustainable Development Goals have been criticized for being too development oriented, hopelessly naive, and weak on implementation mechanisms.

Sustainable Development at the National and Local Level

To what extent has the concept of sustainable development influenced domestic politics and policies? For those who have struggled with these issues, some among the developed countries usually stand out. Apart from European Union as a whole, which embraced sustainable development in the Amsterdam Treaty, Germany, the Nordic countries, the United Kingdom, the Netherlands, Japan, Australia, and Canada (in the early days) are frequently mentioned, while the United States has been described for the most part as disinterested (Dryzek 1997, 2013; Lafferty and Meadowcroft 2000). There are fewer studies of developing countries' engagement with sustainable development, but there are success stories in China, Brazil, Uganda, Kenya, Ecuador, and elsewhere (UN Environment Program 2010).

Sustainable development policies have materialized most markedly in the comprehensive plans and strategy documents as well as measurement and monitoring procedures called for in Agenda 21 (Lafferty and Meadowcroft 2000, 434; Meadowcroft 2007; United Nations 2013a). Many countries have created sustainable development strategies, plans, indicators, and more to monitor and track progress. There are numerous examples of attempts to implement sustainable development policies by merging environment and economics in decision making through environmental policy integration and other measures (Nilsson and Eckerberg 2007). Sustainable development has also been written into constitutions, and is increasingly seen as relevant in international courts and tribunals, especially in relation to the sustainable use of natural resources, precaution, and integration (Schrijver et al. 2010).

Experiences are mixed, however. Most national plans are detached from political decision making, and there are enormous differences in thematic scope, intra- and intergovernmental coordination, participatory engagement, goals and targets, monitoring mechanisms, and political salience (Meadowcroft 2007). This is also reflected in the review of the UNCSD, which was replaced by a high-level political forum at the UNCSD Rio+20. Here it was argued that the UNCSD guidelines for reporting had been "very loose," reporting had been "uneven," and "little support was provided to build capacity for undertaking such reporting in developing countries" (United Nations 2013a, 10).

Despite this, in some countries, and in many cities and local communities, there are identifiable sustainable development policies and attempts to implement them. In some areas, there is also noticeable progress in poverty reduction and health improvements globally (United Nations 2014, 3). At the same time, though, the "major trends that threaten environmental sustainability continue." Global emissions of carbon dioxide are on an "upward trend"; in 2011, they were " almost 50 percent above their 1990 level." In addition, "millions of hectares of forest are lost every year, many species are being driven closer to extinction and renewable water resources are becoming scarcer" (ibid., 4), and "birds, mammals and other species are heading for extinction" (United Nations 2013b, 46). In other words, development is going forward, but at the expense of environmental sustainability (Meadowcroft et al. 2012).

The Inescapable Challenges of Sustainable Development

What, then, can be said about the concept of sustainable development in terms of its significance for environmental policy? How productive or fruitful has the concept been? The answer depends on whom you ask and which version of sustainable development one subscribes to. What is certain is that—as Bill Hopwood, Mary Mellor, and Geoff O'Brian (2005, 47) contend—there is no "single unified philosophy of sustainable development," and "no sustainable development 'ism.'" In most cases, people bring their "already existing political and philosophical outlooks" to the debate. Seeing sustainable development as a "macro concept" on par with notions such as democracy and freedom, and a "fundamental normative idea" (Meadowcroft 2000, 371), it is hardly surprising that sustainable development is contested. There are profound "ideological" conflicts over the core issues of the concept, although more on the means than the objectives.

The issues of environmental limits and economic growth illustrate the competing worldviews. The perspectives expressed in *Our Common Future* concerning limits have been conceived as either implying "no limits" (Dryzek 1997) or at best reflecting "ambiguity concerning existence of limits" (Dryzek 2013). But can limits be both flexible and absolute? Even the most cited source on limits to growth operated with a similar ambiguity. As Donella H. Meadows and her coauthors (1972, 81, 84) write in *Limits to Growth*,

> It is not known how much CO_2 or thermal pollution can be released without causing irreversible changes in the earth's climate. ... We do not know the precise upper limit of the earth's ability to absorb any single kind of pollution, much less its ability to absorb the combination of all kinds of pollution. We do know however that there *is* an upper limit.

Different views on environmental limits exist today. The planetary boundaries approach of Johan Rockström and his coauthors (2009) on defining a safe operating space for humanity has been heavily criticized by Ted Nordhaus, Michael Shellenberger, and Linus Blomquist (2012) for mixing global with regional or local limits, and ignoring the unavoidable political aspects of defining these limits. Apart from the ozone layer, there is only one environmental limit that has been politically recognized internationally—the target adopted in the Paris Agreement: holding the increase in the global average temperature to well below 2 degrees above preindustrial levels and pursue efforts to limit the increase to 1.5 degrees above preindustrial levels.

Underlying the discussions on environmental limits are also widely different views on progress that are necessarily linked to optimism and pessimism (Ekins 1993; Meadowcroft 2012). For the believers in human ingenuity, limits, if they exist, will be overcome. A good example is William

A. Gale, who literarily reached into space to refute *Limits to Growth*. Gale's (1979, iix) edited book explored the "prospects for extraterrestrial settlement" and argued that "limits to growth would be encountered only when we reach volumes of space already developed by other beings."

The relationship between economic growth and environmental sustainability is equally contested, as the debate between "weak" and "strong" sustainability reveals. Is it enough that only the aggregate value of the total capital stock (of natural, physical, and human capital) is nondeclining over time as advocated by weak sustainability? Or is it required that human-made and natural capital each be maintained intact separately? In Neumayer's (2013, 3, 192) analysis, neither weak nor strong sustainability is falsifiable. Support for one or the other depends on basic beliefs and claims about "*future*" possibilities of substitution and technical progress—issues that are essentially "nonfalsifiable," except perhaps in retrospect. It is therefore far less clear what is necessary to provide nondeclining future utility "than either paradigm would want us to believe" (193).

The future is precisely what sustainable development addresses. Hence, there is something inescapable in sustainable development. As asserted by Daniel J. Fiorino (2010, 582, 583), one "does not have to be an environmental doomsdayer or ecological pessimist to recognize the many threats to the environmental system and the implications for the social/political and economic systems"—what he refers to as an "inescapable logic of environmental sustainability." Arguably, the challenges at the core of sustainable development—to reconcile ecological sustainability (the environment) with intra- and intergenerational justice (development) at the global level—will not go away, and there is probably no escape from the sustainable development equation. Development and justice are already intertwined in environmental problems and issues (Meadowcroft 2000), and limits to growth and the need for development are probably both "real" (Mitcham 1995, 324). As such, increasing inequality and the growth of limits will eventually force sustainable development up on the political agenda (Hopwood, Mellor, and O'Brian 2005, 47).

For some, of course, sustainable development remains a sham, co-opted and misused beyond recognition and rescuing. For others, it constitutes an important (global) discourse, a conceptual meeting place for different actors, and a common ground for discussion among a range of developmental and environmental actors "who are frequently at odds" (Sneddon et al. 2006, 259). For others, it is more than that. For Jeffrey D. Sachs (2015, 3), sustainable development is both a normative outlook on the world and an analytic concept, "an intellectual pursuit" that "tries to make sense of the interactions of three complex systems: the world economy, the global society and the Earth's physical environment." Fiorino (2010, 583), writing from a public administration perspective, argues in the same manner. The sustainability concept—being comprehensive, analytic, normative, and flexible—has much to offer by defining a framework for structuring, evaluating, and making choices. It encompasses the major policy issues and serves as an integrating concept, "because it requires that choices be made in the context of interrelationships and interaction" among the environmental, social, and economic dimensions of sustainability. It is analytic in providing a "framework for framing, evaluating and making choices." It also provides a set of normative principles and is flexible—it can be applied usefully at any scale of governance.

Hopefully the world will manage to achieve an absolute decoupling of economic goods from environmental bads. Many studies maintain that it is theoretically possible to stay within the two-degree target—through substantial energy efficiency, a huge growth in renewables, and/or a massive use of carbon capture and storage worldwide. Hence, the proponents of strong sustainability may be theoretically wrong. But so far, few countries have managed to realize an absolute decoupling of greenhouse gas emissions from economic growth, and deforestation and loss of biodiversity continues (United Nations 2014, 2013b). So it seems that while proponents of strong sustainability may be theoretically wrong, they are nonetheless empirically correct. It is an open question whether the future will prove them empirically wrong.

Concluding Remarks

Sustainable development is both simple and notoriously complex. It is simple in the sense that it builds on intuitive ideas that merge environmental and developmental concerns. It assumes that there are no valid moral reasons for refusing poor people their legitimate aspirations for a better life (Lafferty and Langhelle 1999). Yet this cannot be accomplished by merely following the growth path already taken by the developed countries since human activities are already "pressing on the frontiers of the environmentally sustainable" (Meadowcroft 2000, 383). Sustainable development is therefore not just a strategy for the future of developing countries but also for developed countries that "must reduce the excessive stress their past economic growth has imposed upon the earth" (Dryzek 1997, 129). All countries should pursue alternative development trajectories that take into account the interests of present and future generations. The sustainable development equation, then, is simple: meeting needs and aspirations without violating the minimum requirement for sustainable development—that of not endangering the natural systems that support life on earth—is essentially what makes up the challenge of sustainable development.

Sustainable development is notoriously complex in the sense that it demands the cooperation of all countries in a globalized, competitive, capitalist system. *Our Common Future* should be read as a massive call for multilateralism under circumstances of increasing global ecological interdependence. As such, it has been argued here that sustainable development is inescapable—there is simply no way out of the sustainable development equation. Sustainable development is a "name for a way that tries to recognize and bridge both realities" of environment and development "and their sometimes conflicting claims." The precise parameters of that bridging, however, "remain to be determined" (Mitcham 1995, 324). For the reasons discussed above, determining the precise parameters will be politically difficult and contestable, but sustainable development will probably continue to be the main framework for the integration of the challenges of environment and development.

References

Adams, William M. 1990. *Green Development: Environment and Sustainability in the Third World*. London: Routledge.

Barry, John. 2002. "Sustainability." In *International Encyclopedia of Environmental Politics*, ed. John Barry and Gene E. Frankland, 438. London: Routledge.

Bernstein, Steven. 2002. *The Compromise of Liberal Environmentalism*. New York: Columbia University Press.

Borowy, Iris. 2014. *Defining Sustainable Development for Our Common Future: A History of the World Commission on Environment and Development (Brundtland Commission)*. London: Earthscan.

Brundtland, Gro H. 1997. "Verdenskommisjonen for miljø og utvikling ti år etter: Hvor står vi i dag?" *ProSus: Tidsskrift for et bærekraftig samfunn* 4:75–85.

Carter, Neil. 2007. *The Politics of the Environment: Ideas, Activism, Policy*. 2nd ed. Cambridge: Cambridge University Press.

Desai, Nitin. 1986. *Note on the Concept of Sustainable Development*. June 27.

Dryzek, John S. 1997. *The Politics of the Earth: Environmental Discourses*. Oxford: Oxford University Press.

Dryzek, John S. 2013. *The Politics of the Earth: Environmental Discourses*, 3rd ed. Oxford: Oxford University Press.

Ekins, Paul. 1993. "'Limits to Growth' and Sustainable Development: Grappling with Ecological Realities." *Ecological Economics* 8:269–288. doi:10.1016/0921-8009(93)90062-b.

Engfeldt, Lars-Göran. 2009. *From Stockholm to Johannesburg and Beyond: The Evolution of the International System for Sustainable Development Governance and Its Implications*. Stockholm: Regeringskansliet.

Fiorino, Daniel J. 2010. "Sustainability as a Conceptual Focus for Public Administration." *Public Administration Review* 70: 578–588. doi:10.1111/j.1540-6210.2010.02249.x.

"The Founex Report on Development and Environment—1971." 1971. Accessed November 17, 2014, http://www.stakeholderforum.org/fileadmin/files/Earth%20Summit%202012new/Publications%20and%20Reports/founex%20report%201972.pdf.

Gale, William A., ed. 1979. *Life in the Universe: The Ultimate Limits to Growth*. Boulder, CO: Westview Press.

Holden, Erling. 2007. *Achieving Sustainable Mobility: Everyday and Leisure-Time Travel in the EU*. Aldershot, UK: Ashgate.

Hopwood, Bill, Mary Mellor, and Geoff O'Brian. 2005. "Sustainable Development: Mapping Different Approaches." *Sustainable Development* 13:38–52. doi:10.1002/sd.244.

International Union for Conservation of Nature and Natural Resources (IUCN). 1980. *World Conservation Strategy: Living Resource Conservation for Sustainable Development*. Gland, Switzerland: IUCN. Accessed December 13, 2016, https://portals.iucn.org/library/efiles/documents/WCs-004.pdf.

Johnson, Stanley. 2012. *UNEP, the First 40 Years: A Narrative*. Nairobi: UN Environment Program. Accessed December 13, 2016, http://www.unep.org/pdf/40thbook.pdf.

Kates, Robert W., Thomas M. Parris, and Anthony A. Leiserowitz. 2005. "What Is Sustainable Development? Goals, Indicators, Values, and Practice." *Environment* 47 (3): 8–21. doi:10.1080/00139157.2005.10524444.

Lafferty, William M., and Katarina Eckerberg, eds. 1998. *From the Earth Summit to Local Agenda 21: Working towards Sustainable Development*. London: Earthscan.

Lafferty, William M., and Oluf Langhelle, eds. 1999. *Towards Sustainable Developmen: On the Goals of Development—and the Conditions of Sustainability*. London: Macmillan.

Lafferty, William M., and James Meadowcroft, eds. 2000. *Implementing Sustainable Development: Strategies and Initiatives in High Consumption Societies*. Oxford: Oxford University Press.

Langhelle, Oluf. 1999. "Sustainable Development: Exploring the Ethics of *Our Common Future*." *International Political Science Review* 20 (2): 129–149. doi:10.1177/0192512199202002.

Langhelle, Oluf. 2000. "Sustainable Development and Social Justice: Expanding the Rawlsian Framework of Global Justice." *Environmental Values* 9:295–323. doi:10.3197/096327100129342074.

Lélé, Sharachchandra M. 1991. "Sustainable Development: A Critical Review." *World Development* 19 (6): 607–621. doi:10.1016/0305-750x(91)90197-p.

Lipschutz, Ronnie D. 2012. "The Sustainability Debate: *Déjà vu* All over Again?" In *Handbook of Global Environmental Politics*, ed. Peter Dauvergne. 2nd ed. Cheltenham, UK: Edward Elgar.

MacNeill, Jim. 1979. "Environmental Policy in Transition." *OECD Observer* 97:40–41. doi:10.1787/observer-v1979-2-en.

MacNeill, Jim. 1990. "Strategies for Sustainable Economic Development." In *Managing Planet Earth: Readings from Scientific American Magazine*, 109–124. New York: W. H. Freeman and Company.

MacNeill, Jim, Pieter Winsemius, and Taizo Yakushiji. 1991. *Beyond Interdependence: The Meshing of the World's Economy and the Earth's Ecology*. Oxford: Oxford University Press.

Meadowcroft, James. 1997. "Planning for Sustainable Development: Insights from the Literatures of Political Science." *European Journal of Political Research* 31:427–454. doi:10.1111/1475-6765.00324.

Meadowcroft, James. 2000. "Sustainable Development: A New(ish) Idea for a New Century?" *Political Studies* 48:370–387. doi:10.1111/1467-9248.00265.

Meadowcroft, James. 2007. "National Sustainable Development Strategies: Features, Challenges, and Reflexivity." *European Environment* 17:152–163. doi:10.1002/eet.450.

Meadowcroft, James. 2012. "Pushing the Boundaries: Governance for Sustainable Development and a Politics of Limits." In *Governance, Democracy, and Sustainable Development: Moving beyond the Impasse*, ed. James Meadowcroft, Oluf Langhelle, and Audun Rudd, 272–296. Cheltenham, UK: Edward Elgar.

Meadowcroft, James, Oluf Langhelle, and Audun Ruud, eds. 2012. *Governance, Democracy, and Sustainable Development: Moving beyond the Impasse*. Cheltenham, UK: Edward Elgar.

Meadows, Donella H., Dennis L. Meadows, Jorgen Randers, and William W. Behrens III. 1972. *The Limits to Growth: A Report for the Club of Rome's Project on the Predicament of Mankind*. London: Pan Books.

Mitcham, Carl. 1995. "The Concept of Sustainable Development: Its Origins and Ambivalence." *Technology in Society* 17 (3): 311–326. doi:10.1016/0160-791x(95)00008-f.

Neumayer, Eric. 2013. *Weak versus Strong Sustainability: Exploring the Limits of Two Opposing Paradigms*. 4th ed. Cheltenham, UK: Edward Elgar.

Nilsson, Måns, and Katarina Eckerberg, eds. 2007. *Environmental Policy Integration in Practice: Shaping Institutions for Learning*. London: Earthscan.

Nordhaus, Ted, Michael Shellenberger, and Linus Blomquist. 2012. *The Planetary Boundaries Hypothesis: A Review of the Evidence*. Oakland, CA: Breakthrough Institute.

Redclift, Michael. 1992. "Sustainable Development and Global Environmental Change: Implications of a Changing Agenda." *Global Environmental Change* 2 (1): 32–42. doi:10.1016/0959-3780(92)90034-5.

Redclift, Michael. 1993. "Sustainable Development: Needs, Values, Rights." *Environmental Values* 2 (1): 3–20. doi:10.3197/096327193776679981.

Robinson, John. 2004. "Squaring the Circle? Some Thoughts on the Idea of Sustainable Development." *Ecological Economics* 48:369–384. doi:10.1016/j.ecolecon.2003.10.017.

Rockström, Johan, Will Steffen, Kevin Noone, Åsa Persson, F. Stuart Chapin III, Eric Lambin, Timothy M. Lenton, et al. 2009. "Planetary Boundaries: Exploring the Safe Operating Space for Humanity." *Ecology and Society* 14 (2): 32. Accessed November 14, 2014, http://www.ecologyandsociety.org/vol14/iss2/art32.

Sachs, Jeffrey D. 2015. *The Age of Sustainable Development*. New York: Columbia University Press.

Schrijver, Nicholaas J., Duncan French, Ximena Fuentes, Donald K. Anton, Karin Arts, Alan Boyle, Tomer Broude, et al. 2010. "International Law on Sustainable Development." International Law Association, the Hague Conference. Accessed November 14, 2014, http://www.ila-hq.org/download.cfm/docid/70A8B19E-2E03-4397-BC7E64186FC5EC8E.

Sneddon, Chris, Richard B. Howarth, and Richard B. Norgaard. 2006. "Sustainable Development in a Post-Brundtland World." *Ecological Economics* 57:253–268. doi:10.1016/j.ecolecon.2005.04.013.

Springett, Delyse, and Michael Redclift. 2015. "Sustainable Development: History and Evolution of the Concept." In *Routledge International Handbook of Sustainable Development*, ed. Michael Redclift and Delyse Springett, 3–38. Abingdon, UK: Routledge.

UN Environment Program. 2010. *Green Economy: Developing Countries Success Stories*. St-Martin-Bellevue, France: United Nations Environment Program.

United Nations. 1992. "United Nations Framework Convention on Climate Change." Accessed November 16, 2014, http://unfccc.int/files/essential_background/background_publications_htmlpdf/application/pdf/conveng.pdf.

United Nations. 1993. *Report of the United Nations Conference on Environment and Development*. Resolutions adopted by the conference, vol. 1, Rio de Janeiro, June 3–14. New York: United Nations.

United Nations. 1997. "Programme for the Further Implementation of Agenda 21." Adopted by the General Assembly at its nineteenth special session, June 23–28. Accessed November 17, 2014, http://www.un.org/documents/ga/res/spec/aress19-2.htm.

United Nations. 2002. "Plan of Implementation of the World Summit on Sustainable Development." Accessed November 17, 2014, http://www.un.org/esa/sustdev/documents/WSSD_POI_PD/English/WSSD_PlanImpl.pdf.

United Nations. 2012. "The Future We Want." Resolution adopted by the General Assembly on July 27. Accessed December 13, 2016, http://www.un.org/disabilities/documents/rio20_outcome_document_complete.pdf.

United Nations. 2013a. "Lessons Learned from the Commission on Sustainable Development: Report of the Secretary-General." Accessed November 14, 2014, https://sustainabledevelopment.un.org/content/documents/1676SG%20report%20on%20CSD%20lessons%20learned_advance%20unedited%20copy_26%20Feb%2013.pdf.

United Nations. 2013b. *The Millennium Development Goals Report 2013*. New York: United Nations. Accessed November 14, 2014, http://www.un.org/millenniumgoals/pdf/report-2013/mdg-report-2013-english.pdf.

United Nations. 2014. *The Millennium Development Goals Report 2014*. New York: United Nations. Accessed November 14, 2014, http://www.un.org/millenniumgoals/2014%20MDG%20report/MDG%202014%20English%20web.pdf.

United Nations. 2015. "Transforming Our World: The 2030 Agenda for Sustainable Development." Resolution adopted by the General Assembly on September 25. Accessed November 17, 2014, http://www.un.org/ga/search/view_doc.asp?symbol=A/RES/70/1&Lang=E.

World Commission on Environment and Development (WCED). 1987. *Our Common Future*. Oxford: Oxford University Press.

9 Biodiversity: Increasing the Political Clout of Nature Conservation

Yrjö Haila

Biodiversity is a new term—a neologism—launched at a conference held at the Smithsonian Institution in Washington, DC, in September 1986; the conference was titled the National Forum on BioDiversity. The organizers were biologists—most of them well-established academics with backgrounds covering all the biological subfields relevant to environmental issues. Biologists had become seriously worried about the rate of extinctions, which was thought to be increasing. This in itself was nothing new; nature conservation had been an important social movement and public concern since the late nineteenth century. What was new was a perception that much more was at play than the extinction of named species: the increasing rate of extinctions was envisaged as a threat to the stability of the biosphere as a whole and a serious threat to the future of humanity.

By adopting the neologism biodiversity, the organizers of the conference aimed at constructing a new and politically potent interpretation of humanity's interactions with the rest of nature. In this they succeeded to an amazing degree, and quickly: loss of biodiversity is nowadays routinely included in the list of the most serious global environmental threats, and the concern has gained extensive attention internationally. The purpose of this chapter is first to describe how the new term was transformed into a concept, and what the elements of the concept are such that it got such a sweeping success among academic biologists. Second, the chapter will explore the development that led to the consolidation of biodiversity as a major issue on the international environmental scene. The success of biodiversity was supported by its close affiliation with several parallel concerns, which are explained next. The story has a subtext, however: specifying what the most critical environmental changes threatening biodiversity actually are, and what to do about them, has turned out to be difficult. Such problems and obscurities in understanding the concept as well as giving it a practically relevant formulation are taken up as the fourth theme of the chapter. Finally, the last section briefly takes stock of the concept as an innovation in environmental policy.

The First Stage: Conquering the Scientific Home Turf

The story of biodiversity has two conspicuous features: it originated as a neologism that nevertheless was derived from elements that had been well known and discussed since Charles Darwin's 1859 *Origin of Species*; and it broke through into public usage remarkably quickly. The concept was adopted during the preparations for the Washington conference mentioned above.[1] The background of the conference is known exceptionally well thanks to the PhD project of David Takacs (1996), who conducted interviews with the main organizers. Some of the interviewees articulated the clearly political goal of the conference; according to one of the participants, the forum was "an explicitly political event, explicitly designed to make Congress aware of this complexity of species that we're losing" (37). As Takacs interpreted it, another aim of the conference was to promote the role of conservation biologists as spokespersons for a new form of nature conservation:

> Biodiversity shines with the gloss of scientific responsibility, while underneath it is kaleidoscopic and all-encompassing: we can find in it what we want, and can justify many courses of action in its name. … If biodiversity is a much more complex and dynamic focus for conservation efforts than endangered species, it likewise offers a much more complex and dynamic role for biologists in society at large. (99)

The motivation of the Washington conference organizers grew from a worry among biologists in the 1970s concerning an increasing extinction threat. Early summaries include the books by Norman Myers (1979) and Paul R. Ehrlich and Anne H. Ehrlich (1981); the current literature is vast. Extinctions were the explicit focus of the BioDiversity forum in Washington. Strong words have been used to characterize the threat; in his popular monograph, Edward O. Wilson (1992, 265, 297) used expressions such as "miniature holocausts" and "a problem of epic dimensions."

Originally, the concern over a human-caused extinction wave was aroused by theoretical predictions derived from the theory of island biogeography that was constructed in the 1960s (MacArthur and Wilson 1967). According to the theory, the number of species in a particular patch of habitat such as an island is determined as the equilibrium between colonization and extinction (or speciation and extinction, in evolutionary time). Hence, if the human modification of natural habitats leads to a diminishing total area, species will vanish—go extinct—from the remaining patches of original habitat. The application of the theory to conservation rests on an analogy between islands and preserves, assumed to be surrounded by habitat types that are inhospitable to the majority of the original species.[2]

The theoretical background makes the success of the term among ecologists understandable for several interconnected reasons. First of all, the island theory was one of the flagships of modern, mathematically sophisticated population ecology that grew in prominence in the 1960s, and the implications of the theory for nature conservation had been already widely discussed. Second, the theory addressed somewhat indirectly but with apparent success a perennial problem in evolutionary biology: Why are there as many species as there are—not fewer, or more? This question was explicitly raised using mathematical models in the 1950s. In other words, irrespective of the direct conservation implications, the theory became one of the key themes in ecological thought at that time. These factors together raised a hope that a predictive theory could be constructed concerning the burden that increasing human influence will cause on the viability of other species.

Further explaining the academic success of the concept was that *diversity* already was a familiar term. Biodiversity is a contraction of *biological diversity*, which had been used with a reference to the general variability of the biological realm. Diversity was adopted into use in the 1950s as a term that combines the number of species in a given ecological community or geographic area and their relative proportions. Whether diversity is important for the stability and "health" of ecological systems was a much-discussed problem, which of course had obvious implications for nature conservation. James Maclaurin and Kim Sterelny (2008) present a good overview of the multidimensional background of biological discourses on diversity. They list and assess different alternative framings that the concept has been given, covering research fields from paleontology and evolutionary biology to population genetics.

Simultaneously with the launching of the term biodiversity, concerned biologists established *conservation biology* as a new biological discipline. One of the originators of the discipline, Michael Soulé (1985, 727), described it as a *mission- or crisis-oriented discipline*, writing, "Its relation to biology, particularly ecology, is analogous to that of surgery to physiology and war to political science." With this analogy, Soulé wanted to underline the urgency and practical relevance of nature conservation.

Research on biodiversity and related themes has increased exponentially in the academic literature since the mid-1980s. Xingjian Liu, Liang Zhang, and Song Hong (2011) conducted a bibliometric analysis, and reported that the number of journal articles on biodiversity broadly hovered around two hundred per year in the mid-1980s, rose to over 1,000 by the mid-1990s, and reached 7,533 in 2009 (the final year of the analysis). Thus, biodiversity became a household word among conservation scientists and activists early on—although the term lacked a practical political framing for a number of years. This lack is reflected in a curious way in the contents of the second anthology put together with the purpose of advertising the term to the academic community (Reaka-Kudla, Wilson, and Wilson 1997). A great majority of the essays specify tasks and challenges for biological research, particularly for taxonomy; there are few direct ideas about policies for protecting biodiversity.

To sum up, the term biodiversity got an enthusiastic reception within academia. It integrated a range of demanding theoretical problems and an ethical passion to address the looming extinction crisis. The second half of the term, diversity, connects to a feature of life with which biologists are familiar: the fundamental importance of heterogeneity on all levels of the biological world. For specialists in the field, the idea carried by the term was immediately palatable, and evoked thoroughly positive connotations by being theoretically pertinent and interesting.

Is It an Innovation?

As a neologism that referred to a well-known concern, biodiversity was an *invention*, in a most literal sense. But in which sense can biodiversity be characterized as an *innovation*?

A formal criterion, as articulated by economic historians in particular, is whether the invention in question is adopted into broad use or not. This formal criterion is fulfilled: the term spread into use remarkably efficiently.

Another, more evaluative benchmark is "being new and better" (Dyke 2004, 12). This qualitative criterion requires a baseline; in this case, an obvious one is the way conservation concerns were articulated previously. The term fulfilled this evaluative standard: protection of biodiversity was immediately accepted as a more powerful goal than traditional nature conservation due to its rhetorical power and broad coverage. As its reference, the term encompasses biological variability in living nature on all levels. This is a huge topic. As a consequence, specialists who got involved in using the term had to interpret it in their own ways. In this sense, the term functioned as a *theory constitutive metaphor*, to use Sandra Mitchell's (2003, 17) expression. A metaphor is a trope that makes a new phenomenon seem familiar by pointing out a similarity with something already known. Biodiversity works as a metaphor to the extent that it brings knowledge of the variability of life into the environmental scene; it works as a theory constitutive metaphor by bringing forth new and theoretically exciting research topics. On the other hand, the reference of the concept is vague in a peculiar way: it does not specify which particular type of variability really matters, and in which sense.

Somewhat paradoxically, such peculiar vagueness has given additional power to the generative nature of the term. The idea is inherently attractive to those feeling strongly about the health of the ecological systems of the earth, but the significance of biological diversity is not a straightforward "observable" that can be documented by looking around. Quite to the contrary, the use of the term calls for and generates theoretical work. And as the field of interest is the protection of nature in most variable guises, the generative dimension inspires also normative thought via questions such as, What is appropriate human behavior toward living nature? How can we give more space to nature? Such questioning brings forth new aspects of understanding the human condition and feels potentially attractive to a wide constituency. The motto for the UN Decade on Biodiversity, 2011–2020 is "living in harmony with nature." On the other hand, difficulties and controversies necessarily follow when such ambitious declarations are translated into specific goals. This theme will be addressed in later sections of this chapter.

Overall, the concept biodiversity brought new elements into thinking about nature conservation (Haila 2012). Critical for the success of the term biodiversity in this sense is the fuzzy but comprehensive and fundamentally positive connotation of the term diversity. The prefix *bio-* links the term to biology, but without any precise reference. Rather, the term is open-ended by leaving the task of giving it a more precise referent to the speaker, or listener, as the case may be. In this sense, Wilson's (1997, 1) note on biodiversity is fitting: "So, what is it? Biologists are inclined to agree that it is, in one sense, everything." Wilson's statement summarizes the metaphoric power of the word.

An important additional factor was that there had already been shifts in thinking about nature conservation among practitioners in that field. Crucial work was started during the 1980s by nature conservation nongovernmental organizations, especially the International Union of the Conservation of Nature (IUCN 1980) in its *World Conservation Strategy* (Adams 2001). In the IUCN's strategy, the developing world, in which a major part of valuable ecological systems such as tropical rain forests are located, received special emphasis. Martin Holdgate (1999), in his inside history of the IUCN, describes the strategy's preparation as a process of education for the conservationists at the IUCN: they learned that to have any chance of success at all in the developing countries, nature conservation has to address the economic and social needs of the people.

This development also paved the way for the Brundtland Report seven years later by acknowledging the importance of human sustenance for environmental concerns. The close conceptual link with sustainable development is demonstrated in the subtitle of the *World Conservation Strategy: Living Resource Conservation for Sustainable Development*. The expression "living resource conservation" is used as an umbrella phrase in the text of the strategy; diversity is used in the context of "genetic diversity."

The strong link with development was criticized in the beginning by conservationists, but such skepticism has all but died off. This is quite remarkable as sustainable development is continuously regarded as inherently developmentalist, at the expense of environmental values. The reason may be that living resource conservation is relatively easily adaptable to green values.

Rio 1992 and the CBD: Claiming Space in International Environmental Politics

Biodiversity entered the sphere of international environmental politics at the Rio Summit in 1992, with the opening for signatures by UN member states of the Convention on Biological Diversity (CBD). By the closing date of the signing period, June 4, 1992, it had received 168 signatures. The convention entered into force in December 1993. What biodiversity is according to the convention is specified in article 2, "Use of Terms":

"*Biological diversity*" means the variability among living organisms from all sources including, *inter alia*, terrestrial, marine and other aquatic ecosystems and the ecological complexes of which they are part: this includes biodiversity within species, between species and of ecosystems.

Relevant in this regard is also annex I, titled "Identification and Monitoring." The text specifies three targets for monitoring: ecosystems and habitats; species and communities; and described genomes and genes.

The broad scope and development-friendly tone of the CBD (1992) is expressed by the two-page long preamble:

The contracting parties,

Conscious of the intrinsic value of biological diversity and of the ecological, genetic, social, economic, scientific, educational, cultural, recreational and aesthetic values of biological diversity and its components.

Conscious also of the importance of biological diversity for evolution and for maintaining life sustaining systems of the biosphere.

Affirming that the conservation of biological diversity is a common concern of mankind. ...

Concerned that biological diversity is being significantly reduced by certain human activities. ... *Aware* that conservation and sustainable use of biological diversity is of critical importance for meeting the food, health and other needs of the growing world population, for which purpose access to and sharing of both genetic resources and technologies is essential. ... *Determined* to conserve and sustainably use biological diversity for the benefit of present and future generations

Have agreed as follows: ...

Following this preamble, there are forty-two articles and a brief annex I. The objectives of the convention are explicated in article 1, "Objectives," which reads:

The objectives of this convention, to be pursued in accordance with its relevant provisions, are the conservation of biological diversity, the sustainable use of its components and the fair and equitable sharing of the benefits arising out of the utilization of genetic resources, including by appropriate access to genetic resources and by appropriate transfer of relevant technologies, taking into account all rights over those resources and to technologies, and by appropriate funding.

During the CBD's preparations, the issue of using and sharing genetic resources and biotechnology was included in the agenda through an initiative of the so-called Group of 77 developing countries (Chatterjee and Finger 1994). The initiative was accepted, and the result is in articles 15 (Access to Genetic Resources), 16 (Access to and Transfer of Technology), and 19 (Handling of Biotechnology and Distribution of Its Benefits), all of them fairly detailed. The issue of genetic resources led in due time to two separate protocols: the Cartagena Protocol on Biosafety to the CBD (adopted in January 2000), and the Nagoya Protocol on Access to Genetic Resources and the Fair and Equitable Sharing of Benefits Arising from Their Utilization to the CBD (adopted in 2010).

William M. Adams (2001) assumes that the inclusion of biotechnology explains the rapid accumulation of signatures to the convention. This is a fair premise: bioprospecting, or the search for commercially valuable biological resources such as medicinal herbs, has been going on for centuries, and as a rule, local people whose knowledge has been exploited in the process have received no share of the economic benefits; the Madagaskar rosy periwinkle is a famous recent case.[3]

Overall, quite soon after the Rio Summit most countries of the world recognized the protection of biodiversity as a governmental duty. The conspicuous exception was the United States. President Bill Clinton signed the convention in June 1993, and transmitted it to the Senate for advice and ratification. The Senate Foreign Relations Committee supported ratification by a bipartisan vote in 1994, but the convention was never submitted for a vote on the Senate floor due to powerful lobbying by antienvironmentalist interest groups. Yet the United States has sent a delegation of "observers" to the CBD's Conferences of the Parties (COP), and the protection of biodiversity is an established household word in domestic environmental politics (Snape 2010).

The rapid success of the concept in international environmental politics is most likely attributable to the role of biodiversity as a generative metaphor in the field of environmental concerns generally. That it can be and has been connected to everything alive on the earth is a major determinant of its success. The basic argument underlying the concept draws on science, even on technically demanding modeling work on the viability of endangered populations. Basically, biodiversity is an overarching good, and humanity is facing the danger of losing something really valuable. This gives good support for arguments in favor of nature conservation, although the setting of specific targets is much more complicated than what it looks like at the outset.

The significance of the CBD has clearly increased over the years, as the imperative of protecting biodiversity has become institutionalized on various levels and in various ways. It seems that there are several pathways of institutionalization that deserve attention. First, a formidable organization has grown up to perform the work required by the original convention. Second, the implementation of the convention demands specific measures on the national level, emphatically in a multilevel setting; developments within the European Union demonstrate the significance of multilevel structures. Another key feature of the institutionalization, cutting through governance levels, is the importance of a broad variety of actors representing, to borrow Ken Conca's (2005) term, transnational civil society.

Efforts to Establish a Biodiversity Regime

The text of the CBD includes articles on how the work is to be continued. The most important step was to institutionalize COP, originally to be held annually, but since the 2000s every two years. The Cartagena Protocol on Biosafety was adopted in 2000 at an extraordinary meeting.

Susan Baker (2003) observes that the regular participation at COPs has created reporting and implementation dynamics that strengthen the political weight of the CBD. The first conferences labored mainly with putting the institutional structures in place; for instance, working groups were established on specific aspects of the convention, such as reviewing implementation, preserving and maintaining traditional and indigenous knowledge, and the management of protected areas. In the more recent COPs, the decisions and declarations agreed on have increased in ambition. The equitable use of genetic resources has been a major theme, as mentioned above. Other important decisions have been made on setting targets and enforcing management by defining the duties of the member states.

The first organizational steps decreed in the convention were the establishment of a secretariat (article 24), located since 1996 in Montreal, and production of a knowledge base and scientific capacity. No new knowledge base had been provided for the Rio Summit, which is somewhat paradoxical as biodiversity is a heavily science-laden concept. By establishing the Subsidiary Body on Scientific, Technical, and Technological Advice (article 25), the CBD's scientific capacity was strengthened. The UN Environment Program took on the task of organizing a scientific overview of the situation, and thus the *Global Biodiversity Assessment* came out in 1995. Later, the CBD secretariat published *Global Biodiversity Outlooks*, the fourth edition of which came out in fall 2014.

The Strategic Plan for Biodiversity 2011–2020, known as the Aichi Declaration and adopted in 2010 at the tenth COP, is probably the most critical step in the implementation of the convention; it replaced the previous one targeted for 2010. The strategy includes wide-ranging but fairly well thought out indicators, and sets relatively specific targets. An important part of the process is the demand for national strategic action plans, already mentioned in the CBD's article 6, "General Measures for Conservation and Sustainable Use." The article includes two paragraphs; the first sets the task to "develop national strategies, plans or programmes for the conservation and sustainable use of biological diversity," and the second to "integrate, as far as possible and as appropriate, the conservation and sustainable use of biological diversity into relevant sectoral or cross-sectoral plans, programmes and policies."

Baker (2003) argued that the CBD has established a novel environmental governance regime: the process of getting the institutions of the convention to work smoothly created administrative pressures that modified the behavior of environmental authorities on various levels. Significantly, CBD gave additional prestige to conservation authorities both within environmental offices and in cross-sectoral interactions with other authorities.

There are several tensions that make the process quite complicated, however. The European Union is a good test case, as it enacts binding directives, but the member states have the task of integrating the directives in their own policy making and legislation. First, the regime is necessarily plagued by tensions across levels: the CBD versus the EU Commission versus member states. Such tensions had already played out in the preconvention era, as the European Union had a strong background in conservation legislation in the shape of the Birds Directive (1979) and, most important, the Habitats Directive (1992). The Endangered Species Act of the United States (1973) as well as the international conservation discourse in general were key models all along (Verschuuren 2003). Baker (2003, 25) characterizes this connection as "international entanglement." Specific international agreements that are binding in the European Union include the Convention on International Trade in Endangered Species of Wild Fauna and Flora (1973), Ramsar Convention (1976), and Berne Convention (1979). In Baker's view, EU compliance with the CBD was lukewarm in the first few years, but since about 1998 the EU Commission has taken a firmer stand, partially to strengthen its position vis-à-vis the member states in a policy field—land use—that the member states were originally reluctant to delegate to the EU level.

Another source of complexity is what Baker (2003, 36) calls "functional" pressures. The units in nature that require attention do not coincide with national borders. River systems and seas divided between the jurisdictions of several different states are paradigmatic examples, but terrestrial ecosystem types are also shaped according to biogeographic regularities that do not respect national borders.

A third source of complexity is cross-sectoral integration. This is the responsibility of all participants in the CBD, but again, the problems are accentuated within the European Union. Particularly after the adoption of the new biodiversity strategy at the tenth COP in 2010, *mainstreaming* biodiversity into the working of all relevant policy sectors has become a dominant emphasis. The term is introduced in a draft document, "Mainstreaming Biodiversity into Sectoral and Cross-Sectoral Strategies, Plans, and Programmes" (CBD 2007). The need for cross-sectoral integration was already firmly articulated in the ministerial declaration at the sixth COP in 2002 (CBD 2002):

> The most important lesson of the last ten years is that the objectives of the Convention [on Biological Diversity] will be impossible to meet until consideration of biodiversity is fully integrated into other sectors. The need to mainstream the conservation and sustainable use of biological resources across all sectors of the national economy, the society and the policy-making framework is a complex challenge at the heart of the Convention.

These are hugely demanding requirements, if taken literally. The reporting and monitoring routines set by the CBD require that the parties actually do so. Yet the practical consequences of these demanding requirements remain to be seen. Quite obviously, practical experiences are still lacking.

National Biodiversity Strategy and Action Plans (NBSAP) comprise the main tool for implementing the CBD on the national level. The COP routines include regular reporting and assessment of the progress of the signature countries with their NBSAPs, which have strengthened the dynamics of regime constitution on national levels, as Baker (2003) suggested. Mainstreaming is the new terminological fad for describing this trend.

A major requirement for the NBSAPs is that conservation and use of biological resources be integrated together. The practices adopted in different countries vary, depending on each country's institutional structures concerning resource use. One may assume, though, that the requirement, if followed at all, brings about another feature of regime constitution that Conca (2005) describes as "norm-disseminating." What precise form specific norms are given varies depending on national histories. Agriculture, rangeland management, forestry, and fisheries all belong under this umbrella, and every one of these fields is infested with turf struggles between different sectors of government. The situation resembles the tensions prevailing almost everywhere between climate policy and the energy sector.

Biodiversity Breathes Life into Several Parallel Concerns

The text of the CBD is explicitly affiliated with several streams of environmental political thought. Its close connection to sustainability was pointed out above: the early stages of both concepts coincided in the preparation of the World Conservation Strategy by the IUCN. Another demand with wide-ranging procedural implications for environmental policy is in article 14, "Impact Assessment and Minimizing Adverse Effects." It applies environmental impact assessment to projects that are likely to have adverse effects on biological diversity. In later decisions the term *strategic impact assessment* is adopted. Environmental Impact Assessment has become a standard procedure used in evaluating land use decisions with conservation implications.

During the years following the Rio 1992 Summit, the notion of biodiversity has inspired other ambitious conceptual novelties, or the launching of "sister concerns," corresponding to its character as a theory and norm-generating metaphor. In other words, biodiversity has had a dual role as a conceptual innovation in creating a novel understanding of the human duty to protect nature and helping to articulate new perspectives on a set of problems that were discussed already before the concept was adopted.

These sister concerns of resilience, ecosystem services, and natural capital are briefly taken up in what follows.

Resilience

C. S. Holling (1973) introduced the term *resilience* into ecological terminology. The word refers to the capacity of ecological systems undergoing external disturbances to retain their integrity. Holling provided a theoretical formulation for what ecologists had suspected for some time: there is a large class of ecological systems that actually retain their integrity as a result of recurring disturbances. Forests undergoing wildfires or storm damages as well as estuaries and shore meadows subject to flooding are paradigmatic examples. Theoretically, this view builds on interactive nonlinear dynamics in ecological relationships. Biological diversity, understood as the richness and variability of the species included in the systems considered, is viewed as a facilitator of ecosystem resilience.

Since something like the early 1990s, resilience has been conceptualized also in a socioecological context. Stockholm's Resilience Center, affiliated with Stockholm University and the Royal Swedish Academy, is the main promulgator of the approach. The research on socioecological resilience aims at detecting congruence between human-induced changes and changes occurring due to other factors in ecological complexes that are subject to human influence. Specific applications of the idea have received some empirical support concerning particular ecological systems—typically those subject to regular disturbances (Walker and Salt 2006).

Ecosystem Services

The term *ecosystem services* covers the view that natural ecosystems provide humanity with both material and cultural services that have to be cherished. The basic aim is to stress the usefulness and necessity of nature's riches for human sustenance. People have known this since time immemorial, but the idea, formalized in ecological terms, was made popular by a collection of articles edited by Gretchen Daily (1997). The public reception of the idea presents another case of the explosive success of a new concept. The support given to the idea of ecosystem services among conservationists rests on an underlying assumption that there is a close correlation between biodiversity and ecosystem functions that supply the services people enjoy.

On the other hand, critical voices are claiming that the term ecosystem services has a strong anthropocentric flavor, and consequently the need to protect biodiversity may be buried under human-centered instrumental goals. Ben Ridder (2008) presents a good summary of this line of argument. Nevertheless, the mainstream view among ecologists is that the ideal of ecosystem services is rhetorically so powerful that it basically supports the aim of protecting biodiversity. Georgina M. Mace, Ken Norris, and Alastair H. Fitter (2012) offer a good recent defense of the confluence of the two parallel aims; they promulgate ecosystem management as a means to catch both flies with one stroke.

The idea of ecosystem services is attractive, but similar to the concept of biodiversity itself, it may be excessively all encompassing: if everything that happens in ecosystems is "service," the idea does not allow for making distinctions that should be made between different alternatives in specific decision-making contexts.

In research aiming at policy and management advice, biodiversity and ecosystem services have been largely considered together. An example is a large international research project, Millennium Ecosystem Assessment, conducted in 2001–2005 on the initiative of the UN general secretary, Kofi Annan. The first product of the project was a framework for assessment titled *Ecosystems and Human Well-being*, published in 2003. As the title indicates, the goal was to underline the dependence of human sustenance on nature's amenities.

Natural Capital

Biodiversity per se is regarded as a critical natural asset, under the general heading *natural capital*. Of course, the idea that nature is an asset in production is old, but in mainstream economics nature is equated with either "Ricardian land"—that is, land that can be used for some productive purpose that facilitates differential rent—or specific natural resources. In the prevailing production function (known as the Cobb-Douglas function), nature's productive power does not get any explicit expression. This feature of the generally accepted model has led some prominent economists to claim that natural capital can be substituted with human capital so that production can get along without using natural resources at all. Nicholas Georgescu-Roegen heavily criticized this idea in the 1970s (see Bonaiuti 2011).

Ecological economists have emphasized the need to include natural capital in economic accounting. The main point of this debate from the perspective of this chapter is that biodiversity has brought new perspectives into the economic valuation of nature; G. Cornelis van Kooten and Erwin H. Bulte (2000) offer an excellent overview of the important issues. Articulating the value of living nature has been a growth industry for the last couple of decades. A recent and important further development is materialized by the international network Economics of Ecosystems and Biodiversity; its first overview volume, *TEEB Ecological and Economic Foundations*, was published in 2010. The main contribution of this process has been to systematically explore problems and controversial issues that show up when elements of biodiversity are valued within an economic framework. At the very least, the process shows how difficult the task of valuation is.

Problems and Prospects

The idea of biodiversity has brought about an upheaval, if not a revolution, in thinking about the conservation of nature. It is not only the preservation of particular species or magnificent sites that is at issue; the main concern of nature conservation is the viability of ecological systems and ultimately the biosphere. A recent definition of biodiversity reads as follows" "Biodiversity in the broad sense is the number, abundance, composition, spatial distribution, and interactions of genotypes, populations, species functional types and traits, and landscape units in a given system" (Díaz et al. 2006, 1300). This formulation shows that the concept has undergone a drastic transformation since the 1980s, when it was articulated as a concern over an increasing global extinction rate.

A similar transformation is observable in the political framing of the biodiversity issue developed at the COPs of CBD. In the Strategic Plan for Biodiversity 2011–2020, approved at the tenth COP in 2010, the significance of biodiversity is articulated as follows (CBD 2010b):

> Biological diversity underpins ecosystem functioning and the provision of ecosystem services essential for human well-being. It provides for food security, human health, the provision of clean air and water; it contributes to local livelihoods, and economic development, and is essential for the achievement of the Millennium Development Goals, including poverty reduction.

The level of ambition expressed by this formulation is huge. Biodiversity is painted as an agent that "underpins," "provides for," "contributes to," and "is essential for." The question that arises is, How do we align biodiversity with realistic and "doable" measures to reach such goals? The next sections bring up specific aspects of this dilemma.

The Global Nature of the Problem Is Overplayed

Since the early worries of a global extinction wave, it has been customary to perceive the loss of biodiversity as a global issue. The recurring reference to global extinction statistics gives the impression that the driving forces of extinctions are also global, but this is not true. It is an example of what might be called *a fallacy of global averages* (Haila 2004). The crux of the matter is that the global biota does not form one functionally unified whole, in which all species would be in an equal position in the way global extinction statistics suggest.

A comparison of biodiversity with climate change is illuminating. The nature of the medium is different. The earth's atmosphere is a unified geophysical system, whereas the biosphere is divisible into different sections or subsystems, geographically, taxonomically, and ecologically. When one is collecting background data for assessing the likelihood of climate change, it is possible to take estimates of greenhouse gas discharges of single countries such as China, Canada, and Guatemala, and add up the effect of each of them for the global atmospheric balance. In the case of biodiversity, no comparable addition is possible. Furthermore, when different alternative criteria are used for dividing a particular section of biodiversity into components, different kinds of patterns are observed.

Consequently, efficient policies to address the deterioration of biodiversity should always be sensitive to the context. The protection of forest biodiversity, for example, sets different tasks in tropical moist forests versus northern coniferous forests. It is deceptively simple, and apparently politically easy, to formulate overarching global goals precisely because these may remain as mere declarations, devoid of specific obligations for any specific polity.

It Is Difficult to Specify Criteria and Set Credible Targets

Every locality is enmeshed in a larger spatial context, and its boundaries can be drawn in many different ways depending on what criteria are adopted. As a result, strict sensitivity to the context is a crucial feature of biological diversity locally: there is no unambiguous way to determine the relevance of a specific locality from the point of biodiversity preservation in a larger geographic and temporal context. The evaluative ambiguity could be lessened if clear criteria could be given for assessment, but unambiguous criteria are not available. A popular solution among conservation biologists has been to take some assumedly untouched state of nature as a reference (Haila 1999). This, however, is deceptive. First of all, the earth's ecosystems have already been modified by human sustenance practices, and hence some conception is needed on what sort of change has been particularly harmful. An additional complicating factor is that nature has changed as a consequence of nonhuman factors, as demonstrated by the dramatic shift in conditions after the retreat of continental glaciers following the latest glaciation, merely ten thousand years ago. What could "pristine" nature be in regions that were crushed by two miles of ice merely ten thousand years ago?

The need is to focus on those aspects of nature on which human existence on the earth depends—that is, on human-modified ecosystems that we depend on and that depend on us in turn. Biodiversity supports the continuous productivity of those systems. A garden provides a small-scale model; coastal polders and other types of reclaimed wetlands supply large-scale models. The critical question is what does the key type of biodiversity look like in each specific case that does not have straightforward "natural" models.

Assessing biodiversity loss also presents serious theoretical problems. This requires a model of what is assessed, and an analysis of the model requires that it be appropriately bounded. Physicist Leo Smith (2007) offers advice on the uncertainties that this requirement brings forth. As to biodiversity, problems begin with the demand to define it as a system. What Sandra Díaz and her colleagues (2006) describe in the citation above cannot be bounded as a system such that one could analyze its "state." Isabelle Stengers (2010) discusses this methodological issue. In practice, biodiversity can only be assessed by using surrogates, such as numbers of species, but then the requirement of context specificity returns with a vengeance. All species are not equal in terms of their significance as indicators of biodiversity. Patterns observed using different surrogates entitle different conclusions. Trends and tendencies can be assessed using specific indicators, quite reliably into a near future, but the task of synthesis based on different indicators is difficult. It is likely that different specific indicators present mutually contradictory trends; furthermore, critical parts of the biosphere such as fish stocks in the oceans and rain forests in the Amazonia are uncoupled from one another.

The CBD organization has addressed the problem of reliable indicators from early on. In 2004, the ninth COP formulated a set of recommendations. As Andrew Balmford and his colleagues (2005) point out, the categories named were overly broad. The need to combine global relevance and sensitivity to an understanding of local ecological contexts is a stumbling block.

Problems with Legitimacy

A positive novelty of the biodiversity issue is that nature gets a political voice. What this means in practical terms is more complicated, though. The complex nature of biodiversity has brought forth new challenges for the legitimacy of nature conservation among the public at large. Top-down initiatives, such as the establishment of Natura 2000 protected areas in the European Union, have given rise to popular opposition in many member countries.[4] Ewald Engelen and his colleagues (2008) note that the nature of the legitimacy of conservation has changed, from substantive to procedural. Procedures for drawing up and implementing conservation plans matter more than before, and public participation is increasingly considered necessary.

On the other hand, the emphasis on procedural legitimacy opens up space for local initiative and public-private partnerships in voluntary conservation. There are conflicting trends in this aspect, too. Positive results have been reached by using innovative methods of providing incentives for landowners to set aside natural areas. "Conservation banking" in the United States is a good example, but there are many variants adopted in different countries in somewhat different ways. On the downside, pressed by formally strict demands set by conservation regulations such as the Habitats Directive of the European Union, environmental authorities may stick to the letter of the directive and lack the courage to address the actual local circumstances.

Problems with Feasibility

The goal of mainstreaming biodiversity protection into actual practices of other administrative sectors in the society is, as mentioned above, hugely demanding. The relative weight of different sectors naturally varies across countries depending on natural conditions and economic history. Professionalism among specialists within the sectors has fueled turf struggles.

It is easier to come up with ambitious political goal setting than to formulate credible policies. It is possible these days to come to agreement about ambitious general declarations on biodiversity preservation, but efficient policy is an entirely different matter. The previous goal to halt the decline by 2010, and the new goal setting agreed on in Nagoya in 2010 as well as the 2020 program of the European Union, published in 2011, serve as illustrations. The problem is that it is difficult to specify policies that could possibly stop the deterioration of biodiversity when it is assessed using the methods that are prevailing today. There simply is no straightforward way of halting the expansion of the material basis of the current world society, which faces the task of providing decent conditions of life for a population soon to reach ten billion members.

Feasibility analysis, as recommended by Giandomenico Majone (1989, 71), is a good first step: "It is often more fruitful to ask what cannot be done and why, rather than what can be done."[5] Feasibility analysis involves charting the constraints and impossibilities that restrict the space of action in a given field. It is not difficult to identify constraints faced by biodiversity conservation: the long temporal horizon required for achieving changes in infrastructure and people's ways of life; sustenance of local populations; conflicts over priorities with other policy fields; and so on.

As Majone stresses, developing good arguments and testing them through practical experience is a precondition of any policy progress. Setting goals and constructing instruments for reaching those goals is a dialectic process, and testing the realism of the goals is essential from the processual point of view. On too ambitious (and ambiguous) goals, Majone (1989, 69) comments, "To try to do something that is inherently impossible is, to borrow from Oakshott, always a corrupting exercise."

A productive pathway may be opened up by various sector-specific sustainability programs in areas in which biological diversity plays a role.

Examples include forestry, agriculture, water management, urban and recreational environments, and so on. In the context of such specified projects, the protection of biodiversity can be given a specific meaning.

Ambiguities

There is no doubt that the concept of biodiversity has been an influential innovation in environmental policy both internationally as well as on national and supranational (such as the EU) levels. It is still premature, however, to predict what specific shape biodiversity politics and policy will take in the future. The concept has come to stay, but there is a basic ambiguity at its core. On the one hand, the idea that biological diversity has to be protected is rhetorically powerful—it demands that nature must be given a voice in how human global sustenance is shaped. On the other hand, formulating credible policies is difficult, for two main reasons. For one, the requirement of context specificity renders general goals vacuous. An even more fundamental reason is that the material pressure exerted by the global human society on the biological systems of the earth is going to grow for the foreseeable future. How could this happen without infringing on the quality of particular ecosystems?

To clarify the ambiguity, it is helpful to revisit the stages of the breakthrough of the concept.

First, the term captured an underlying worry—humanity has the potential to inflict deadly harm on nature. It succeeded as a generative metaphor and in two dimensions. The term generated theoretical and intellectually challenging work within a broad range of biological disciplines, but it also gave strength to the normative demand of respecting nonhuman nature. The enthusiastic reception of the term among both academics and conservationists indicates that biodiversity was adopted as a theory- and norm-generating metaphor.

Second, the concern was firmly incorporated into international environmental policy at Rio 1992; this was a key event. A favorable political conjuncture had been created by a new understanding of the importance of living resources for development ("stars in lucky positions"), as Adams (2001) observes. It was important that genetic resources and bioprospecting were included in the convention. A fairly convincing governance regime got established, as Baker has argued, first on the international level (the secretariat and regular COPs), and then, gathering strength, on national and supranational (the EU) levels. A critical contributing factor was that environmental agencies and civil servants adopted biodiversity protection as a powerful assertion, or tool, or perhaps weapon, to use in turf struggles with other sectors of domestic administration representing more narrow economic and industrial interests.

The strength of the concept is its overarching scope. To take one more example, one of the targets included in the Aichi Declaration is formulated as follows (CBD 2010a):

> ***Target 14***: By 2020, ecosystems that provide essential services, including services related to water, and contribute to health, livelihoods and well-being, are restored and safeguarded, taking into account the needs of women, indigenous and local communities, and the poor and vulnerable.

But such strengths also have weaknesses. With the goals listed in the Aichi target 14, biodiversity is broadened way beyond its original reference to biological diversity. These targets could not actually be reached without a complete, even revolutionary turn in the direction of global social development. Who believes such a turn will take place within the next couple of years? How does one assess success versus failure concerning goals articulated in such a universalizing language? There is no differentiation between what can and what cannot be done. Consequently, credible criteria on what kind of measures pave the way for future success are lacking.

To draw these strands together, biodiversity brings into the open the fact that the human future is tied to the continuous functioning of the life-support systems of the earth's biosphere. This requires compassionate care and stewardship; goal-driven and instrumental management is not sufficient. At the same time, the expertise-driven, top-down political style, attenuated by a globalist overreach, makes it difficult to create space for compassionate stewardship. The gap between these polarities cannot be easily bridged, but that is what is needed.

Notes

1. The proceedings are documented in Wilson and Peter 1988.
2. For a summary of the argument, see Wilson 1992, 208–216.

3. The literature on this theme is huge. For good overviews, see Karasov 2001; van Kooten and Bulte 2000.

4. For examples, see Keulartz and Leistra 2008.

5. See also Haila 2012.

References

Adams, William M. 2001. *Green Development: Environment and Sustainability in the Third World*. 2nd ed. London: Routledge.

Baker, Susan. 2003. "The Dynamics of European Union Biodiversity Policy: Interactive, Functional, and Institutional Logics." *Environmental Politics* 12:23–41. doi:10.1080/09644010412331308264.

Balmford, Andrew, Peter Crane, Andy Dobson, Rhys E. Green, and Georgina M. Mace. 2005. "The 2010 Challenge: Data Availability, Information Needs, and Extraterrestrial Insights." *Philosophical Transactions of the Royal Society B: Biological Sciences* 360:221–228. doi:10.1098/rstb.2004.1599.

Bonaiuti, Mauro, ed. 2011. *From Bioeconomics to Degrowth: Geogescu-Roegen's "New Economics" in Eight Essays*. New York: Routledge.

Chatterjee, Pratap, and Matthias Finger. 1994. *The Earth Brokers: Power, Politics, and World Development*. London: Routledge.

Conca, Ken. 2005. "Old States in New Bottles? The Hybridization of Authority in Global Environmental Governance." In *The State and the Global Ecological Crisis*, ed. John Barry and Robyn Eckersley, 181–205. Cambridge MA: MIT Press.

Convention on Biological Diversity (CBD). 2002. "Annex to The Hague Ministerial Declaration of the Conference of the Parties to the Convention on Biological Diversity. Accessed January 15, 2017, https://www.cbd.int/decision/cop/?id=7195.

Convention on Biological Diversity (CBD). 2007. "Mainstreaming Biodiversity into Sectoral and Cross-Sectoral Strategies, Plans, and Programmes." Module B-3, July. Accessed December 16, 2016, https://www.cbd.int/doc/training/nbsap/b3-train-mainstream-en.pdf.

Convention on Biological Diversity (CBD). 1992. "Text of the Convention." Accessed January 15, 2017, https://www.cbd.int/convention/text.

Convention on Biological Diversity (CBD). 2010a. "Aichi Biodiversity Targets." Accessed January 15, 2017, https://www.cbd.int/sp/targets.

Convention on Biological Diversity (CBD). 2010b. "Strategic Plan for Biodiversity 2011–2020." Accessed January 15, 2017, https://www.cbd.int/decision/cop/?id=12268.

Daily, Gretchen C., ed. 1997. *Nature's Services: Social Dependence on Natural Ecosystems*. Washington, DC: Island Press.

Díaz, Sandra, Joseph Fargione, F. Stuart Chapin III, and David Tilman. 2006. "Biodiversity Loss Threatens Human Well-being." *PLoS Biology* 4 (8): 1300–1305. doi:10.1371/journal.pbio.0040277.

Dyke, Chuck. 2004. "Open Source Art and Distributed Genius: Reflections on Tuomi's *Networks of Innovation*." *Framework: The Finnish Art Review* 2:12–15.

Ehrlich, Paul R., and Anne H. Ehrlich. 1981. *Extinction: The Causes and Consequences of the Disappearance of Species*. New York: Ballantine Books.

Engelen, Ewald, Jozef Keulartz, and Gilbert Leistra. 2008. "European Nature Conservation Policy Making." In *Legitimacy in European Nature Conservation Policy: Case Studies on Multilevel Governance*, ed. Jozef Keulartz and Gilbert Leistra, 3–21. New York: Springer.

Haila, Yrjö. 1999. Socioecologies. *Ecography* 22:337–348. doi:10.1111/j.1600-0587.1999.tb00571.x.

Haila, Yrjö. 2004. "Making Sense of the Biodiversity Crisis: A Process Perspective." In *Philosophy of Biodiversity*, ed. Markku Oksanen and Juhani Pietarinen, 54–82. Cambridge: Cambridge University Press.

Haila, Yrjö. 2012. "Genealogy of Nature Conservation: A Political Perspective." *Nature Conservation* 1:27–52. doi:10.3897/natureconservation.1.2107.

Holdgate, Martin. 1999. *The Green Web: A Union for World Conservation*. London: Earthscan.

Holling, C. S. 1973. "Resilience and Stability in Ecological Systems." *Annual Review of Ecology and Systematics* 4:1–24. doi:10.1146/annurev.es.04.110173.000245.

International Union for Conservation of Nature and Natural Resources (IUCN). 1980. *World Conservation Strategy: Living Resource Conservation for Sustainable Development*. Gland, Switzerland: IUCN. Accessed December 13, 2016, https://portals.iucn.org/library/efiles/documents/WCs-004.pdf.

Karasov, Corliss. 2001. "Who Reaps the Benefits of Biodiversity." *Environmental Health Perspectives* 109:A582–A587. doi:10.1289/ehp.109-a582.

Keulartz, Jozef, and Gilbert Leistra, eds. 2008. *Legitimacy in European Nature Conservation Policy: Case Studies on Multilevel Governance*. New York: Springer.

Liu, Xingjian, Liang Zhang, and Song Hong. 2011. "Global Biodiversity Research during 1900–2009: A Bibliometric Analysis." *Biodiversity and Conservation* 20:807–826. doi:10.1007/s10531-010-9981-z.

MacArthur, Robert H., and Edward O. Wilson. 1967. *The Theory of Island Biogeography*. Princeton, NJ: Princeton University Press.

Mace, Georgina M, Ken Norris, and Alastair H. Fitter. 2012. "Biodiversity and Ecosystem Services: A Multilayered Relationship." *Trends in Ecology and Evolution* 27:19–26. doi:10.1016/j.tree.2011.08.006.

Maclaurin, James, and Kim Sterelny. 2008. *What Is Biodiversity?* Chicago: University of Chicago Press.

Majone, Giandomenico. 1989. *Evidence, Argument, and Persuasion in the Policy Process*. New Haven, CT: Yale University Press.

Mitchell, Sandra D. 2003. *Biological Complexity and Integrative Pluralism*. Cambridge: Cambridge University Press.

Myers, Norman. 1979. *The Sinking Ark*. Oxford: Pergamon Press.

Reaka-Kudla, Marjorie L., Don E. Wilson, and Edward O. Wilson, eds. 1997. *Biodiversity II: Understanding and Protecting Our Biological Resources*. Washington, DC: Joseph Henry Press.

Ridder, Ben. 2008. "Questioning the Ecosystem Services Argument for Biodiversity Conservation." *Biodiversity and Conservation* 17: 781–790. doi: 10.1007/s10531-008-9316-5.

Smith, Leo. 2007. *Chaos: A Very Short Introduction*. Oxford: Oxford University Press.

Snape, William J., III. 2010. "Joining the Convention on Biological Diversity: A Legal and Scientific Overview of Why the United States Must Wake Up." *Sustainable Development Law and Policy* (Spring): 6–47.

Soulé, Michael E. 1985. "What Is Conservation Biology?" *Bioscience* 11:727–734. doi:10.2307/1310054.

Stengers, Isabelle. 2010. *Cosmopolitics I*. Minneapolis: University of Minnesota Press.

Takacs, David. 1996. *The Idea of Biodiversity: Philosophies of Paradise*. Baltimore: Johns Hopkins University Press.

van Kooten, G. Cornelis, and Erwin H. Bulte. 2000. *The Economics of Nature: Managing Biological Assets*. Malden, MA: Blackwell.

Verschuuren, Jonathan. 2003. "Effectiveness of Nature Protection Legislation in the EU and the US: The Birds and Habitats Directives and the Endangered Species Act." *Yearbook of European Environmental Law* 3:305–328. doi:10.1007/1-4020-2105-4_4.

Walker, Brian, and David Salt. 2006. *Resilience Thinking: Sustaining Ecosystems and People in a Changing World*. Washington, DC: Island Press.

Wilson, Edward O. 1992. *The Diversity of Life*. London: Penguin.

Wilson, Edward O. 1997. Introduction to *Biodiversity II: Understanding and Protecting Our Biological Resources*. Ed. Marjorie L. Reaka-Kudla, Don E. Wilson, and Edward O. Wilson, 1–3. Washington, DC: Joseph Henry Press.

Wilson, Edward O., and Frances M. Peter, eds. 1988. *Biodiversity*. Washington, DC: National Academy Press.

10 Environmental Justice: Making Policy, One Skirmish at a Time

Karen Baehler

Social justice advocates long have worried that environmental agendas would crowd out the concerns of the poor and other marginalized populations. At the 1972 UN Conference on the Human Environment in Stockholm, Indian prime minister Indira Gandhi (1972, 3) warned of "grave misgivings" among people in the developing world "that the discussion on ecology may be designed to distract attention from the problems of war and poverty." She exhorted the environmental community "to prove to the disinherited majority of the world that ecology and conservation will not work against their interests, but will bring an improvement in their lives." Two years before that speech, *Time* magazine reported on a growing backlash against environmentalism in the United States: "Cleveland Mayor Carl Stokes has said that providing housing, clothing and food for the poor should take precedence over finding ways to combat air and water pollution" ("The Rise of Anti-Ecology" 1970, 44).

Such statements sound dated to contemporary ears because it is now well understood that environmental burdens play a significant role in perpetuating cycles of disadvantage, and that environmental activism can support social policy priorities rather than competing with them. Thanks to several decades of accumulated evidence, we now know that members of vulnerable population groups tend to experience disproportionately higher levels of exposure to environmental hazards, less access to clean air and water, and fewer real opportunities to have their environmental concerns heard and remedied than their wealthier and whiter counterparts. The concept of environmental justice captures not only the sea change in understanding that occurred when these patterns of maldistribution were first recognized but also the advances in grassroots politics that developed alongside the conceptual innovation.

Environmental justice as we know it today arose within a particular historical setting in the United States. Given those origins, the concept acquired associations with race and place (i.e., residential segregation and racial politics) that subsequently limited its international diffusion somewhat, at least until recently. Other, related concepts that deal with equity, broadly conceived, include *environmental equity*, *climate justice*, and *environmentalism of the poor*. In the quarter century since environmental justice emerged as a concept, the environmental justice lexicon has penetrated official environmental discourse and research agendas around the globe, and the overarching idea has established itself as a focus of environmental governance and activism.

Beyond conceptual take-up, the movement's impact on the actual distribution of environmental benefits and burdens has been mixed. While robust, enforceable environmental justice policies at larger scales are hard to find, many local environmental justice coalitions have notched victories against specific abuses, and the cumulative effect of these wins may amount to policy making by other means. Further progress depends on whether the diverse and fluid networks that have emerged from local initiatives accumulate enough collective power to generate systemic policy change. If environmental justice becomes a force for larger-scale reforms, then a robust menu of environmental justice–specific policy levers will need to be developed, tested, and refined.

Origins and Evidence: The Big Bang

The concept of environmental justice arose cumulatively from local struggles in the United States. By the late 1970s, both the antipoverty and civil rights movements were in need of fresh initiatives to engage their supporters, and the antitoxics campaign was hitting its stride following the revelations in Love Canal, New York. Local environmental battles in predominantly African American communities attracted less national media attention than Love Canal, but eventually sparked a coordinated campaign. The first of these battles began in 1979 with a public interest lawsuit that alleged a pattern of discrimination in the siting of Houston's waste sites. Although the Houston plaintiffs lost their bid for relief, the case paved the way for future court challenges and generated the first empirical study of waste siting relative to residential demographics, authored by sociologist

Robert Bullard (1990). Just as the Houston case was wrapping up, housewife and mother Hazel Johnson started organizing her mostly African American neighbors to assert their rights to the remediation of toxic contamination in the Altgeld Gardens neighborhood of Chicago, where, incidentally, a young community activist named Barack Obama was launching his career.[1]

Finally, in 1982, when the mostly African American residents of Warren County, North Carolina, lay down in front of trucks carrying PCB-laden soil to a local hazardous waste facility, the civil rights establishment took note. In her detailed history of the Warren County protests, Eileen Maura McGurty (1997) notes that local residents first staked their case on standard not-in-my-backyard (NIMBY) grounds: they argued that the technical specifications of their local facility were inadequate to handle PCBs, and recommended that the waste be shipped to a more suitable facility in Alabama. Only when that assertion failed in court did the local organizers reframe their problem as racial discrimination and reach out to the civil rights establishment for support. The larger dimensions of the issue then became clear when it was learned that the Alabama facility also sat in the middle of a majority poor and black community, as did many other such facilities in the US South.

Although the Warren County protesters had to wait twenty years for remediation, their larger cause set off a chain of historically significant events. The Congressional Black Caucus immediately instructed the US Government Accounting Office to undertake the first official government study of demographic patterns in hazardous waste siting. The resulting report was narrow in scope and thin on methodology, but raised suspicions about race and income-based discrimination (US Government Accounting Office 1983). A subsequent study by the United Church of Christ's Committee on Racial Justice (Lee 1987) found race to be a more powerful predictor of hazardous waste location than income, property values, or proximity to waste producers. The Reverend Benjamin Chavis, director of that study and a veteran of the Warren County struggle, memorably observed that the United States produces a ton of hazardous waste annually for every resident, and "blacks are toting more than their ton" (Darst 1987).

These two reports launched a twenty-year surge of research into environmental discrimination by both race and class. The methodological rigor of that research has improved considerably over time, thanks in large part to a steady flow of critiques by thoughtful skeptics. Academic consensus now generally affirms the correlation between race, income, and exposure to environmental risks in the United States (Ringquist 2005), so much so that the inequitable distribution of environmental benefits and burdens has achieved the status of "a stylized fact" (Banzhaf 2012, 2).

The US environmental justice movement reached maturity quickly once the problem was articulated. At a cluster of national conferences in the late 1980s and early 1990s, a larger agenda for change was built and a cadre of skilled, dedicated leaders arose to champion the new cause. During this time, the original term, *environmental racism*, quietly faded from general use, while *environmental equity* rose to prominence, especially within government circles. The leaders of the nascent movement ultimately rejected the environmental equity label for reasons explained by Beverly Wright, an early movement leader:

> I remember an old lady standing up, and she said, "What is equity? What does that mean? Does that mean that we want equity in poisoning people?" And we're like, "Well, no," and she said, "What we want is justice. We don't want anybody to be exposed to toxic chemicals." (US EPA 2014)

Shifting from *equity* to *justice* signaled an important clarification of purpose for the nascent social movement: the environmental justice ideal was to reduce hazards for everyone while ensuring that environmental progress did not leave minorities and the poor behind. That inclusive orientation quickly became codified in seventeen principles of environmental justice, which were adopted in 1991 at the First National People of Color Environmental Leadership Summit in Washington, DC, where 650 grassroots leaders gathered to share experiences and coalesce as a movement.[2] In addition to the siting of hazardous waste facilities, other issues that came under the environmental justice banner during this period included strip-mining in Appalachia, uranium mining on Indian lands, and pesticide hazards faced by migrant farmworkers.

Ultimately, what happened in the United States in the 1980s and early 1990s defined environmental justice as a new, grassroots concept inextricably intertwined with a new, US-accented social movement. In the words of one longtime environmental justice leader, Vernice Miller Travis,

> We gave birth to a conversation that people would recognize as their own. We gave it a language, we gave it words, we gave it a science base, we gave it a public policy base, and we gave it a base that was rooted in the power and mobilization of people on the ground so it couldn't be denied. (US EPA 2014)

The Innovation: Making and Breaking a False Dichotomy

Environmental justice exposes a previously unnamed vector of distributive injustices—namely, the piling of environmental burdens on top of existing economic and social disadvantages in poor and minority communities. Of course, such piling on has been occurring nearly everywhere for centuries, if not millennia, and social reformers in previous eras worked hard to expose and remedy the most blatant practices. Perhaps best known among these precursors to the environmental justice movement were the settlement houses, whose residents fought back against uncollected trash, spewing factories, contaminated water, and lack of green space in the tenement districts of late nineteenth-century London and Chicago (Gottlieb 2005). Disconnected moments of local awareness such as these could not coalesce into a distinct concept of environmental justice, however, until the master concept—*the environment*—arose to describe what the various maldistributions had in common. Likewise, environmental policies and laws had to be conceived as well as adopted before people could demand equal protection under them or fair participation in making them.

The idea of fairness in environmental distribution follows almost automatically from the emergence of the master concept of environment. Insofar as the environment refers to a set of highly valued goods and services associated with human relationships to air, water, and land, it is hardly surprising that people would take an interest in who gets what, and who makes the decisions, within that newly identified space. It was just a matter of time before fairness in environmental distribution joined the larger policy conversation about fairness in housing, transportation, education, employment, health care, and so on. Given the apparent inevitability of its arrival, a question arises: How can the concept of environmental justice qualify as an innovation?

The answer is twofold. First, environmental justice does more than just give a name to the inevitable, predestined topic of environmental distribution. It also places that topic against the backdrop of a particular historical and geographic setting—the US environmental justice movement, as described in the previous section's potted history. The US experience and US definition of environmental justice thus provide a benchmark of comparison and contrast for subsequent developments in both thinking and practice within the broad field of environmental distribution. As the next section will discuss, environmental justice's US overtones aided the diffusion of the concept in some directions and hindered it in others. The US backstory also opened the door to the invention of additional terms and concepts to identify different experiences and approaches in different settings, all of which have enriched the field with local nuances and cultural variants. Second, environmental justice qualifies as an innovation because it resolves a set of long-standing antagonisms between the environmental and social justice movements. Without the invention of environmental justice and related concepts, the discourse about environmental distribution might never have moved beyond the false dichotomy between helping marginalized populations and protecting the environment.

As a US-born innovation, environmental justice carries a strong emotional charge. The language of the US civil rights movement links justice to righteousness, liberation, and moral struggle. It evokes the cry of the poor and plight of the oppressed. Likewise, environmental injustices tend to produce powerful, gut-level reactions that correspond with philosopher Judith Shklar's depiction of injustice as a primary rather than secondary phenomenon. According to Shklar (1989, 1135), instances of injustice "have an immediate claim on our attention" and may serve as flash points for social change if they are vivid enough. By contrast, the reverse of injustice—justice—has less of that self-evident quality. Justness is often in the eye of the beholder. To illustrate this point, take a hypothetical scenario in which a chemical company openly states its intention to emit potentially cancer-causing chemicals only from its plants in low-income neighborhoods. Most citizens would react to this proposal with instantaneous outrage and protest. The injustice of it would be self-evident and undeniable. But what if such releases happened to be occurring already in poor neighborhoods due to the legacy of earlier industrial patterns? Would those same outraged citizens agree on what to do? Would they agree to prohibit all releases from the offending plants if it meant shutdowns and large-scale job losses, or would they agree to allow some releases in order to preserve jobs? Would they all immediately recognize the just solution in the same way that they immediately spotted the injustice?

If not, and if Shklar was right about the primacy of injustice, why, then, did the positive variant of this chapter's concept (environmental *justice*) rise to prominence instead of the negative (environmental *injustice*)? The answer has to do with the internal logic of political mobilization. Although horror stories about community poisoning and references to environmental degradation as "slow violence" grab people's attention, sustaining public interest often requires more hopeful framings that inspire and activate. Strong social movements seem to understand the benefits of positive framing—for example, prolife, prochoice, marriage equality, civil rights, gun rights, animal rights, the Arab Spring, the Green Belt Movement, greening the ghetto, the peace movement, and so forth. At the same time, the affirmative environmental justice label works because it reveals no particulars about exactly how justice should be achieved. Thanks to the pluralism inherent in both environment and justice, the composite term has proven capacious enough to accommodate a wide range of policy proposals and visions of reform.[3] The concept's longevity and global reach flow from its ability to fill a diversity of political niches while upholding the core universal insight that environmental injustice in any form demands immediate redress. This balancing act is part of the concept's innovation.

The capacity of environmental justice to accommodate competing interests also describes the practical side of the innovation. Environmental agenda setters long have faced pressure from factions inside and outside the environmental movement pulling in different directions. Two important meetings held in the United States in the 1970s—the Conference on Environmental Quality and Social Justice in Woodstock, Illinois, in 1972; and in 1976, the Working for Economic and Environmental Justice and Jobs summit in Black Lake, Michigan—focused on what many at the time saw as unavoidable trade-offs: humans versus nature; jobs versus environmental protection; and minority, low-income versus rich, white definitions of what counts as the environment. In order to depolarize the debate and develop inclusive solutions, environmental harms and benefits first needed to be seen as relevant to the lives of marginalized populations, here and now, at home, school, and work (Bullard 1990). The dialectic of environment as luxury versus environment as survival needed a synthesizing concept before the other points of friction could be addressed constructively. Environmental justice filled that need.

Gaps: A Need for Other Concepts

Given the importance of racial segregation and disadvantage as a vector of injustice in the United States, it is not surprising that the US brand of environmental justice was taken up in other countries where race, ethnicity, or indigeneity figure prominently in policy discourse. Ties to the US movement can be traced through the history of environmental justice movements in South Africa in the wake of apartheid; Brazil on behalf of exploited workers and peasants; Israel, with a focus on "minorities and residents of the periphery"; central and eastern Europe, where the Roma people can be found at ground zero of nearly every environmental disaster left from the Soviet era; and most recently, Australia, Canada, and China.[4] Although collaboration across these networks is growing, progress to date toward a genuinely transnational environmental justice movement has been patchy.

Issues of residential segregation are less prominent in Europe than the United States. Where race- and ethnicity-based segregation occurs in European cities, it tends to be a relatively recent phenomenon tied to late twentieth-century immigration from Asia, Africa, and the Middle East rather than a deeply embedded historical pattern (Iceland 2014).[5] Although some European social problems concentrate in areas of high poverty, European social policy agendas typically focus less on geography and more on aggregate populations with demographic characteristics tied to vulnerability, especially income, gender, and age, alongside race and ethnicity. Such an orientation makes sense given the nature of European environmental policy making, which generally emanates from the national and EU levels in a process better characterized as top-down than bottom-up. Within that social and political context, a US-accented concept of environmental justice closely tied to residential segregation and grassroots organizing seems out of place. Other concepts are needed to capture the full range of European perspectives on fairness in environmental distribution.

One such concept is environmental rights. There was widespread support in the 1970s for including a right to environmental quality in the European Convention on Human Rights, but the drafters could not agree on the wording. In lieu of substantive environmental rights, reformers pushed for procedural rights, such as the rights to be informed about environmental matters, participate in environmental decision making, and seek remedies in court. These procedural rights were realized in the UN Economic Commission for Europe's 1989 Aarhus Convention, which defines environmental justice as access to justice systems and legal remedies, not the fair distribution of environmental goods and bads themselves.[6] The term *environmental democracy* has gained some traction when advocates and policy makers want to concentrate on issues of popular participation in environmental policy making.[7]

When European organizations use the environmental justice label to refer to environmental distribution as opposed to legal access, they often make a point of associating it with the US movement. The related but distinct term environmental equity thus came into use in Europe for the very reasons that the first US environmental justice leaders rejected it. *Equity* carries connotations of impartiality that help European researchers and their audiences separate the science of environmental distribution from the political discourse. It also connotes procedural integrity, which accommodates both environmental democracy and rights to a clean environment (Organisation for Economic Co-operation and Development 2012). In the European context, environmental equity, or (increasingly) inequality, has come to refer mostly to aggregate patterns of distribution of environmental burdens and benefits across large population groups defined by income, class, age, and gender, rather than case-specific instances of environmental injustice in precise places tied to race and ethnicity. European research on environmental health inequity has expanded recently with initiatives by both the Organisation for Economic Co-operation and Development and the World Health Organization along with increasing interest on the part of universities.

In contrast to environmental justice's grassroots identity as a resistance movement, environmental equity framing appears most frequently in public health research (Braubach and Fairburn 2010). Some of those public health researchers have stated frankly that environmental equity framing helps them depoliticize their work and separate it from the work of US scholar-advocates (Northridge et al. 2003). To date, European studies of environmental equity/inequality also have paid more attention than environmental justice studies generally to the question of who bears the costs of environmental regulations (European Environment Agency 2011; Organisation for Economic Co-operation and Development 2003; Serret and Johnstone 2006). Other recent studies have stretched beyond a traditional environmental health focus to consider a diverse array of outcomes, including comparative rates of death by automobile accident in richer and poorer neighborhoods (Stephens et al. 2001), and unequal distribution of the "well-being benefits" associated with "viewing, engaging with, and accessing woodlands and forests in Britain" (O'Brien and Morris 2014, 356).

References to both environmental justice and environmental equity have increased in frequency in Canada and Australia since the mid-2000s, with lawyers and health researchers playing central roles. Likewise, Swedish academics have invoked both environmental justice and social equity when reporting the larger ecological footprints of middle-class, car-dependent, ethnic Swedes compared with their lower-income, immigrant neighbors, more of whom live in multifamily buildings and use public transportation heavily (Bradley et al. 2008). The environmental justice concept fits the place-based nature of that Swedish study and the importance of ethnicity in explaining its findings. As the authors point out, it seems both unfair and inefficient that some Swedish cities have centered their sustainability strategies on changing behaviors in low-income, immigrant-rich neighborhoods—the very places where negative environmental impacts are already smaller than elsewhere.

Similar social justice issues can be found in recent debates over how countries should share the burden of reducing greenhouse gases in order to slow climate change. Should less developed countries with small carbon footprints be expected to cut emissions at a rate below that of the more developed, higher-emitting countries? This is one of several dilemmas highlighted by the concept of *climate justice*. Resolutions 7/23 (2008) and 10/4 (2009) of the United Nation's Human Rights Council express concern "that climate change poses an immediate and far-reaching threat to people and communities around the world and has implications for the full enjoyment of human rights" (Res. 7/23). Both resolutions also acknowledge that "the world's poor are especially vulnerable to the effects of climate change," and particularly those living in fragile areas (especially small island states and developing countries) as well as women, indigenous people, members of minorities, and the disabled. Climate justice therefore encompasses not only the more familiar question of how the costs of reducing greenhouse gases should be distributed around the globe but other questions, too: Who suffers the impacts of climate change most immediately and intensely? Who will sit at the table when international agreements are struck? Who will ensure that the resulting agreements are enforced fairly and effectively everywhere?

The issue of climate change raises additional overlapping themes of sustainability and sustainable development, which tie directly to well-established policy frameworks in Europe and internationally. Julian Agyeman and Bob Evans (2004) coined the term *just sustainability* to capture the fusion of these discourses with that of environmental justice, while avoiding some of the traps associated with the high-visibility concept of sustainable development. Sustainable development famously emphasizes intertwining dependencies between ecological, social, and economic systems, thereby reinforcing a central message of environmental justice: that environmental protection and social protection are mutually dependent. Perhaps the clearest point of difference between environmental justice and sustainable development is the latter's "basic belief that the interests of future generations should receive the same kind of attention that those in the present generation get" (Anand and Sen 2000, 2030). The environmental justice community takes a relatively clear stand on the much-debated question of whether intra- and intergenerational justice are deserving of equal weight. Even the exceedingly broad "Seventeen Principles of Environmental Justice" do not mention any allowances for future people to make claims against current people. Principle 16 calls for future generations to be educated about social and environmental issues, but that falls far short of equating the needs of current and future generations. Just sustainability offers a mental model for addressing these conflicts.

Another recent addition to the lexicon, environmentalism of the poor, describes the community-level, grassroots component of sustainable development (Martinez-Alier 2002). Whereas the language of sustainable development has circulated mostly at the national and international levels of policy discourse, spontaneous manifestations of the core ideas behind sustainable development have sprung up in villages and regions throughout the Global South, often in protest against the exploitation of natural resources by large corporations partnering with national governments. The leaders of these community-based resistance movements typically adopt names that capture local resonances. India's Chipko movement, for example, began in 1973 as a series of localized protests against harmful and discriminatory forest management policies. Chipko—meaning "stick to"—activists became famous for the practice of physically surrounding trees to block corporate loggers' access. Three years after the Chipko movement started, Brazilian rubber tappers and their families used similar tree-hugging techniques to prevent ranchers from converting rain forest—the rubber tappers' sole source of subsistence—into pastureland. A year after the rubber tappers kicked off their campaign, future Nobel Prize winner Wangari Maathai initiated Kenya's tree-planting initiative, known as the Green Belt Movement. Environmentalism of the poor fuses environmental awareness with the international human rights agenda.

Finally, *ecojustice* deserves mention as a near synonym for environmental justice in many parts of the world. The ecojustice brand has proven especially popular with faith-based organizations in the United States whose environmental advocacy is rooted in ideas about God's creation and humanity's role as stewards of creation. Critics of this school of thought have noted that distinguishing starkly between people and the rest of nature can lead not only to neglect of human-built environments but also to support for conservation-oriented policies that unwittingly deprive current people of access to natural resources critical to their livelihoods (as addressed by environmentalism of the poor). Many faith-based ecojustice organizations work hard to avoid these pitfalls by defining creation as inclusive of both human and nonhuman values. The resulting, modified version of ecojustice overlaps extensively with environmental justice, and ecojustice groups of various stripes often collaborate with environmental justice coalitions around specific issues and local campaigns.

What Counts as Environmental Justice?

The early leaders of the US environmental justice movement saw environmental justice as a highly inclusive concept that should and would spread horizontally to other countries, and vertically to international agendas. But they could not control the concept's evolution.

As it happened, the concept diffused somewhat, as described in the previous section, but evidence of a large-scale international movement remains sparse. Gordon Walker (2009) found references to environmental justice in documents from thirty-seven countries as of 2008. What this number means is unclear, however, because separating casual, token references to environmental justice from genuine discourse about identifiable environmental injustices can be difficult. Clearly, awareness of the environmental justice concept is now widespread, and the term has become familiar to many. Still, the number of verified cases of active uptake—in the form of advocacy, policy making, and/or engaged debate—remains fairly small, and most of those cases (i.e., Israel, central and eastern Europe, Canada, Australia, and China) arose recently, within the last ten years. Even in academia, where conventional wisdom holds that the environmental justice field is now globalized, nearly half the academic articles published in 2009 with environmental justice as a keyword were written by authors located in the United States (another 20 percent were located in the United Kingdom), and 60 percent of the articles referred only to US cases (Reed and George 2011). These empirical findings, combined with the modest number of verified cases of environmental justice uptake outside the United States, suggest that direct diffusion of the concept has been limited.

What about initiatives focused on remedying skewed distributions of environmental benefits, burdens, and authority that do not use the term environmental justice? Can they be classified under the environmental justice rubric if the people and organizations involved in those initiatives do not use the term? Many academics would say yes, because environmental justice is defined not by the verbatim phrase but instead by the core ideas behind the phrase. Surveys of environmental justice thus commonly gather historical and contemporary local cases of grassroots action in response to exploitative and discriminatory environmental practices and package them under the environmental justice label to identify sociopolitical patterns (Carruthers 2008; Claudio 2007; Environmental Justice Atlas, n.d.). Such retrospective labeling may be undertaken by academics and advocates who become attached to particular terminology and want to see it used more widely to facilitate horizontal, transnational networks. Gathering heterogeneous projects under a single, umbrella label is one way of building solidarity and strengthening the larger cause.

Behind the question of what counts as environmental justice lies the unfilled need for a master frame that can accommodate the multiplicity of terms, concepts, and experiences that have accumulated to date. The only remaining question, then, is what to call the master frame. History suggests that repurposing existing terms will not work perfectly, because terms tend to become freighted with the baggage of their origins. Although environmental justice has been declared a "paradigm" (Taylor 2000), both environmental justice and environmental equity fall short as master labels because of their hot/cold, populist/analytic, and US/European associations. David Pellow (2000, 581) has proposed "environmental inequality" as an alternative overarching concept. His approach is attractive on several fronts. *Equality* avoids the bloodless connotations of equity, while *inequality* captures the insight that this issue area is driven first and foremost by problems, which tend to present obviously and urgently, and not by any clear vision of an ideal outcome, which remains as elusive as pure justice.

There is a simple and direct answer to the central concern posed by this section: What counts as environmental justice is anything that calls itself environmental justice, and in most cases, that involves either theoretical or practical connections to the US movement.

Practical Consequences: Policy Making by Other Means

The proliferation of terms and concepts described in the previous sections complicates the academic task of mapping the larger space in which environmental justice sits. From a practical perspective, however, the multiplicity of terms clearly has not prevented various groups from working together toward common purposes in both translocal and transnational networks. Environmental justice victories on the ground often owe their success to the capacity of heterogeneous groups to join forces around a particular cause, thereby building support and marshaling resources from a diverse set of publics.

To take one example, consider Chicago's decade-long Quit Coal campaign to close two of the United States' oldest and dirtiest power plants on the borders of two low-income, majority Latino and African American neighborhoods. The coalition of environmental, health, faith, labor, and environmental justice organizations that eventually coalesced around this cause undertook a wide array of actions, including marches, press conferences, opinion polls, letter-writing campaigns, lawsuits, events at city hall, and a people's hearing. One group of protesters rappelled off the Pulaski Bridge and dangled by ropes over the Chicago Canal to block a coal barge destined for the Crawford plant, while others scaled the Fisk plant's 450-foot smokestack to paint "Quit Coal" in large block letters in Spanish and English. Each phase of action was led by a slightly different subset of groups that came together for specific purposes. The cumulative effort eventually achieved its goal of retiring the two plants in 2012 and 2014. Similar Quit Coal collaborations operate in other locations around the world.

The first directory of environmental justice groups was published in 1991. As of 2014, that directory included more than eight hundred environmental justice and collaborating organizations, many of them community based.[8] Viewed from above, the community-by-community approach to environmental justice amounts to a war-of-attrition strategy based on the theory that if all vulnerable communities fight back against excessive environmental burdens, then all previous paths of least resistance will be blocked and status quo practices will have to change. In a world where rich and poor equally assert their NIMBY opposition to toxic waste dumps and related "bads," and where marginalized communities demand green amenities equivalent to their wealthy counterparts, the problem of environmental injustice will be all but solved. To the extent that this bottom-up, war-of-attrition approach works, it may be seen as commensurate with, if not superior to, strictly enforced executive or legislative rules against environmental maldistribution.

What about policy making by conventional means? Efforts in the name of environmental justice and environmental equity have left their marks on law, regulation, and public administration. Several European countries actively consider the financial impact of environmental rules on low-income households when designing policy details (Organisation for Economic Co-operation and Development 2003); the United Kingdom, for instance, imposes a lower value-added tax rate on household energy bills compared to other taxable items. Germany provides a 50 percent rebate on electricity taxes for specific types of heaters used most frequently by low-income families. The Netherlands exempts certain low-income households from fees associated with waste collection and sewerage.

A parallel judicial approach to environmental justice operates in some countries with the goal of ensuring fair access to courts and equal participation in environmental proceedings. Toward that end, the European Union has made serious efforts to implement the Aarhus Convention among member states (European Commission 2007). In one recent case, the European Court of Justice found the United Kingdom to be in violation of the Aarhus Convention because of the high costs associated with bringing environment-related suits in English and Welsh courts. In the United States, access to judicial remedies faces significant constraints. Although Title VI of the Civil Rights Act prohibits discrimination by race, it applies only to activities that receive federal funding, and a recent precedent (*Alexander v.*

Sandoval 532 US 275 [2001]) requires plaintiffs to prove not only disparate impacts but also intent to discriminate.

Many of the public policy outputs associated with environmental justice and environmental equity have taken the form of official studies, requirements for studies and assessments, and statements of intentions rather than active interventions. As of the early 2000s, a catalog of environmental justice initiatives in the United Kingdom consisted of eight studies or reports, one high-profile nongovernmental organization campaign in Scotland, the establishment of one new nongovernmental organization and one new network of organizations, one pro–environmental justice statement by a quasi-governmental body, and one genuine policy change involving the establishment of a dedicated tribunal to hear environmental cases (Agyeman and Evans 2004). In the United States, federal agencies must incorporate environmental justice awareness into their core business and consider the distributional effects of their environmental decisions on vulnerable communities. The US Environmental Protection Agency (EPA) offers trainings, technical assistance, and grants to communities engaged in environmental justice activities. Most state governments in the United States operate similar environmental justice programs.

This rather-thin inventory of official actions may seem surprising given the speed with which the environmental justice movement achieved official recognition at its start. In 1987, an EPA spokesperson expressed bewilderment when asked about race and hazardous waste siting: "We don't have any data on the sociology of these things," he was quoted as saying (Darst 1987). Barely three years later, EPA administrator William K. Reilly established the Environmental Equity Working Group, which became the Office of Environmental Equity in 1992. In 1994, President Bill Clinton signed Executive Order 12898, "Federal Actions to Address Environmental Justice in Minority Populations and Low-Income Populations," and Clark Atlanta University opened the nation's first Environmental Justice Resource Center under Bullard's leadership.

That was the golden era, according to most US historians of environmental justice, some of whom were also leading activists of the time. But the Clinton Executive Order's policy impact depended on robust implementation, and this has rarely been seen. Efforts to help federal departments implement the Executive Order in the 1990s generated little more than cursory attention to environmental justice impacts in draft environmental impact statements (Rose et al. 2005). Even the EPA itself was not conducting environmental justice reviews in accordance with the Executive Order ten years after the order was signed (Murray and Hertko 2011). As a result, hazardous waste sites in minority and low-income areas are less likely to be given cleanup priority than sites elsewhere, other things being equal, and this skewing actually appears to have worsened since the Executive Order was signed in 1994 (O'Neil 2007). As federal government attention to environmental justice waxed and waned in the post-Clinton period, state governments often stepped in to the void, but rarely with enforceable rules or remedies.

Under President Obama, environmental justice became a priority again. His first EPA administrator, Lisa Jackson, revitalized the environmental justice office and emphasized its role as a support organization for the multitude of community-level initiatives operating across the United States. The idea of policy making by other means—that is, through community action—thereby received endorsement at the highest levels of national government.

Significance, Impact, and Prospects

In the early days of environmentalism, advocates for low-income communities and disadvantaged minorities raised concerns about whether environmental progress might come at the expense of their agendas. At the first Earth Day gathering in the United States in 1970, the director of the National Welfare Rights Organization, George Wiley, queried,

> Are you going to ask the poor people of this country to bear the cost of cleaning up air pollution and doing something about other environmental problems? In all likelihood, a good many of the approaches that you are likely to take are going to be paid for directly at the expense of the poorest people in this country. ... Is the ecology movement planning to place any serious priority on the problems of the environment of the ghetto and the barrio, or our urban areas where pollution is worse? (quoted in Henderson 1974)

Since then, the environmental justice movement has worked to ensure that the answer to Wiley's last question would be yes. Activist Majora Carter (2006) called this "greening the ghetto" and linked it to sustainable urban development in her TED talk. As a result of the movement's efforts, many large, well-known environmental groups have incorporated environmental justice principles into their missions, and reoriented their agendas to address a broader range of issues and constituencies. Ad hoc coalitions of social welfare, civil rights, human rights, and environmental groups have formed to fight regressive practices at the local, national, and international levels. Although different factions within the environmental community still compete fiercely over resources and priorities (and always will), hardly anyone still contends that the environment consists of natural areas apart from cityscapes, or that environmental action is anything but a necessity for the poor and people of color.

The significance of environmental justice as a concept and organizing theme flows out of that fundamental change in orientation from environment as luxury to environment as fundamental for all. Thus anchored conceptually, environmental justice makes practical demands on the mainstream environmental, conservation, sustainability, civil rights, human rights, anticolonialist, and antipoverty movements to accommodate each other's concerns and support each other's projects. The clarity of the central idea—the self-evident injustice of environmental injustice and the moral imperative to correct it—has motivated important advances in cross-sectoral community organizing and networking. The integrated nature of these efforts makes it difficult to determine where the environmental justice work ends and the general socioeconomic development work begins. Take Spartanburg, South Carolina, for example, where a small community organization was established in 1997 to address health problems associated with hazardous waste sites. As the intertwining nature of the community's environmental and social problems became clearer, the organization gradually expanded its vision to become the hub of a comprehensive revitalization plan fueled by $250 million in public and private funding as well as partnerships with 120 organizations in 2014 (Fields 2014).

The vitality of environmental justice comes largely from communities like Spartanburg and the South Bronx, where Carter is greening the ghetto, and the fluid partnerships that often grow out of community action. Local victories on behalf of specific communities nearly always can be attributed to some coalition or another of small, scrappy, grassroots organizations that came together to fight each other's battles, with support from the big-name organizations that have incorporated environmental justice into their core business (e.g., Friends of the Earth, Oxfam, Greenpeace, World Wildlife

Fund, and others). Some of those coalitions also have scaled up to work at the national and international levels, where efforts are under way to link environmental and human rights policies (Khoday and Perch 2012). The US EPA in recent years has stitched together a model of how governments can support community-level action for environmental justice through technical assistance and networking support.

Perhaps the environmental justice label will gain more adherents in Europe now that increasing attention is being paid to cycles of disadvantage within the immigrant enclaves of European cities, particularly those formed by Muslim minorities. If and when immigrant neighborhoods within Europe are found to have greater exposure to environmental risks and lower access to environmental services, the environmental justice frame with its US civil rights resonances may catch on, especially if locally based community organizations arise to protest and seek change. A short-lived experiment with environmental justice–style organizing in the United Kingdom in the mid-1990s suggests that the success of such efforts will depend on the ability of organizers to adapt the larger concept to local realities, including political timing (Agyeman and Evans 2004). That experiment also suggests that community organizing has a better chance of gaining long-term traction if the impetus comes from inside rather than outside. The UK initiative was launched by a local franchise of Friends of the Earth, a large, multinational environmental organization, which may provide a partial explanation for its lack of staying power (Walker 2012).

Swedish research points the way toward a European variant of environmental justice that fuses sustainability concerns with distributive issues. The newer term, just sustainability, captures this line of inquiry nicely and includes high-profile European issues such as fuel poverty. As this and other applications of the overarching idea of environmental maldistribution proliferate, more terms and concepts surely will be invented, while the demand for a single, encompassing term to describe all the heterogeneity is unlikely to abate.[9]

Could environmental justice fill that broader niche eventually? Perhaps so. New generations that enter the field will have less direct knowledge of the concept's early history and associations. If they develop the habit of using environmental justice as a broader label, it might break free from its cultural and political ties. Evidence of this second-generation usage can be found here and there (Mohai et al. 2009; Sze and London 2008). The words themselves, taken separately, accommodate an inclusive reading: justice naturally covers distributive justice as well as retributive, procedural, and participatory justice. Likewise, the word environmental on its own can encompass a broad range of ecological and natural resource–based concerns in urban and rural areas, among rich and poor, and in developed, developing, and middle-income countries. Environmental can even be taken as implying environmental policy, the impacts of which (including incidence of costs) are also subject to evaluation on the grounds of fairness and justice.

One important test of a second-generation rebranding of environmental justice would be its uptake at the local, national, transnational, and international levels outside the United States. Apart from retrospective labeling, if organizations, projects, and policies classify themselves as examples of environmental justice at work, then that is what they are. And to the extent that this occurs more broadly in the future, the concept of environmental justice will have evolved, a new conceptual innovation will have emerged, and this chapter will need updating.

Could the broader adoption of environmental justice by activists and policy makers lead to stronger collective action along with greater real progress against environmental maldistribution? To date, environmental justice has tended to be strong on problem identification and direct local action, but weak on systemic solutions. The rhetoric of environmental justice and related concepts has penetrated official agendas in conventional policy-making spheres, but the resulting impact has been less than might be expected from a well-established movement. Although changes in the aggregate environmental burdens borne by vulnerable populations are difficult to measure precisely, the accumulating literature suggests that breakthroughs remain elusive.

Faster progress will require innovations in prevention, enforcement, and remediation where the needs are greatest, combined with expanded roles for vulnerable communities in decision-making processes. Beyond embracing these goals, however, it is not entirely clear how to redesign environmental policies to ensure that they also meet social justice objectives, or how to redesign social welfare policies to ensure that they also address environmental concerns. Recognition of environmental rights accompanied by effective means of enforcing them will contribute to meeting environmental justice goals, but will not do the whole job. Creative, intersectoral policy thinking is sorely needed. Sustainable development may offer a model of such thinking, with its emphasis on changing the way that we "do development" to account for both environmental and equity dimensions.

The greatest advances toward environmental justice probably depend on reducing income and wealth disparities, full stop. History has demonstrated repeatedly that wealthier, higher-status communities have the motivation and means to do whatever it takes to ensure access to environmental benefits as well as freedom from environmental burdens for their members. When more communities have such means, more instances of environmental injustice will be remedied. In the final analysis, the environmental justice movement's greatest achievement may consist of building capacity for sustainable, inclusive, grassroots development that narrows the income, wealth, and efficacy gaps, one community at a time.

Notes

Thanks go to Andrew Breza and Rachel Nusbaum for research assistance.

1. Johnson and Bullard are often referred to in the United States as the mother and father of environmental justice.

2. For the seventeen principles, see www.ejnet.org/ej.

3. For a good discussion of environmental justice as a fundamentally pluralist concept, see Schlosberg 2009.

4. On Israel, see http://www.aeji.org.il/en/background.

5. The Roma are a notable exception.

6. Its formal name is the Convention on Access to Information, Public Participation in Decision-Making, and Access to Justice in Environmental Matters.

7. The US Environmental Protection Agency's definition of environmental justice as "the fair treatment and meaningful involvement of all people regardless of race, color, national origin, or income with respect to the development, implementation, and enforcement of environmental laws, regulations, and policies" also conveys a commitment to fair participation.

8. The University of Michigan's Multicultural Environmental Leadership Development Initiative maintains the directory.

9. For example, note the variant *green equity* (Khoday and Perch 2012).

References

Agyeman, Julian, and Bob Evans. 2004. "Just Sustainability: The Emerging Discourse of Environmental Justice in Britain?" *Geographical Journal* 170:155–164. doi:10.1111/j.0016-7398.2004.00117.x.

Anand, Sudhir, and Amartya Sen. 2000. "Human Development and Economic Sustainability." *World Development* 28:2029–2049. doi:10.1016/s0305-750x(00)00071-1.

Banzhaf, Spencer. 2012. "Regulatory Impact Analyses of Environmental Justice Effects." *Journal of Land Use and Environmental Law* 27:1–30.

Bradley, Karin, Ulrika Gunnarsson-Ostling, and Karolina Isaksson. 2008. "Exploring Environmental Justice in Sweden: How to Improve Planning for Environmental Sustainability and Social Equity in an 'Eco-Friendly' Context." *MIT Journal of Planning—Projections* 8:68–81.

Braubach, Matthias, and Jon Fairburn. 2010. "Social Inequities in Environmental Risks Associated with Housing and Residential Location: A Review of Evidence." *European Journal of Public Health* 20:36–42. doi:10.1093/eurpub/ckp221.

Bullard, Robert. 1990. *Dumping in Dixie: Race, Class, and Environmental Quality*. Boulder, CO: Westview Press.

Carruthers, David V. 2008. *Environmental Justice in Latin America: Problems, Promise, and Practice*. Cambridge, MA: MIT Press.

Carter, Majora. 2006. "Greening the Ghetto." February. Accessed March 18, 2015, http://www.ted.com/talks/majora_carter_s_tale_of_urban_renewal.

Claudio, Luz. 2007. "Standing on Principle: The Global Push for Environmental Justice." *Environmental Health Perspectives* 115:A500–A503. doi:10.1289/ehp.115-a500.

Darst, Guy. 1987. "Church Group Sees Racist Pattern in Placement of Landfills." Associated Press, April 15. Accessed March 18, 2015, http://www.apnewsarchive.com/1987/Church-Group-Sees-Racist-Pattern-in-Placement-of-Landfills/id-373fce512fa51a9cd74bc7e11debb299.

Environmental Justice Atlas. n.d. "World Map." Accessed December 19, 2016, http://ejatlas.org.

European Commission, Directorate-General for the Environment. 2007. *Summary Report on the Inventory of Member States' Measures on Access to Justice in Environmental Matters*. Brussels: Milieu Ltd.

European Environment Agency. 2011. *Environmental Tax Reform in Europe: Implications for Income Distribution*. Luxembourg: Publications Office of the European Union.

Fields, Timothy, Jr. 2014. "A Dream Realized: Community-Driven Revitalization in Spartanburg, South Carolina." EPA Blog, August 26. Accessed March 18, 2015, https://blog.epa.gov/ej/2014/08/a-dream-realized.

Gandhi, Indira. 1972. "Man and Environment." Speech to the plenary session of the UN Conference on the Human Environment, Stockholm, June 14.

Gottlieb, Robert. 2005. *Forcing the Spring: The Transformation of the American Environmental Movement*. Washington, DC: Island Press.

Henderson, Hazel. 1974. "Redefining Economic Growth." In *Environmental Quality and Social Justice in Urban America*, ed. James Noel Smith, 123–145. Washington, DC: Conservation Foundation.

Iceland, John. 2014. *Residential Segregation: A Transatlantic Analysis*. Washington, DC: Migration Policy Institute.

Khoday, Kishan, and Leisa Perch. 2012. "Green Equity: Environmental Justice for More Inclusive Growth." In *Research Brief 19*. Brasília: International Policy Centre for Inclusive Growth.

Lee, Charles. 1987. *Toxic Wastes and Race in the United States: A National Report on the Racial and Socio-economic Characteristics of Communities with Hazardous Waste Sites*. New York: United Church of Christ Commission for Racial Justice.

Martinez-Alier, Joan. 2002. *The Environmentalism of the Poor: A Study of Ecological Conflicts and Valuation*. Cheltenham, UK: Edward Elgar.

McGurty, Eileen Maura. 1997. "From NIMBY to Civil Rights: The Origins of the Environmental Justice Movement." *Environmental History* 2:301–323. doi:10.2307/3985352.

Mohai, Paul, David Pellow, and J. Timmons Roberts. 2009. "Environmental Justice." *Annual Review of Environment and Resources* 34:405–430. doi:10.1146/annurev-environ-082508-094348.

Murray, Sylvester, and Mark D. Hertko. 2011. "Environmental Justice and Land Use Planning." In *Justice for All: Promoting Social Equity in Public Administration*, ed. Norman J. Johnson and James H. Svara, 192–206. Armonk, NY: M. E. Sharpe.

Northridge, Mary E., Gabriel N. Stover, Joyce E. Rosenthal, and Donna Sherard. 2003. "Environmental Equity and Health: Understanding Complexity and Moving Forward." *American Journal of Public Health* 93:209–214. doi:10.2105/ajph.93.2.209.

O'Brien, Liz, and Jake Morris. 2014. Well-Being for All? The Social Distribution of Benefits Gained from Woodlands and Forests in Britain." *Local Environment: The International Journal of Justice and Sustainability* 19:356–383. doi:10.1080/13549839.2013.790354.

O'Neil, Sandra George. 2007. "Superfund: Evaluating the Impact of Executive Order 12898." *Environmental Health Perspectives* 115:1087–1093. doi:10.1289/ehp.9903.

Organisation for Economic Co-operation and Development. 2003. *Conceptual Framework for Analysing the Distributive Impacts of Environmental Policies*. Paris: Organisation for Economic Co-operation and Development.

Organisation for Economic Co-operation and Development. 2012. *Review of the Implementation of the OECD Strategy for the First Decade of the 21st Century*. Paris: Organisation for Economic Co-operation and Development.

Pellow, David Naguib. 2000. "Environmental Inequality Formation." *American Behavioral Scientist* 43:581–601. doi:10.1177/0002764200043004004.

Reed, Maureen G., and Colleen George. 2011. "Where in the World Is Environmental Justice?" *Progress in Human Geography* 35:835–842. doi:10.1177/0309132510388384.

Ringquist, Evan J. 2005. Assessing Evidence of Environmental Inequities: A Meta-Analysis." *Journal of Policy Analysis and Management* 24:223–247. doi:10.1002/pam.20088.

"The Rise of Anti-Ecology." 1970. *Time* 96 (August 3): 44–46. Accessed March 18, 2015, http://content.time.com/time/magazine/article/0,9171,876696,00.html.

Rose, Linda, Natalie A. Davila, Kelly A. Tzoumis, and Daniel J. Doenges. 2005. "Environmental Justice Analysis: How Has It Been Implemented in Draft Environmental Impact Statements?" *Environmental Practice* 7:235–245. doi:10.1017/s1466046605050374.

Schlosberg, David. 2009. *Defining Environmental Justice: Theories, Movements, and Nature*. New York: Oxford University Press.

Serret, Ysé, and Nick Johnstone, eds. 2006. *The Distributional Effects of Environmental Policy*. Cheltenham, UK: Edward Elgar.

Shklar, Judith. 1989. "Giving Injustice Its Due." *Yale Law Journal* 98:1135–1151. doi:10.2307/796574.

Stephens, Carolyn, Simon Bullock, and Allister Scott. 2001. *Environmental Justice: Rights and Means to a Healthy Environment for All*. Special Briefing No. 7. London: Economic and Social Research Council.

Sze, Julie, and Jonathan K. London. 2008. "Environmental Justice at the Crossroads." *Sociology Compass* 2:1331–1354. doi:10.1111/j.1751-9020.2008.00131.x.

Taylor, Dorceta E. 2000. "The Rise of the Environmental Justice Paradigm." *American Behavioral Scientist* 43:508–580. doi:10.1177/0002764200043004003.

US Environmental Protection Agency (EPA). 2014. "Environmental Justice: The Road to Executive Order 12898." In *Celebrating 20 Years of Making a Visible*

Difference in Communities. Accessed March 18, 2015, https://www.epa.gov/environmentaljustice/events/20th-anniversary.html.

US Government Accounting Office. 1983. *Siting of Hazardous Waste Landfills and Their Correlation with Racial and Economic Status of Surrounding Communities*. Accessed March 18, 2015, http://www.gao.gov/products/RCED-83-168.

Walker, Gordon. 2009. "Globalizing Environmental Justice." *Global Social Policy* 9:355–382. doi:10.1177/1468018109343640.

Walker, Gordon. 2012. *Environmental Justice: Concepts, Evidence, and Politics*. Abingdon, UK: Routledge.

11 Environmental Security: Policy within a Violent Imaginary

Johannes Stripple

The concept of environmental security revolves around the idea that there is a connection between the health of the natural environment and the security of persons, states, cultures, ecosystems, or the biosphere. While sustainability has come to evoke a cooperative and peaceful imaginary, environmental security articulates a violent and conflict-ridden future. For instance, one of the current claims is that global warming increases the risk of civil war in Africa, and political instability will follow when subsistence farmers are forced to leave their livelihoods because of drought, flooding, and famine. Other projections depict a series of "regional hot spots"—multiple conflict constellations—around the globe due to the current rates of environmental change (German Advisory Council on Global Change 2008). The concept of national security, as it has been understood within world politics since the 1920s, brings drama, urgency, and priority to environmental issues. The temptation for activists, international organizations, scholars, and media to situate the environment within a violent imagery is understandable. It is meant as a call to take immediate action and put in the resources needed to avoid future violence.

What are we to make of these "discourses of fear" (Hulme 2008, 5) that circulate among the media, scholars, think tanks, international organizations, and the military establishment? How does the concept of environmental security enable us to apprehend the nexus between the environment, conflict, and human suffering? What does it imply to say that the Syrian war was caused by climate change? And what does it imply to approach and govern the environment through the lens and practices of security? What kind of policy, politics, and governance is then legitimated?

Security is a derivate concept. It necessarily presupposes something to be secured (Krause and Williams 1997, x). For many years during the twentieth century, the concept of security was simply taken for granted, and questions about what security is or what it is that needs securing were never asked. Security was understood to be about the security of the state in relation to the military aggression of other states. Nature, or "environmental change," was not linked with insecurity, instability, and violent conflicts. The environment was neither seen as a threat nor a value to be secured, though this changed in the early 1990s. The Cold War order waned, and the new order had yet to be identified. At this time, the journalist Robert Kaplan traveled to West Africa, understanding what he saw in Sierra Leone as being a microcosm of where our world was heading. Kaplan's (1994) essay, "The Coming Anarchy: How Scarcity, Crime, Overpopulation, Tribalism, and Disease Are Rapidly Destroying the Social Fabric of Our Planet," which was published as a cover story in the *Atlantic Monthly*, captured the mood of the day. It is time, he wrote, to "understand the environment for what it is: the national-security issue of the early 21st century" (58). Kaplan saw chaos and mass migration as a result of the combination of rapid population growth, the spread of disease, deforestation and soil erosion, air pollution, water depletion, and so on, given that these issues were set to be the defining foreign policy issues in the post–Cold War security order. And he concluded, "West Africa's future, eventually, will also be that of most of the rest of the world" (48). Thus, the concept of environmental security was born as the Berlin Wall came down, being launched as an overarching violent characteristic of the times to come.

While writings on security and the environment existed before the 1990s, it was the end of the Cold War that unlocked and energized a range of new discussions. It became a topic among policy makers, the media started to pay attention, and scholars initiated empirical investigations to understand under what conditions environmental change might give rise to violent conflicts. Interestingly, the ways in which Kaplan penned his essay—the limits/scarcities, urgencies, geographic tropes of we/them, and failed forms of statehood—tell us much about the context in which claims to environment and security *can* be made. As Jon Barnett (2013, 204) notes, "The most influential interpretations of environmental security are those that fit well the orthodox security paradigm." Security is a powerful word, so when a problem is identified as a security problem, it becomes "high politics"—an issue of highest priority that can help mobilize significant resources and support. This is why it has been tempting for those who would like to see environmental issues rise on the political agenda to invoke the language of security.

From Science to Policy

If the concept of environmental security took off in the early 1990s, what were its precedents? Rita Floyd and Richard A. Matthew (2013, 3) go all the way back to Thucydides's *The Peloponnesian War* and Plato's *Republic*. Sparta, a society living within its limits, is compared to Athens, a society heavily reliant on imports. Another precedent is Thomas Malthus (1798), whose *An Essay on On the Principle of Population* made the case that if human populations grow faster than their agricultural output, the resulting gap between supply and demand will result in famines, epidemics, and wars. The term *Malthusian* is now part of the English language, and the idea of scarcity as a limit on humanity's potential has been reiterated over the years, from *Limits to Growth* (Meadows et al. 1972) to *Planetary Boundaries* (Rockström et al. 2009).

Apart from Malthus, environmental security has its conceptual prehistory in an important set of writings from the 1960s. Generally speaking, these were books that discussed the negative outcomes of a deteriorating environment for humanity and/or the planet (Carson 1962; White 1967; Ehrlich 1968; Meadows et al. 1972). In the 1970s, progressive scholars started to directly engage the prevalent security establishment. "We need to revamp our entire concept of 'national security' and 'economic growth' if we are to solve the problems of environmental decay," as Richard A. Falk (1971, 185) wrote in *This Endangered Planet*. Or as Harold H. Sprout and Margaret Sprout (1971, 406) put it in *Toward a Politics of Planet Earth*, "The thrust of evidence is simply that the goal of national security as traditionally conceived—and still very much alive—presents problems that are becoming increasingly resistant to military solutions." A few years later, Lester Brown (1977, 5), then the president of the World Watch Institute, penned the brief report "Redefining National Security," which directly criticized the military supremacy and state-based definition of security: "The overwhelmingly military approach to national security is based on the assumption that the principal threat to security comes from other nations. But, the threats to security may now arise less from the relationship from nation to nation and more from the relationship of man to nature." The report urges countries to confront these threats in a cooperative manner, but the conclusion raises severe concerns over the security establishment's response to the issue at hand: "National defense establishments are useless against these new threats. Neither bloated military budgets, nor highly sophisticated weapons systems can halt the deforestation or solve the firewood crises now affecting so many Third World countries" (37).

With remarkable foresight, these writings identified the need for the international system to respond because environmental problems pose threats to international stability and well-being. According to Barnett (2013, 192), the idea that environmental change could cause *war* was first suggested when Richard Ullman (1983), a Princeton professor of international affairs, wrote the article "Redefining Security," which appeared in the influential US journal *International Security*. A few years later, environmental scientists such as Norman Myers (1986) and Arthur H. Westing (1986) argued that environmental degradation would cause violent conflict. By 1989–1990, we find numerous attempts to "redefine security," especially among established journals. Jessica Tuchman Mathews (1989) had a piece in *Foreign Affairs*, contending that resources had to be seen as a new national security priority. Nevertheless, we should remember that writings up to 1990 on environmental security were by and large a "peripheral concern to Western security institutions occupied with the hard business of winning the Cold War" (Barnett 2013, 193). The writings by Brown, Myers, Ullman, and others were all attempts to mobilize the security establishment and bring in a new constituency, but the window of opportunity was not yet open, and the arguments did not resonate among practitioners.

The end of the Cold War created new intellectual spaces, and environmental security started to make its way into concrete security policy agendas. From around the 1990s, the concept of environmental security began to "feature regularly in academic journals, in the speeches of politicians and security bureaucrats, and in the work of environmental organizations" (ibid., 205). Calls for "common security" were made in *Our Common Future* (Brundtland 1987), and security surfaced in the preparations for the Rio conference in 1992. The UN secretary-general's *Agenda for Peace* identified "ecological damage" as a new risk for stability (Boutros-Ghali 1995, 5). Within the United States, the Bush administration incorporated the concept of environmental security into the National Security Strategy in

1991–1992, advancing the notion that in concert with its allies, the United States will "achieve cooperative international solutions to key environmental challenges, assuring the sustainability and environmental security for the planet as well as growth and opportunity for all" (Fiksel and Hecht 2012). A report done by the US Army War College claimed that international environmental issues could lead to instability and conflict (Butts 1993), while US president Barack Obama made the link between climate change and national security when he gave his acceptance speech for the Nobel Peace Prize in December 2009. In 2010, the US Pentagon included climate change for the first time in the important strategy document *Quadrennial Defense Review* (US Department of Defense 2010).

Climate change is understood to play a significant role in shaping the future security environment. It may act as an accelerant of instability or conflict that, in combination with other issues, could contribute to failed-state scenarios that demand military intervention. In the recently published second *Quadrennial Defense Review* (US Department of Defense 2014), climate change gets a bit less attention, but the approach is the same. Climate change is a potential threat multiplier that could enable terrorist activity and other forms of violence. The Department of Defense thus needs to be ready for a security environment in which climate impacts affect the operating environment.

In an overview of the growing military interest in climate change, Emily Gilbert (2012) shows that these contentions reverberate all around the world in security establishments. For example, the Australian Department of Defence (2007) argues for a new role for the military in resource protection, whereas in the United Kingdom the Ministry of Defense's Global Strategic Trends Program 2007–2036 (DCDC 2006) asserts a potential role for the military not just in relation to climate impact and disaster relief but also in geoengineering interventions in outer space. The American Security Project has surveyed all countries to determine to what extent governments worldwide consider climate change to be a security issue. It found that 70 percent of the nations in the world state that climate change is a security concern, but it does not provide any numbers of what that translates into in terms of planning and budget allocations (Holland and DeGarmo 2014). In 2007, the environment finally made it to the UN Security Council, which hosted its first debate on climate change, and in July 2011 it again considered the impact of climate change on international peace and security.

The deliberations did not lead to a resolution, although the final statement expressed a "concern that possible adverse effects of climate change may, in the long run, aggravate certain existing threats to international peace and security" (UN Security Council 2011, 1).

Innovation and Contestation

We usually carry out conceptual innovation by inventing new words or combining two existing ones. Linguistically, environmental security is a "single-word modifier," in which one word (environmental) modifies the meaning of another word (security). If we talk about environmental security (and not about environment and security), we understand it as a compound—the process of combining two words to create a new one. It is therefore crucial that we are clear about the *meaning of security* in order to understand what it is that the notion of *environmental* attempts to modify. While security has meant many different things over the years, it was a particular meaning of security that came to be dominant in the post–World War II era within academia, policy making circles, and think tanks. Security has historically been understood in various ways, and the very idea of what it means to be secure has fluctuated. The word *security* comes from the Latin phrase *sine cura*, which means "without worries." In the early days of the modern state system, security started to mean being safe, protected and free from danger. In the twentieth century, security gained a particular meaning in world politics, with connotations of survival, urgency, and priority. By labeling an issue as a security issue, "an agent claims a need for and a right to treat it by extraordinary means" (Buzan et al. 1998, 26). While concerns (e.g., the territorial integrity of the state) established as security issues receive attention, priority, and resources, they are also removed from the public as well as processes of democratic deliberation and contestation.

The word environmental has primarily attempted to modify the meaning of security that emerged in the 1940s. At this time *national security* entered our vocabulary and quickly became an organizing concept—a comprehensive label for a range of phenomena that were previously discussed as war, defense, military, and foreign policy (Yergin 1977). The early Cold War saw the establishment of the UN Security Council along with US agencies such as the National Security Agency and National Security

Council. From the 1940s onward, the *subject* of security was the modern state, and *the main threat* to the survival of the state was military aggression from other states. When the Cold War ended, organizations such as the UN Development Program (UNDP) attempted to change the terms of the security discourse and launched the concept of *human security*. "The concept of security has for too long been interpreted narrowly: as security of territory from external aggression, or as protection of national interests in foreign policy or as global security from the threat of a nuclear holocaust. It has been related more to nation-states than to people" (UNDP 1994, 22). Hence, the reformulation by the UNDP puts individuals at the center and acknowledges a range of threats to people's security. The environment is one of the seven categories of threats identified by the UNDP, while the other threats are economic, food, health, personal, community, and political. The UNDP's legacy in focusing on development meant that human security came to be associated with the vulnerability of the most marginal and poor. As Simon Dalby (2013, 122) puts it, "Human security is formulated by the UNDP precisely in terms of providing the conditions that make development possible."

The original publication by the UNDP also resonates well with a core principle of environmental thought—namely, precaution. Human security is about anticipating dangers and acting on them before they occur. Since human security shifts the focus to the conditions that make people insecure in particular places, the concept also brings the larger picture into view. It turns the attention to modernity, trade, and the expansion of the global economy—the "appropriation of rural natural resources to supply commodities to the metropoles" (ibid., 126). While the academic community, international organizations, and nongovernmental organizations have heavily endorsed aspirations to human security, the concept has not been much adopted in policy by states (Kerr 2013). The assessment *so far* has been that it is "a stalled initiative" (Suhrke 2004), but this might change—as practices of security are not set in stone.

When considering the efforts (in the 1980s and early 1990s) to innovate conceptually around environment and security, it is important to see that these attempts were of different kinds and potentially incompatible with each other. The environment was treated as both a new type of threat and something to be secured. The "environment as threat" framing was about the ecological crises being increasingly understood in security terms, in which environmental threats were added onto a traditional, geopolitical national security agenda. The "securing the environment" framing emphasizes the integrity of the environment, and therefore becomes disruptive toward the state-centric and militaristic mind-set of the mainstream security discourse (Elliott 1998).

Another line of argument was put forward by US academic Dan Deudney, who thought that environmentalists should stay away from trying to come up with new conceptual innovations at the interface of environment and security. He contends that it is analytically misleading to think of environmental degradation as a national security threat because the traditional focus of national security has little in common with both environmental problems and solutions. Deudney (1990, 461) is particularly skeptical about the different values, connotations, and "mind-sets" inherent in environmentalism vis-à-vis security institutions: "For environmentalism to dress their programmes in the blood-soaked garments of the war system betrays their core values and creates confusion about the real tasks at hand."

Looking back at twenty-five years of discussions around environmental security within academia, international organizations, think tanks, and defense planners, it is striking that despite a lot of contestation, there is an underlying commitment to a materialist ontology. This assumption leads to recurrent conversations as to whether the environment "really" is a security issue or not, and what type of approach would be "best" or most "desirable." The concept of environmental security is taken for granted as something that the expert can and should define objectively, and compare to a certain state of affairs in the world. The main form of knowledge has consequently been the quest for empirically validated generalizations about cause-and-effect relationships. The environment becomes a variable, a cause whose effect (e.g., the incidence of conflict or violence) will be measured on the basis of case studies or large-N statistical data.

An Increasingly Violent World?

A key characteristic of the concept of environmental security is how it draws attention to a violent future. It is intended to instigate action today that could avoid the problems of tomorrow. This could be illustrated with the recent Group of 7 report, *A New Climate for Peace* (Rüttinger et al. 2015), which highlights that climate change heightens the risk for instability and conflict, and hence must be a foreign policy priority. Of course, the overarching questions looming are whether environmental change will actually cause violent conflict (within or between states). If so, what are the mechanisms and processes through which this will occur? For this reason, academic research has been devoted to specifying "the environment" as a variable (often in terms of the scarcity of a certain natural "resource") and measuring its influence on "security," operationalized, for example, as violent conflict within or between states, or (sometimes) as human well-being.

While the writings on these issues in the early 1990s were sweeping and anecdotal, the latter part of the 1990s saw the emergence of more empirically driven research projects. Thomas Homer-Dixon (1991, 1994, 1999) put forward several hypotheses on the probable linkages between environmental change and acute conflict, underscoring the role of the scarcity of renewable resources that places stress on sociopolitical systems, and when given other variables (conditions), will erupt into subnational violence and strife. Based in Zurich around Guenther Baechler, another research group conducted similar research on the linkages between environmental degradation and violence. These two groups carried out numerous case studies in which they attempted to determine the influence of environmental factors in generating specific conflicts. According to Barnett (2013, 197), the "common findings of both projects are that: unequal consumption of scarce resources contributes to violent conflicts; violent conflicts where environmental scarcity is a factor are more likely in low-income resource dependent societies; and, when mechanisms that enable adaptation to environmental scarcity fail, violent conflict is a more likely outcome." Environmental change never "causes" conflict in a direct sense; it instead works as an "exacerbating factor" or "threat multiplier," to use a word frequently invoked in policy discussions on a climate-changed world.

An important methodological development has been to use more aggregated and detailed data. There has been a range of new studies on climate change and conflict based on statistical methods, in which large-scale data sets on, for instance, temperature and precipitation are correlated with data sets such as the Armed Conflict Dataset of the Uppsala Conflict Data Program and the Peace Research Institute Oslo. Interestingly, this research and debate has taken place in journals such as the US-based *Science* and

Proceedings of the National Academy of Sciences as well as in op-eds in the *New York Times* and articles circulated in the blogosphere. That the concept of environmental security from 2009 has moved to *Science* and *Proceedings of the National Academy of Sciences* indicates that it is seen to be "maturing," and it is assumed that the earlier questions and arguments can be settled through the application of statistical methods. Historically, compared to the situation in general in political science and international relations, security studies in the United States are not as hard-core rational choice or about number crunching; "the typical article in *International Security* uses historical case studies—maybe one in-depth historical case study—to examine a hypothesis framed as a cause-effect relationship and very much tied into general debates" (Waever and Buzan 2013, 403). But this new research is certainly not just about settling "old disputes." It is more about forecasting what might come tomorrow. This is the type of knowledge that actors such as the US Department of Defense hope the academic community will be able to provide insights about: Will there be more violence in a climate-changed world? In this respect, an interesting report is *Climate and Social Stress: Implications for Security Analysis*, prepared by the US National Research Council, which set up a specific committee for the task. The task was to "evaluate the evidence on possible connections between climate change and U.S. national security concerns and to identify ways to increase the ability of the intelligence community to take climate change into account in assessing political and social stresses with implications for U.S. national security" (Steinbruner et al. 2013, 1). The report clearly indicates that environmental security has now become rather mainstream.

Preparations for the effects of climate change seem to be under way within militaries and national security communities (Holland and DeGarmo 2014). The US television network NBC did a news report about the Weather Warriors, the Pentagon's Combat Climatologists, which is a frontline force monitoring the apocalypse (Dokoupil 2014). Obama announced a new set of tools to help vulnerable populations around the world to strengthen their climate resilience as well as a new executive order requiring all federal agencies to factor climate resilience into the design of their international development programs and investments (White House 2014).

The task of the intelligence community is to provide indicators and warnings about a wide variety of threats that might pose, whether directly or indirectly, significant risks to national security. The climate issue is one of those concerns that need to be taken into account. The report highlights climate surprises—for example, disruptive events that might affect integrated systems such as the grain market—but also puts a lot of emphasis on the need to understand adaptation and changes in vulnerability that could create or alleviate social as well as political stresses.

On the science side, one well-known article of the statistically driven wave of scholarship is the Marshall B. Burke and colleagues (2009) piece "Warming Increases the Risk of Civil War in Africa," published in the *Proceedings of the National Academy of Sciences*. They find "strong historical linkages between civil war and temperature in Africa, with warmer years leading to significant increases in the likelihood of war. When combined with climate model projections of future temperature trends, this historical response to temperature suggests a roughly 54% increase in armed conflict incidence by 2030, or an additional 393,000 battle deaths if future wars are as deadly as recent wars" (20670). Halvard Buhaug responded in the same journal with an article titled "Climate Not to Blame for African Civil Wars." Buhaug (2010) is discomforted that "vocal actors within policy and practice contend that environmental variability and shocks, such as drought and prolonged heat waves, drive civil wars in Africa." His model, which uses measures of drought, heat, and civil war, arrives at the opposite conclusion: "Climate variability is a poor predictor of armed conflict. Instead, African civil wars can be explained by generic structural and contextual conditions: prevalent ethno-political exclusion, poor national economy, and the collapse of the Cold War system."

The journal *Science* ran a piece by Jürgen Scheffran and his colleagues, who note that some quantitative empirical studies support a link between climate change and violent conflict, whereas others find no connection or only weak evidence. Overall, "current debates over the relation between climate change and conflict originate in a lack of data" (Scheffran et al. 2012, 869). Hence, while there is a community of scholars who agree that statistical methods can settle disputes around climate-induced violence, the community neither agrees on how to measure the independent variable (the climate) or how the dependent variable should be conceptualized (how many battle deaths constitute a civil war?).

A recent widely circulated study has used a radically different conceptualization of the dependent variable, from "civil war" to "interpersonal conflict." Solomon M. Hsiang and his colleagues published "Quantifying the Influence of Climate on Human Conflict" in *Science*, in which they attempted to demonstrate a correlation between climate extremes and violence across a range of time periods, countries, and different levels of conflict. It is a kind of "meta-analysis" of sixty studies that compared levels of violence in a given population, during periods of normal climate, with levels of violence during periods of extreme climate. The studies came from fields as diverse as archaeology, criminology, economics, geography, history, political science, and psychology. They found that higher temperatures and extreme rainfall led to large increases in conflict: "The magnitude of climate's influence is substantial: for each one standard deviation (1s) change in climate toward warmer temperatures or more extreme rainfall, median estimates indicate that the frequency of interpersonal violence rises 4% and the frequency of intergroup conflict rises 14 % (Hsiang et al. 2013, 1235367-1). They followed up with a piece in the *New York Times* titled "Weather and Violence," in which they wrote that our children and grandchildren will face an increasingly hot and angry planet (Burke et al. 2013), while the *Atlantic* ran the story "Hotter Weather Actually Makes Us Want to Kill Each Other." On the New Security Beat, the blog of the Woodrow Wilson International Center for Scholars' Environmental Change and Security Program, Joshua Busby (2013) suggests that we now need to find out the causal processes that connect climate effects and patterns of conflict. Despite the different data sets and methodologies involved, the last years of writings have nurtured the idea that we are going to live in an increasingly violent world—a world of large-scale geopolitical tensions related to both intimate and interpersonal violence.

Environmental Security as Ambiguous Concept

Arnold Wolfers (1952) noted as early as 1952 that national security was an ambiguous symbol, and Karen Litfin (1999, 360) reiterated that environmental security functions as "an ambiguous symbol" for a wide array of policy and analytic positions. Philosophically speaking, ambiguity is an attribute of a concept whose meaning cannot be resolved, and it is not the same as vagueness. While philosophers often work hard to reduce the ambiguity of a concept, sentence, or argument, ambiguity has a productive side that enables the concept to travel among different constituencies. According to Litfin, "Because it functions as an ambiguous symbol, environmental security has attracted a remarkable array of proponents, ranging from environmentalists to Western military institutions" (361). Some of the issues that can be put under the term environmental security might seem rather disparate, ranging from warfare as a major cause of environmental destruction, renewable energy deployment in the military, large-scale environmental changes along with their impact on national and international security, and the possibility of interstate war as a result of resource abundance or scarcities. People from different communities of practice (e.g., military officials, think tanks, the media, and scholars) use environmental security as a point of reference for conversations, even though different meanings and interpretations have been circulated. Environmental security is not imposed by just one community, and simultaneously satisfies different concerns. For the activist, environmental security could indicate a radical priority to preserve the biosphere, while for a traditional security institution it is a discourse of fear that legitimates and maintains military spending as preparation for warfare.

The ambiguity and abstraction of environmental security allows for different communities of security practice to engage with it, from the military and international development organizations to reinsurance companies and academic research departments. The concept can accommodate different activities, such as the modeling of climate variability and conflict risk in East Africa, while being an organizing metaphor for a range of discussions that have to do with warfare, and preparations for warfare, as a major cause of environmental destruction. The violent imaginary that is brought in is likely to stay for the time being. It is continuously invoked in the media through news and commentaries on environmental impacts (flooding and heat waves) in both metropolitan areas and the developing world. The recent surge in the advanced statistical modeling of civil wars and violence in a warming world, and the discussion of this research in prestigious natural science journals and high-profile media outlets, will keep environmental security on the agenda. Its place on the agenda, however, is largely influenced by nonenvironmental events such as Cold War dynamics, the war on terror, or financial crises.

When considering the place of environmental security in contemporary discussions, we should be careful not to idealize security as the desired end goal. There is a "productive" aspect of security that needs to be understood. Security "is not a noun that names something, it is a principle of formation that does things" (Dillon 1996, 16). Mike Hulme (2008) warns that the current discourse of fear about future climate catastrophes could legitimize attempts to conquer our climate futures through ideas of control and mastery of the planet, or individual and collective behavior. Gilbert (2012) is afraid that the current bringing together of climate change and security does not lead to any reconsideration of questions around security, as one might have hoped for, but rather to the militarization of climate change, in addition to a nationalist and defensive strategy modeled on future disaster scenarios of resource conflicts.

Recent writings and articulations around climate change and the Anthropocene give reasons for pause. When climate change is situated in the earth system science imagination of nonlinearity within complex systems, and flips between multiple unstable equilibriums and physically chaotic behavior (Lövbrand et al. 2009), we risk being paralyzed and ambivalent instead of being motivated to take action. Gerda Reith (2004, 393) argues that "the profound uncertainty generated within a globalized, indeterministic world erodes the basis for decision making, freezes action, and ultimately blocks the possibility of forward movement into the future." The future becomes a site of anxiety and unknowns out of reach of human intervention. Looking across different domains of life (terrorism, transspecies epidemics, and climate change), Ben Anderson (2010) notes the commonalities around how these domains have been enacted as threats, and how the future is now problematized as being indeterminate and uncertain, which thus has to be met with an extraordinary proliferation of anticipatory action. Mark Duffield (2011, 763), a development scholar, argues that this condition turns climate change into an "environmental terror"—an environment that because it operates through uncertainty and surprise, has itself become terroristic. Complex adaptive systems require different strategies of securing; if uncertainty prevails, if we cannot predict and prevent, then we have to invest in "preparedness" and "resilience" (Methmann and Oels 2015). And as it turns out, the resilience paradigm individualizes the problem. Environmental governance becomes a question of fostering the resilience of various populations (mostly southern) to make them fit for survival in times of uncertain and apocalyptic climate change (ibid.). From a Foucauldian perspective, security can be understood as a specific mode of governing. Hence, as climate security as apocalypse enables a new set of practices for fostering resilience, the structural causes of climate change become depoliticized and naturalized (Oels 2014, 212). Climate politics is left with a politically debased subject that "accepts the disastrousness of the world it lives in as a condition for partaking of that world, and which accepts the necessity of the injunction to change itself in correspondence with the threats and dangers now presupposed as endemic" (Reid 2012, 74). Political life in an increasingly violent climate-changed world consequently becomes about embracing insecurity as the new normal. Environmental security, institutionalized as a matter of fostering resilience, seems to lead to the depoliticization (less politics, or fewer claims about, say, justice or democracy) instead of the politicization of the causes and impacts of climate change that was hoped for when the concept was originally innovated.

Conclusion

The concept of environmental security has been significant in nurturing a violent imagination of environmental and climate futures. It has been taken up and circulated across a range of political, social, and cultural realms. The environment is now well established as part of an imagery of a world that is becoming more violent, more conflict ridden, and less secure for many people, from those in the Sahel to those who took refuge in the Superdome in New Orleans during Hurricane Katrina. In particular, a climate-changed world feeds into a horizon of the future that is increasingly understood as indeterminate and uncertain, thereby requiring new modes of preparedness and precaution.

As environmental and climate impacts become ever more serious and known about, the relation to security was bound to be made. From the early writings by environmental nongovernmental organizations such as the World Watch Institute, to recent reports by mainstream military organizations like the Central Intelligence Agency or Pentagon, we learn that the environment is not a romantic and peaceful place but rather a source of conflict, violence, and human suffering. As discussed throughout this chapter, the uptake of the environment within the security establishment has been problematized and criticized on several grounds. As a discourse of fear, it potentially legitimizes military spending, encourages policy paradigms of control and mastery, and articulates a depoliticized idea of who is to blame for a life rendered insecure in a climate-changed world. Because the concept of security has referred to notions of urgency and priority, it has been tempting to use it in order to raise awareness and support for environmental issues.

Perhaps for both good and bad, the concept of environmental security has engaged new constituencies such as defense departments, military planners, and development agencies. The debate around environmental security has rearticulated fundamental questions of what it might mean to be secure today in relation to how modern dangers can be approached—although the answers to such questions are anything but straightforward. On the one hand, we seem to have come a long way since Brown of the World Watch Institute suggested the need to think about the environment in terms of security in 1977, while on the other hand, many other things remain the same.

It should be kept in mind that security is a powerful set of practices at the heart of the modern state. In his book *Leviathan*, the British philosopher Thomas Hobbes (1651) wrote that without security, life becomes solitary, poor, nasty, brutish, and short. In this influential (liberal) tradition of thought, solidified during the Cold War aspirations to national security, it is the state as our primary political community that has made us secure, and it is the state that is the entity to be secured, often through the use of military power.

The waning of the Cold War did open a space for reconsidering security in new ways. The environment was understood to be a new type of threat, but it was also invoked as a value to be secured. The academic research frontier has moved on from single case studies of sites in which environmental scarcities potentially contribute to violent conflicts, to large-scale statistical analyses of the links between climate change and violence. Despite the increased precision of the environment and security nexus, the problems of situating and linking the environment within the imagery and practices of security has been raised by numerous scholars and activists. It has been pointed out to be the wrong mind-set, the wrong institutions, and a mode of governing that leads to the displacement of politics. Environmental security remains an ambiguous concept with many fault lines among and within academia, think tanks, environmental organizations, and the military establishment. There are always repeated calls for deciding how environment and security *should* be defined and understood; if only a particular definition were adopted, the reasoning goes, everything would be fine. But the calls for straddling this contested terrain through adopting a particular definition are largely unfruitful. We could wish the dense practices of international security to be different (less militaristic, less state-centric, more human, etc.), but they won't go away that easily. While claims about environment and security shed light on the assumptions of international security (which threats, priorities, and subjects should be made secure), a violent imagination of environmental and climate futures does also reproduce this very same security discourse. There is no need for more normative writings on how environmental and climate security *should* be defined; rather, attention should be on understanding the consequences when environment and security is invoked. What difference does it make, for example, to link the war in Syria to climate change? What is gained or lost, illuminated or obscured, empowered or disempowered?

Security, understood as characteristic of world politics or a characteristic of the modern welfare state, is best understood as a verb and not as a noun. It is a mode of governing that does things and needs to be approached in terms of its effects. Hence, what kind of new political practices become legitimized when climate change is increasingly governed as an emergency?

References

Anderson, Ben. 2010. "Preemption, Precaution, Preparedness: Anticipatory Action and Future Geographies." *Progress in Human Geography* 34:777–798. doi: 10.1177/0309132510362600.

Australian Department of Defence. 2007. *Defence Annual Report*. Accessed June 5, 2015, http://www.defence.gov.au/annualreports.

Barnett, Jon. 2013. "Environmental Security." In *Contemporary Security Studies*, ed. Alan Collins, 190–205. Oxford: Oxford University Press.

Boutros-Ghali, Boutros. 1995. *An Agenda for Peace*. 2nd ed. New York: United Nations.

Brown, Lester. 1977. "Redefining National Security." In *Paper 14*. Washington, DC: World Watch Institute.

Brundtland, Gro Harlem, and the World Commission on Environment and Development. 1987. *Our Common Future: Report of the World Commission On Environment and Development*. Oxford: Oxford University Press.

Buhaug, Halvard. 2010. "Climate Not to Blame for African Civil Wars." *Proceedings of the National Academy of Sciences of the United States of America* 107 (38): 16477–16482. doi: 10.2307/20779692.

Burke, Marshall, Solomon Hsiang, and Edward Miguel. 2013. "Weather and Violence." *New York Times*, August 30. Accessed February 23, 2017, http://www.nytimes.com/2013/09/01/opinion/sunday/weather-and-violence.html.

Burke, Marshall B., Edward Miguel, Shankar Satyanath, John A. Dykema, and David B. Lobell. 2009. "Warming Increases the Risk of Civil War in Africa." *Proceedings of the National Academy of Sciences of the United States of America* 106 (49): 20670–20674. doi: 10.1073/pnas.0907998106.

Busby, Joshua. 2013. "Why Do Climate Changes Lead to Conflict? Provocative New Study Leaves Questions." New Security Beat, September 12. Accessed September 29, 2014, https://www.newsecuritybeat.org/2013/09/climate-lead-conflict-provocative-study-leaves-questions.

Butts, Kent H. 1993. *Environmental Security: What Is DOD's Role?* Carlisle, PA: Strategic Studies Institute, US Army War College.

Buzan, Barry, Ole Waever, and Jaap de Wilde. 1998. *Security: A New Framework for Analysis*. London: Lynne Rienner Publishers.

Carson, Rachel. 1962. *Silent Spring*. New York: Houghton Mifflin.

Dalby, Simon. 2013. "Environmental Dimensions of Human Security." In *Environmental Security: Approaches and Issues*, ed. Rita Floyd and Richard Matthew, 121–138. New York: Routledge.

DCDC. 2006. *The DCDC Global Strategic Trends Programme, 2007–2036*. Development, Concepts, and Doctrine Centre, Ministry of Defense. Accessed June 5, 2015, http://www.cuttingthroughthematrix.com/articles/strat_trends_23jan07.pdf.

Deudney, Dan. 1990. "The Case against Linking Environmental Degradation and National-Security." *Millennium: Journal of International Studies* 19 (3): 461–476. doi: 10.1177/03058298900190031001.

Dillon, Michael. 1996. *Politics of Security: Towards a Political Philosophy of Continental Thought*. New York: Routledge.

Dokoupil, Tony. 2014. "Weather Warriors: Meet the Pentagon's Combat Climatologists." NBC, September 20. Accessed October 1, 2014, http://www.nbcnews.com/news/military/weather-warriors-meet-pentagons-combat-climatologists-n206451.

Duffield, Mark. 2011. "Total War as Environmental Terror: Linking Liberalism, Resilience, and the Bunker." *South Atlantic Quarterly* 110 (3): 757–769. doi: 10.1215/00382876-1275779.

Elliott, Lorraine. 1998. *The Global Politics of the Environment*. New York: New York University Press.

Ehrlich, Paul R. 1968. *The Population Bomb*. New York: Ballantine Books.

Falk, Richard A. 1971. *This Endangered Planet: Prospects and Proposals for Human Survival*. 1st ed. New York: Random House.

Fiksel, Joseph, and Alan Hecht. 2012. "Environment and Security." *Encyclopedia of Earth*. Accessed December 5, 2012, http://www.eoearth.org/view/article/51cbf3307896bb431f6ac12b.

Floyd, Rita, and Richard A. Matthew. 2013. "Environmental Security Studies: An Introduction." In *Environmental Security: Approaches and Issues*, ed. Rita Floyd and Richard A. Matthew, 1–20. New York: Routledge.

German Advisory Council on Global Change. 2008. *Climate Change as a Security Risk*. London: Earthscan.

Gilbert, Emily. 2012. "The Militarization of Climate Change." *ACME: An International E-Journal for Critical Geographies* 11 (1): 1–14.

Hobbes, Thomas. 1651. *Leviathan, or, The Matter, Forme, and Power of a Commonwealth, Ecclesiastical and Civil*. London: Printed for Andrew Crooke.

Holland, Andrew, and Albert James DeGarmo. 2014. "Global Security Defense Index on Climate Change." American Security Project. Accessed October 1, 2014, www.americansecurityproject.org/climate-energy-and-security/climate-change/gsdicc.

Homer-Dixon, Thomas. 1991. "On the Threshold: Environmental Changes as Causes of Acute Conflict." *International Security* 16:76–116. doi:10.2307/2539061.

Homer-Dixon, Thomas. 1994. "Environmental Scarcities and Violent Conflict." *International Security* 19:5–40. doi:10.2307/2539147.

Homer-Dixon, Thomas. 1999. *Environment, Scarcity, and Violence*. Princeton, NJ: Princeton University Press.

Hsiang, Solomon M., Marshall Burke, and Edward Miguel. 2013. "Quantifying the Influence of Climate on Human Conflict." *Science* 341: 1235367. doi: 10.1126/science.1235367.

Hulme, Mike. 2008. "The Conquering of Climate: Discourses of Fear and Their Dissolution." *Geographical Journal* 174 (1): 5–16. doi:10.1111/j.1475-4959.2008.00266.x.

Kaplan, Robert D. 1994. "The Coming Anarchy: How Scarcity, Crime, Overpopulation, Tribalism, and Disease Are Rapidly Destroying the Social Fabric of Our Planet." *Atlantic Monthly* 273 (2): 44–76. Accessed December 20, 2016, http://www.theatlantic.com/magazine/archive/1994/02/the-coming-anarchy/304670.

Kerr, Pauline. 2013. "Human Security." In *Contemporary Security Studies*, ed. Alan Collins, 121–135. Oxford: Oxford University Press.

Krause, Keith, and Michael C. Williams. 1997. "Preface: Towards Critical Security Studies." In *Critical Security Studies*, ed. Keith Krause and Mark C. Williams, vii–xxi. Minneapolis: University of Minnesota Press.

Litfin, Karen. 1999. "Environmental Security in the Coming Century." In *International Order and the Future of World Politics*, ed. T. V. Paul and John A. Hall, 328–351. Cambridge: Cambridge University Press.

Lövbrand, Eva, Johannes Stripple, and Bo Wiman. 2009. "Earth System Governmentality: Reflections on Science in the Anthropocene." *Global Environmental Change* 19 (1): 7–13. doi: 10.1016/j.gloenvcha.2008.10.002.

Malthus, Thomas Robert. 1798. *An Essay on the Principle of Population.* London: J. Johnson.

Mathews, Jessica Tuchman. 1989. "Redefining Security." *Foreign Affairs* 68 (2): 162–177. doi: 10.2307/20043906.

Meadows, Donella H., Dennis L. Meadows, Jorgen Randers, and William W. Behrens III. 1972. *The Limits to Growth.* New York: Universe Books.

Methmann, Chris, and Angela Oels. 2015. "From 'Fearing' to 'Empowering' Climate Refugees: Governing Climate-Induced Migration in the Name of Resilience." *Security Dialogue* 46 (1): 51–68. doi:10.1177/0967010614552548.

Myers, Norman. 1986. "The Environmental Dimension of Security Issues." *Environmentalist* 6:251–257. doi:10.1007/bf02238056.

Oels, Angela. 2014. "Climate Security as Governmentality: From Precaution to Preparedness." In *Governing the Global Climate: Rationality, Practice, and Power*, ed. H. Bulkeley and Johannes Stripple, 197–216. Cambridge: Cambridge University Press.

Reid, Julian. 2012. "The Disastrous and Politically Debased Subject of Resilience." *Development Dialogue* 58:67–80.

Reith, Gerda. 2004. "Uncertain Times: The Notion of 'Risk' and the Development of Modernity." *Time and Society* 13 (2–3): 383–402. doi: 10.1177/0961463x04045672.

Rockström, Johan, Will Steffen, Kevin Noone, Åsa Persson, F. Stuart Chapin III, Eric F. Lambin, Timothy M. Lenton, et al. 2009. "A Safe Operating Space for Humanity." *Nature* no. 461 (7263): 472–475.

Rüttinger, Lukas, Dan Smith, Gerald Stang, Dennis Tänzler, and Janani Vivekananda. 2015. *A New Climate for Peace: Taking Action on Climate and Fragility Risks*. Accessed June 5, 2015, http://newsroom.unfccc.int/media/252731/newclimateforpeace.pdf.

Scheffran, Jürgen, Michael Brzoska, Jasmin Kominek, P. Michael Link, and Janpeter Schilling. 2012. "Climate Change and Violent Conflict." *Science* 336 (6083): 869–871. doi: 10.1126/science.1221339.

Sprout, Harold H., and Margaret Sprout. 1971. *Toward a Politics of the Planet Earth* . New York: Van Nostrand Reinhold Co.

Steinbruner, John D., Paul C. Stern, and Jo L. Husbands, eds. 2013. *Climate and Social Stress: Implications for Security Analysis*. Washington, DC: National Academies Press.

Suhrke, Astri. 2004. "A Stalled Initiative." *Security Dialogue* 35 (3): 365. doi: 10.1177/096701060403500322.

Ullman, Richard. 1983. "Redefining Security." *International Security* 8 (1): 129–153. doi: 10.2307/2538489.

UN Development Program (UNDP). 1994. *Human Development Report*. New York: Oxford University Press.

UN Security Council. 2011. "'Contextual Information' on Possible Security Implications of Climate Change Important When Climate Impacts Drive Conflict." 6587th Meeting of the UN Security Council, July 20. Accessed December 21, 20216, http://www.un.org/press/en/2011/sc10332.doc.htm.

US Department of Defense. 2010. *Quadrennial Defense Review*. Accessed June 5, 2015, http://www.defense.gov/News/Special-Reports/QDR.

US Department of Defense. 2014. *Quadrennial Defense Review*. Accessed June 5, 2015, http://www.defense.gov/News/Special-Reports/QDR.

Waever, Ole, and Barry Buzan. 2013. "After the Return to Theory: Past, Present, and Future of Security Studies." In *Contemporary Security Studies*, ed. Alan Collins, 393–409. 3rd ed. Oxford: Oxford University Press.

Westing, Arthur H. 1986. *Global Resources and International Conflict: Environmental Factors in Strategic Policy and Action*. New York: Oxford University Press.

White, Lynn. 1967. "The Historical Roots of Our Ecologic Crisis." *Science* no. 155 (3767): 1203–1207. doi: 10.1126/science.155.3767.1203.

White House, Office of the Press Secretary. 2014. "Fact Sheet: President Obama Announces New Actions to Strengthen Global Resilience to Climate Change and Launches Partnerships to Cut Carbon Pollution." Accessed September 23, 2014, https://www.whitehouse.gov/the-press-office/2014/09/23/fact-sheet-president-obama-announces-new-actions-strengthen-global-resil.

Wolfers, Arnold. 1952. "'National Security' as an Ambiguous Symbol." *Political Science Quarterly* 67 (4): 481–502. doi: 10.2307/2145138.

Yergin, Daniel. 1977. *Shattered Peace: The Origins of the Cold War and the National Security State*. Boston: Houghton Mifflin.

12 Green Economy: Reframing Ecology, Economics, and Equity

Daniel J. Fiorino

The green economy is the idea that a society's ecological and economic goals can be pursued as a mutually reinforcing, positive sum. It accepts that economies will increase in scale and efficiency, but argues that economic growth may occur in less harmful ways ecologically through the use of new policies, patterns of investment, technology innovation, and behavioral change. The ultimate goal is that of a green economic transition, in which ecological objectives and policies are effectively integrated with many others—energy, transportation, manufacturing, and infrastructure, to name a few—and all sectors of society work more collaboratively to maximize the opportunities for positive-sum solutions.

The concept of a green economy offers both a means of reframing conventional ecology-economy relationships and defining a more pragmatic framework for making as well as implementing policy choices. As a reframing concept, it turns old assumptions about conflicts among ecology and economy into a set of propositions about complementarity and synergy. In the past, ecology-economy linkages have been portrayed as a series of zero-sum choices: actions to reduce pollution, lessen resource use, or preserve ecosystems would, it was maintained, subtract from economic progress by reducing competitiveness, increasing costs to businesses and consumers, pushing domestic manufacturing offshore, and increasing unemployment. Conversely, economic growth and prosperity were assumed to lead inexorably to many forms of ecological degradation. In a green economy framing, these assumptions are seen as a false choice.

The innovations embodied in the green economy concept are, first, this *reframing* of presumed zero-sum relationship into a potentially positive sum, and second, providing a mechanism for organizing and integrating a diverse set of public and private actions into a coherent, pragmatic *policy framework*.

This chapter examines the green economy as a conceptual innovation in environmental policy. It begins with a discussion of the concept itself; its intellectual origins in political science, economics, and business; and its recent rise to political relevance. It then turns to the concept in practice along with its political and policy impact. The chapter concludes by assessing the concept's future.

What Is the Green Economy?

To one line of thinking, the green economy is potentially, if not inevitably, an oxymoron (Brand 2012). That economies can expand in production and consumption while not causing ecological degradation is a proposition that defies common sense and a great deal of evidence. Economic expansion leads to more production and consumption, energy use, urbanization, land and water use, and other pressures. Although economies may become more ecoefficient as they mature, growth in terms of increases in economic scale will overwhelm this effect.

Yet the relationships among economic growth and ecological protection are complex; ecological stresses do not necessarily increase linearly with higher incomes. First, empirical evidence suggests that some forms of health-related air and water pollution decline absolutely as economies and incomes grow (Fiorino 2011; Scruggs 2003). This is due to demands for government action, policy interventions, and improved governance. Second, growth is associated with slower rates of population growth and better social conditions, such as the improved status of women, thereby offsetting many causes of health and ecological damage. Third, policy choices enable societies to decouple, to a degree, economic growth from ecological damages by guiding growth away from dirty and toward green economic sectors (Dellink et al. 1999).

The green economy concept recognizes these factors in an effort to reframe economy-ecology relationships. It goes further, however, in seeking out not only complementary but also synergistic linkages through new patterns of investment; institutions that integrate across policy sectors; methods for valuing ecosystem services and incorporating that into decisions; policies that deliver ecological and social as well as economic benefits like ecotaxes; technologies that improve well-being while using resources efficiently or protecting ecosystems; incentives that shift consumption in ecologically benign directions via tax or pricing policies; and regulation that promotes technology innovation and diffusion.

The Green Economy as Policy Framing

A major theme in policy studies is that the ways in which issues are expressed and understood influences political debates and coalitions, proposed solutions, and outcomes. If concepts matter in environmental policy, it is often because they are used to frame issues in meaningful ways. Framing issues with concepts like the green economy provide interpretative story lines that give meaning to a complicated world by "weaving an intricate web of cause and effect that can be used to define problems, diagnose cause, attribute blame and responsibility, make moral judgments, and suggest remedies" (Guber and Bosso 2013, 439–440). The typical depiction of ecology-economy relationship as a zero sum is replaced by a positive framing that is designed to build political support by reconciling what were perceived as incompatible goals.

The use of the green economy as a framing concept is suggested in Hajer's work on ecological modernization. In this view (Hajer 1996, 247), social change occurs as a result of "discourse coalitions" that form around shared concepts and "dominate the perception of the nature of the ecological dilemma at a specific moment in time." Policies are determined by the "outcome of struggles among political decision-makers to achieve 'discursive hegemony' ... to get their construction of the environmental problem and its proper solution adopted as public policy" (Lundqvist 2000, 21). An updated and more politically accessible version of ecological modernization, the green economy concept fits this description; its advocates seek to have the green economy version of the world adopted as public policy. To be sure, ecological modernization was an academic concept coming out of the social sciences, and its direct role in the emergence of the green economy is limited, but it does define the intellectual roots of the newer concept.

As a framing notion, the green economy concept serves several purposes. It bridges the main source of political conflict on environmental issues—the ecology-economy trade-off. Beyond this, it provides an analytic foundation for attaching economic value to ecological resources not appreciated in markets, and facilitates the formation of political coalitions for challenging opponents of greening, such as fossil fuel or large-scale agricultural interests. Indeed, fossil fuel interests in the United States base their opposition to a green energy transition on the argument that it leads to job losses, high prices for energy, and reduced competitiveness—in effect, on the ecology-economy zero sum. In response, the Obama administration stressed the win-win benefits of clean energy as an alternative framing of economy-ecology issues.

There is empirical support for this issue framing. An analysis of public opinion on climate change finds that "people's willingness to change their behavior or support a measure is inversely related to its cost or inconvenience" (Chong 2015, 120). Economic conditions influence public opinion. In the United States in recent years, for example, higher unemployment was associated with reduced public perceptions of threats from climate change, and renewed GDP growth helped restore public concerns about the climate issue. In summary, "there is evidence the economy-versus-environment dichotomy can be overcome with the right plausible story" (132).

The Green Economy as Policy Framework

As policy framework, a green economy agenda differs somewhat from a conventional one. This is not to say that many existing strategies and tools are not used in a green economy framework. Technology standards, for instance, are used in green economic strategies, although their form and the institutional context in which they are applied may differ. Similarly, such tools as emissions and effluent trading, information disclosure, subsidy reform, and ecotaxes appear in the green economic repertoire. Still, a green economy framework differs in scope, objectives, policy tools, and context from past ones.

At a strategic level, policies are focused more on the sources than the effects of economic activity. An emphasis on sector-based policies and investments is an illustration. Policies are aimed at sectors that create environmental stress—agriculture, buildings, energy, manufacturing, tourism, or transport—and the sources of natural capital that make economic activity possible, such as forests, habitats, or fisheries. Green economy strategies also are more likely to integrate across standard policy domains; ecological issues are handled in conjunction with economic, housing, energy, water, and others. The scope of many green economic plans, such as those of South Korean and Ethiopia discussed later, constitute nothing less than an industrial policy. Ideally, a search for synergy leads to higher-level, longer-term strategies.

Tactically, a green economic agenda stresses some policy tools over others. Given the focus on integrating ecological and economic objectives as well as strategies, there is a premium on economic incentive tools, such as carbon taxes, water effluent trading, subsidy reform, or payments for ecosystem services. A carbon tax, for example, can increase the costs of fossil fuels relative to renewables, encourage clean energy investments, drive energy efficiency, promote employment by shifting taxation from labor to resources, and fund green investment. Payments for ecosystems services, a relatively new tool, compensate landowners who maintain ecosystems, such as wetlands or riparian buffers, by sustaining valuable critical ecological services and natural assets through the use of financial incentives.

Critics of the green economy agenda often assert that it relies entirely on technology innovation. While it is true that technology plays a central role—consider renewable energy, water reuse, irrigation, or monitoring technologies—the green economic agenda is much broader. Energy conservation and efficiency involve changes in behavior, or modifications in lighting, insulation, and building design that are low on the technology scale. Reducing deforestation or overfishing, paying landowners to maintain critical ecosystem services, shifting to low-tillage farming, cutting fertilizer use, and striving for denser urban development depend not as much on new technologies as on changes in behavior, investments, and policies. Technology is central to but not the entire policy agenda.

Intellectual Antecedents of the Green Economy

Concepts come from somewhere. They may begin with the writing of an economist or ecologist, the musings of a politician looking for a new policy pitch, or a chance encounter at a conference. If concepts matter, it is useful to ask where they come from and how they evolve. In the case of the green economy, the intellectual origins lie in innovative thinking among select groups of political scientists and economists in the 1980s and 1990s as well as business scholars at about the same time. Over time, these ideas have been incorporated into public debates, political advocacy, national and global economic discourse, and in some cases, national economic or environmental policy.

In political science, the origins of the concept may be traced to ecological modernization theory in the 1980s. A theme in some of the early writing on environmental politics had been the need for fundamental transformation in existing political and economic systems in developed countries. Liberal democracy was viewed as being inadequate for forcing difficult choices of lost growth, reduced consumption, and less material gratification. Capitalist systems, it was contended, were hardwired to take society through inevitable cycles of investment, growth, and expansion. Only basic, systemic political and economic change, including a transition to more authoritarian government and centralized economies, could avert long-term ecological catastrophe (Ophuls 1977; summarized in Dryzek 2013). Ecological modernization presented an alternative vision by rejecting the idea that ecological protection and economic success were inherently in conflict. The field is defined by such assertions as "economic development and ecological crisis can be reconciled to form a new model of development for capitalist economies" (Gibbs 2000, 10).

Competing versions of ecological modernization emerged over the years. For early proponents like Joseph Huber, it meant a private sector process of technology innovation, industrial transformation, and ecoefficiency. This version is still evident in industrial greening, decoupling, and ecoefficiency. Michael E. Porter and Claas van der Linde's (1995) work on the benefits to business of environmental innovation expresses this, as do Amory Lovins and others in ecoefficiency through their factor-four and factor-ten analysis (e.g., Hawken et al. 1999). In contrast to this narrow, business stream of thought is writing by Martin Jänicke and others on the role of the state and public policy in green economic transitions. For them, ecological modernization is a macrolevel process of political development; specific governance capacities are critical and a focus on business too narrow (Jänicke 1996, 2008, 2012).

Arthur P. J. Mol and Gert Spaargaren (2000) define five aspects of ecomodernization: an emphasis on the changing role of science and technology as not only a cause but also a solution to ecological problems; an appreciation of the role of private sector firms and markets in combination with public policy; a new role for the state, stressing decentralized, flexible, and consensual governance; different roles for social movements, which become partners and innovators as well as critics; and new discourses that reframe ecology-economy issues in ways that may generate more widespread political support. Other studies examine the political, social, and economic conditions under which a transition to a more ecologically advanced society develop. Among them are economic development; cognitive capacities in science and technology; effective, accountable governance with low corruption; an active civil society that both criticizes and collaborates; and participatory and policy-integrating capacities (Weidner 2002; Fiorino 2011).

Ecological modernization theory has been lauded and criticized. Joseph Murphy (2000, 5) views it as a major contribution, "given that most work in environmental social science starts by assuming the inability of industry and the state to do anything other than create such problems" as ecological harm. Criticism of ecological modernization over the years reflects the same kinds of issues critics raise today with respect to the green economy. They see it as perpetuating the rich-country, corporate dominance that is the source of our ecological crises. Of course, the validity of these criticisms depends on which version of modernization is on the table—the one focused on ecoefficiency and innovation in business, or the more systemic, institutional change version. Even advocates of the latter, however, recognize that "despite its impressive potential," ecological modernization is "not sufficient to ensure a long-term stabilization of the environment" (Jänicke 2008, 563). Many issues require lifestyle and perhaps more basic social change; dramatic ecoefficiency gains will be insufficient in the face of exponential growth.

Ecological modernization is a line of academic argument and research, and it would be difficult to trace its direct influence on recent expressions of the green economy concept. Still, the academic case foreshadowed recent policy expressions of the concept and criticisms it has drawn.

A more significant and long-term influence on the green economy concept is ecological economics as it developed in the 1980s and 1990s. Building on the work of Herman Daly on limiting throughput in a steady-state economy, Nicholas Georgescu-Roegen on thermodynamics and entropy, and C. S. Holling on ecosystem stability and resilience, this school offered a critique of the limitations in conventional, welfare economics as well as a new vision, concepts, and methods (van den Bergh 2001, 14). The goal is "to provide an integrated and biophysical perspective on environment-economy interactions" (13) as well as to make economists more aware of "ecological impacts and dependencies," and ecologists "more sensitive to economic forces, incentives, and constraints" (Costanza 1989, 1). It was in many ways a striking departure from mainstream welfare economics.

Ecological economists seek to expand and reorient conventional welfare economics in several ways. One is by placing natural capital—clean water, ecosystems and their services, forests, and the like—on not just an equal but also at times a preferential footing relative to other forms of capital. Natural capital differs from financial or physical capital because many forms of critical natural capital have no substitutes in sustaining well-being and due to the existence of thresholds, beyond which ecosystems may flip into an alternative state. Natural and other forms of capital are complementary, though rarely substitutable.

A second difference is the recognition that economies should be conceptualized as subsystems of the ecological system and thus subject to biophysical, planetary constraints. "Continuous exponential growth of any physical sub-system of a finite system is impossible" (Farley 2012, 43). The near-exclusive focus on growth (defined as increasing real GDP) is a recipe for disaster given the "overwhelming theoretical and empirical support for the existence of ecological and biological thresholds" (42). Ecological economics demonstrates a concern with the effects of growth, development over growth, distributions of income and wealth, and alternative measures of progress. Where ecological economists differ among themselves is in their solutions to the growth issue. A central concept has been Daly's (1991) formulation of the steady-state economy, in which the size of the economy does not expand beyond what is sustainable under ecological and biophysical constraints. The exact prescriptions for dealing with the growth issue vary, from managed, moderate growth (Victor 2008; Jackson 2011), to selective reductions in brown sectors and expansions in green ones (van den Bergh 2011), to degrowth through a deliberate contraction in the size of the economy (Alexander 2012).

A major concern of ecological economics is to value ecosystems and their services appropriately. They are undervalued because "they are not adequately quantified in terms comparable with economic services and manufactured capital" (Bratt and de Groot 2012, 6). In economic terms, ecosystems serve as a fund that is "capable of generating a flux of ecosystem services over time" (Farley 2012, 40). Among these ecosystem services are provisioning (food, freshwater, and raw materials), regulatory (carbon storage and moderation of extreme weather), habitat for species, and cultural services (recreation and a sense of place).

Many of these ideas were captured in one of the first books to clearly use the term green economy, *Blueprint for a Green Economy* (Pearce et al. 1989). This work served as a bridge between standard welfare and the newer ecological economics. It also made a case for recognizing the interdependence among ecology and economics, and viewing ecology as the foundation of economic success. This and a later volume (Pearce and Barbier 2000) argued that ecological valuation puts nature on the same level as economics. Many ecological economists, however, would depart from the *Blueprint* analysis because it accepts the desirability of economic growth.

The Green Economy Concept in Practice

Conceptual innovations rarely break through to acquire policy and political significance in one step or a single event, as this book demonstrates. They emerge through predecessor concepts, like ecological economics; are applied narrowly and later expand, like sustainability; or become relevant when used as an organizing concept for a movement, as has been the case with environmental justice.

The green economy broke through as a concept in global discourse, national planning, and local development late in the first decade of this century. It took center stage on the policy agendas of such influential organizations as the World Bank, UN Environment Program (UNEP), Organisation for Economic Co-operation and Development, (OECD), and Association of Academies and Societies of Sciences in Asia. It defines a foundation concept for many countries and major policy reports.

This section considers three aspects of the concept in practice: Why did it break through when it did? How has it been expressed in practice? What political and policy impacts has it had?

Why Did the Green Economy Concept Take Hold When It Did?

The intellectual origins of the green economy concept go back decades. It was only in the last decade ago or so, however, that it became influential at a policy level. Certainly one factor explaining its rise was the global financial crisis of 2008–2009 and the recession that followed. Typically described as the greatest economic crisis since the 1930s, the recession offered a challenge and opportunity. The challenge was to restore confidence in financial markets and reinvigorate growth. The opportunity was to channel economic stimulus spending toward green economic sectors. Reflecting this view, a 2009 report from the OECD (2009, 7, 8) asserted not only that the "current economic crisis is not an excuse to weaken long-term efforts to achieve low-carbon growth" but that recession "can also open new opportunities—don't waste a crisis!"

A second factor was the accumulating evidence on the effects of climate change. The fact of global ecological limits, boundaries, or thresholds took on renewed significance (Meadows et al. 2004). The assessments since the 1970s on the limits to growth now had a real-world, global validation. Although the threats to local and regional ecosystems had been evident for years—think of the Chesapeake Bay or Great Lakes—the global effects of climate change on sea levels, water resources, food production, human health, habitat, and species brought the macrolevel conflicts between growth and ecological well-being into sharp focus. Add to this the results of large-scale analyses of ecological threats, such as the Millennium Ecosystem Assessment (2005) and others (Boston 2011), and the time was right for concluding that the world had to stop growing or grow in fundamentally different ways.

A third factor in the rise of the green economy concept could well have been frustration with progress on sustainable development (Brand 2012). Despite incremental changes, the world had not changed dramatically since the report of the World Commission on Environment and Development in 1987. Nationally, sustainable development has been nearly invisible in the United States, although it is used locally. Internationally, despite many sustainability initiatives, the concept has not delivered the transformational change many had hoped for. It may be that the concept of the green economy offers a more politically accessible framing of the ecology-economy relationship than does sustainable development.

Expressions of the Green Economy

The emergence of the green economy concept in recent years has been impressive. Its current relevance is due in part to the process leading up to the Rio+20 Earth Summit in 2012. The "green economy in the context of sustainable development and poverty eradication" served as one of two conference themes. Consistent with the use of the concept as it developed, the emphasis was on a search for synergies and win-win strategies, with green economy as "a lens for focusing on and seizing opportunities to advance economic and environmental goals simultaneously" (United Nations 2010, 4). Green economy proponents claim it is not meant to replace sustainable development but rather to help achieve the same long-term goals (5).

Reflecting this framing is UNEP's *Towards a Green Economy*. For UNEP (2011, 9), a green economy is "one that results in improved well-being and social equity, while significantly reducing environmental risks and economic scarcities." Such an economy "is not generally a drag on growth but a new engine of growth"; it is a "net generator of decent jobs," and "vital strategy for the elimination of persistent poverty" (10). Investing 2 percent of annual, global GDP, about US$1.3 trillion, could create an economy that grows, but within ecological limits. This may be achieved by sustainably managing and restoring such natural capital sectors as water, forestry, agriculture, and fisheries, while increasing the efficiency and reducing the ecological impacts of transport, energy, manufacturing, and buildings. UNEP's macroeconomic model projects that after a brief period of slightly slower growth to replenish renewable resources, a green strategy would deliver more growth, reduce poverty, and create more jobs than business as usual. Whether these projections are empirically valid, of course, is a matter of debate.

Another prominent expression is the OECD's *Towards Green Growth*. Green growth is "fostering economic growth and development while ensuring that natural assets continue to provide the resources and environmental services on which our well-being depends" (OECD 2011, 9). It is based on the principles of sustainable natural resource use, energy efficiency, and a fair valuation of ecosystem services. Like UNEP, this report provides an optimistic scenario of green economic prospects, if needed incentives and policies are adopted. Among these, aside from investments in green economic sectors, are ecotaxes such as a carbon tax; well-designed regulation to promote innovation; energy, water, and materials efficiency; and cuts in subsidies for unsustainable activity related to fossil fuels, irrigation, mining, and so on. The OECD report also stresses the economic valuation of natural capital and critical ecosystem services.

A third example is the World Business Council for Sustainable Development's (2012) *Vision 2050: The New Agenda for Business*. The good news for business, it states, is that "growth will deliver billions of new consumers who want homes and cars and television sets." The bad news is that "shrinking resources and potentially changing climate will limit the ability of all 9 billion of us to attain or maintain the consumptive lifestyle that is commensurate with wealth in today's markets." The report offers a two-part vision for midcentury. The first vision aims for "a standard of living where people have access to and the ability to afford education, healthcare, mobility, the basics of food, water, energy and shelter, and consumer goods." The second envisions "a standard of living [that] can be sustained with the available natural resources and without further harm to biodiversity, climate, and other ecosystems." The report explicitly recognizes the existence of planetary limits of some kind and sets out a strategy for fulfilling its two-part vision.

For the World Bank (2012, 2), green growth is "efficient in its use of natural resources, clean in that it minimizes pollution and environmental impacts, and resilient in that it accounts for natural hazards and the role of environmental management and natural capital in preventing physical disasters." A premise of the analysis in *Inclusive Green Growth* is that "sustained growth is necessary to achieve the urgent development of the world's poor and that there is substantial scope for growing cleaner with growing slower" (xi). The World Bank places a premium on inclusive growth that is sensitive to the needs of the poor; growth drives poverty reduction, but it must be equitable. Like UNEP and the OECD, the World Bank views environmental damages as "reaching a scale where they are beginning to threaten both growth prospects and the progress achieved in social indicators" (3). Such growth is affordable: "many green policies pay for themselves directly, and the others make economic sense once externalities are priced and eco-system services are valued" (4). Like its counterparts, the World Bank rests its case on the costs of environmental degradation to economic and social well-being rather than the intrinsic worth of nature.

Several aspects of these reports are worth noting. One is that they are aimed at presenting both a policy framing (or in this case, a reframing) and policy framework. All imply the existence of some kind of ecological limits, boundaries, or thresholds, but are vague on just where or when they take hold. All accept the necessity for growth, making the case that it will accelerate, continue, or at worst decline marginally. All express a central tenet of green economy thinking—that business as usual growth at some point degrades the natural capital that makes human well-being possible. All imply or suggest it may be necessary to rethink growth as an overriding goal. None call for an end to or deliberate slowing down of growth or incomes; that would undermine the use of the green economy as a framing concept. Yet there is at least the possibility that a green economy could be a bridge toward a society in which growth is not the most relied on measure of well-being, although that view is not prominent in the literature.

In summary, the following are associated with recent expressions of the concept:

- appreciation of the existence of ecological limits (or boundaries, or thresholds) of some kind, with a lack of specifics on what they are and when they could be tested or breached
- recognition of the inevitability and desirability of growth, but also of a need (in varying degrees) to transform institutions, policies, and behavior along with the composition of growth
- agreement that a green or at least a far greener economy is feasible within existing economic and political systems, while recognizing that the barriers to change are large
- recognition that economic growth is needed to improve human well-being, especially in poor nations, but that equity issues should be considered and addressed
- agreement that at the core of a green economy is explicit valuation of ecological assets and services, making ecology more relevant to political and economic decision making
- a view that although economic prosperity generally serves to increase ecological pressures, it provides the human, scientific, technological, and financial means for managing them

Clearly, the green economy is closely related to many other influential concepts. Such concepts as natural capital and ecosystem services are closely related. Decoupling is relevant; it is a common means of expressing the delinking of the ecology-economy relationship. Concepts like sustainable development and growth, even smart growth as applied to urban policy, are near siblings of green economy. These and many others are part of the conceptual landscape for integrating and reconciling ecological with economic goals in modern societies.

Assessing the Impact of the Concept

As one of the newer concepts in this book, the long-term impact of the green economy is yet to be established. It certainly has been adopted in international political discourse, and exerts influence at a national level in many countries. It has become influential as a means of reframing ecology-economy relationships and competes with sustainable development as a framing concept in many settings. As a policy framework, particularly one that is implemented on any scale, its impact is still undetermined.

With major organizations like UNEP, the OECD, and the World Bank behind it, the green economy does not lack international presence. These and other organizations created the Global Green Growth Institute, "founded on the belief that economic growth and environmental sustainability are not only compatible objectives; their integration is essential to the future of mankind."[1] It develops resources and supports green economic planning in Indonesia, Brazil, Ethiopia, and elsewhere. A range of organizations, including the World Wildlife Fund, International Union for the Conservation of Nature, and New Economy Foundation make up the Green Economy Coalition, focused on a "resilient economy that provides a better quality of life for all within the ecological limits of the planet."[2]

A country where the concept has had a clear policy impact is the Republic of Korea, which has been using a low-carbon, green growth strategy since 2008 as "the pillar of a new vision for the economy" (Matthews 2012, 761). It began as a Green New Deal to promote a recovery from the 2009 financial crisis and recession, was enacted by the legislature in 2009, and defines an industrial strategy for infrastructure investments in energy, buildings, water, transport, and other sectors. A transition to a low-carbon economy is central to the plan, as is the emphasis on energy security, green cities, technology, industries, consumption, and climate adaptation. With strong executive support, and guided by the Presidential Committee on Green Growth with cross-industry representation (including the crucial finance and industry/resources ministries), it demonstrates, according to one observer, that "a strong role for government is needed if a country is to shift off its 'business as usual' pathway," and that "enormous changes can be effected by strong and committed state action," even in a democracy (767–768). The issues here, as in most democratic systems, is whether the plan will extend beyond the current government, and if it ultimately will be sufficient to offset the effects of high growth rates.

Another country that has embraced the concept is Ethiopia. Adopted in the wake of the policy reports discussed above, *Ethiopia's Climate Resilient Green Economy Strategy* is framed as an effort to "unlock economic growth, create jobs for the growing population, and deliver wider socio-economic benefits" (Federal Democratic Republic of Ethiopia 2011, 11). As a policy framework, it is aimed at the energy, agriculture, forestry, buildings, and water sectors. It was developed as a strategy not only to avoid the costs of environmental degradation but also to take advantage of market opportunities by, for example, generating renewable energy for export. The strategy includes measures for leapfrogging to new, energy-efficient technologies in multiple sectors as well as protecting and restoring forests. It includes a mix of policy measures, but it is largely a long-term investment plan organized around green economic principles and objectives led by the prime minister's office along with a ministerial committee. With high annual growth rates, Ethiopia tests the efficacy and durability of a green growth framework while serving as one of many laboratories for evaluating it.

Within the United States nationally, the green economy is most visible as a reframing device. The Obama administration embraced the "clean energy highway" as a way to create jobs, promote durable growth, and open up business opportunities in energy efficiency and renewables. The opposition, in turn, has called on the old ecology-economy zero-sum framing as a rhetorical strategy for protecting fossil fuel interests against such a transition. Green economic issue framing is apparent in other actions as well. For example, the US Environmental Protection Agency (2013) asserts that "water is vital to a productive and growing economy in the United States, and directly affects the production of many goods and services." Coalitions like the BlueGreen Alliance of environmentalists and unions stress the positive ecology-environment linkages of job creation and infrastructure spending. Green jobs are used as an organizing principle for new economic strategies (Jones 2008). Whether or not this will provide a successful long-term framing device in the context of US politics is still an open question.

Domestically, an explicit application as policy framework was in the American Reinvestment and Recovery Act of 2009, which allocated some $80 billion to environmental technology and infrastructure projects. If this had been part of a long-term, stable investment plan, it would have had an obvious application of the concept as a strategic policy blueprint. The more visible uses of the concept are local, where many cities increasingly frame economic and ecological goals in complementary as well as synergistic terms. They draw on elements of the green economy framing by searching for a "peaceful coexistence between economic development and the environment" (Portney 2013, 9), while accepting that "sustainability and climate change strategies could also be engines of economic development" (Fitzgerald 2010, 1).

Evaluating the Green Economy

Like many concepts that seek to reconcile perceived conflicts, the green economy has generated criticism as well as praise. For many critics, it is a cynical effort to justify continued, rapid, and ultimately ecologically and socially destructive economic growth. For others, it is an agenda to limit growth, expand government, undermine markets, and harm vital industries like energy and agriculture. At the same time, for its advocates, the green economy offers a politically defensible and economically pragmatic path for reconciling the inevitability of and need for some kind of growth with ecological limits (Fiorino 2014). Because it is contested politically, views on the concept are influenced by where one stands on issues of economic growth and inequality, the government's role in society, the legitimacy of business engagement, the intrinsic relative to instrumental value of nature, and even definitions of well-being and happiness. Debates over the concept of the green economy reflect views about society, economy, ecology, and politics.

From the Left, criticisms turn on its validity as a policy framework, ecological and social impacts, and preference for humans over nature. These criticisms generally fall into four categories:

• *It perpetuates the capitalist, neoliberal values and institutions at the root of our ecological and social dilemma.* Capitalism, industrialism, and liberal democracy are incapable of the needed transformations. One group predicts its failure "due to the prevailing capitalist and imperialistic conditions as well as the unquestioned faith in progress" (Federal Coordination of Internationalismus 2012, 1).

- *It lacks empirical validity and promises far more than it can deliver.* The "you can have your cake and eat it too" aspect is viewed as impractical. Any gains from decoupling ecological degradation from growth will be overwhelmed by the sheer scale of the latter: "it is impossible for the world economy to grow its way out of poverty and environmental degradation" (Daly 1998, 285).
- *It reinforces rich nation and corporate power.* Emphasis on technology, business, and economic expansion ignores as well as reinforces past patterns of inequity and exploitation. Debates over equity and justice would, it is argued, "close permanently with the sealant of the *Green Economy*" (Federal Coordination of Internationalismus 2012, 4). It could further empower rich nations to use trade policies, intellectual property rights, or foreign aid to block economic development in poor nations (Khor 2011).
- *It makes ecology a commodity by placing it on the market exchange table.* Ecological economists developed methods for attaching economic value to ecological assets and services to justify their preservation. Critics view valuation as making nature yet another commodity that subjects it to market trade-offs through cost-benefit and other analyses (McCauley 2006).

Opposition from the Right is based on claims that it will undermine the capacity for growth by increasing energy and other resources costs, put national economies at a competitive disadvantage, and justify larger and more interventionist roles for governments. Such positions are motivated in part by the likelihood that such economic sectors as fossil fuels, industrial agriculture, and commercial developers will be losers in a green economic transition. Some on the Right charge that the green economy, like sustainability, is merely a vehicle for realizing the long-standing liberal goals of income redistribution, centrally planned economies, higher taxes and more regulation, and heavy-handed global governance.

The core issue for many critics is economic growth. As a political reframing concept, the green economy promises growth advocates that living standards will continue to increase due to expanding GDP. UNEP (2011), for example, estimates in *Towards a Green Economy* that if its 2 percent investment plan is adopted, a global increase in per capita income of 14 percent beyond a business-as-usual scenario is possible by 2050. At the same time, its model projects (beyond business as usual) a 21 percent increase in forestland, a nearly 22 percent decrease in water demand, a 40 percent fall in primary energy demand, and a nearly 50 percent reduction in the ratio of global footprint to biocapacity. Indeed, UNEP projects that its scenario enables even *more* growth than a business-as-usual path.

Such assertions fly in the face of decades of arguments by growth critics. Population growth and affluence in rich nations caused the water, climate, and other ecological crises; the solution is to reverse the global growth machine, reduce physical scale, and bring the economies in line with planetary limits. One prescription is a steady-state economy, "with constant stocks of people and artifacts, maintained at some desired, sufficient levels of maintenance 'throughput' ... by the lowest feasible flows of matter and energy from the first stage of production ... to the last stage of consumption" (Daly 1991, 17). What this means in practical terms varies. For some growth critics, the appropriate solution is deliberate, managed degrowth, "an equitable downscaling of production and consumption that increases human well-being and enhances ecological conditions" (Alexander 2012, 351). Poorer nations would continue to grow to a reasonable level while richer ones would carry out a deliberate economic contraction.

Others, skeptical of the practical and political feasibility of actual degrowth, propose carefully managed, moderate growth along with a rethinking of progress and human well-being. In *Prosperity without Growth*, Tim Jackson challenges the notion that a bigger economy is a better one beyond some level. Constant growth leads not to a better life but instead to a more competitive, unequal, stressful, and ecologically damaging one—a view with some support in research on happiness (Graham 2011). "An unequal society is an anxious society," Jackson (2011, 117) writes, "one given too readily to 'positional consumption' that adds little to overall happiness but contributes significantly to unsustainable resource throughput." He is critical of "the myth of decoupling," in which green economy proponents argue that falling ecointensity per unit of economic output will keep economies within planetary limits at high growth rates. Similarly, a macroeconomic analysis (Victor 2008) concludes that slower, well-managed growth is superior to current policy in reducing unemployment and poverty while avoiding ecological degradation.

Green economy advocates may offer several responses to these criticisms. The critique from the Right may be dismissed because it lacks real-world validity. Green economic investments and policies do create winners and losers, and the opposition largely comes from the likely losers. The overall effects on growth and employment range from slightly negative in the near term to positive in the long haul. Indeed, clean energy investments yield more jobs per unit than do those in fossil fuel sectors (Pollin et al. 2014).

The other criticisms are not easily dismissed. The green economy does rely on existing economic and political institutions to a large degree to achieve ecological goals. Markets, capital, and democracy are central. At the same time, proponents make the case for major changes in policies and investments. Rather than risk uncertain, radical change that could make a green transition even less likely, they call for "a restructuring of the capitalist political economy along more environmentally sound lines, but not in a way that requires an altogether different kind of political economic system" (Dryzek 2013, 170).

The most telling criticism is that the framework will not deliver a global green economy. Part of this argument is that too little of the green agenda will be adopted. Still, what are the odds of even more radical change occurring? The other part of this contention is that even with a robust pursuit of a green agenda, the sheer scale of growth will at some point stretch ecological limits. On the other hand, proposals for deliberate degrowth may be a recipe for failure: they fly in the face of political logic; assume an unrealistic mastery of economic policy; may actually undermine support and capacities for a green transition by strengthening the arguments regarding economy-ecology trade-offs; and are unlikely to deliver the needed changes. Even a smaller, brown economy does not do the job; whether a green economy does or not is another question. The more promising path for green economy advocates is to consider the more politically realistic and likely more effective calls for guided, ecologically sound growth from writers like Jackson (2011) and Peter Victor (2008).

What should receive more attention in debates over the green economy is economic inequity. Ecological economists make a telling point in their emphasis on economic distribution. A great deal of evidence suggests that many problems are a product of inequality, independent of income levels

(Wilkinson and Pickett 2009). Inequality may lead to more unsustainable consumption, more status competition, and economic insecurity that reinforces a growth-at-all-costs mind-set along with an erosion of social capital, an underappreciation of public goods, and less capacity for collective action, all of which arguably are central to a green transition (Uslaner 2002; Holland et al. 2009; Magnani 2000).

The Future of the Green Economy Concept

There is ample evidence that a far greener economy is entirely possible under the reframing and policy framework expressed in the concept. Renewable energy, resource efficiency, smart urban growth, new models for agriculture or transportation—the list goes on. The question, of course, is whether or not this is sufficient to remain within planetary limits given the anticipated increases in economic scale in the coming decades and beyond. If the green economy is to be effective as a framing concept and policy framework, modifications are necessary. All are reflected to varying degrees in recent expressions of the concept, but should be addressed more directly. These include:

- Be realistic in accepting, as some green economy proponents do, that growth will continue, but rethinking and redefining it is essential. This means, for example, recognizing that richer is not always happier; that status competition can be socially and ecologically harmful; and that GDP and income are narrow measures of human well-being (Kubiszewski et al. 2013).
- Recognize the role of economic and political inequality. Equitable growth is less damaging and more ecologically responsible. Inequality can lead to more consumption, more economic insecurity, erosion of social trust, and less capacity for collective action. Central to the concept of a green economy should be an emphasis on quality of life as consisting of more than simply acquiring more wealth, as the happiness research suggests (Graham 2011).
- Beyond rethinking what growth means, be "indifferent or neutral about it" in evaluating goals and developing strategies (van den Bergh 2011, 882). Rather than call for overall degrowth that may not even address the problems, focus on economic activities and sectors such as fossil fuels, urban

sprawl, and others with *selective degrowth*. More targeted policy and investment strategies could deliver significant ecological benefits.

• Build the concept more explicitly on the irreplaceability of ecological assets and services, for their own sake as well as their economic value. Ecological economics provides an analytic base and evidence for doing this, but it would help to allay the concerns of many critics if there were more appreciation of ecological goods on their own and not just for their instrumental value.

• Recognize that business greening through ecoefficiency and technology innovation is necessary but insufficient. Government must play a role by putting a price on carbon, reforming subsidies, linking policy across sectors, reorienting investments, promoting technologies, pricing water and other resources effectively, and other measures.

Like any macrolevel concept that aims to reconstruct how we think about issues and solutions, the concept of a green economy offers ample grounds for criticism. It could be used to justify marginal adjustments from business as usual as a way of sustaining a growth-at-all-costs mindset. It also could be the basis for a politically effective reframing of ecological and economic issues as well as a pragmatic blueprint for change. For proponents, it sets out a middle ground between the politically unrealistic, risky view that economies need to stop growing or actually contract, and the ecologically irresponsible view that ecological limits do not exist, or will be overcome simply through technology, innovation, and the benefits of greater wealth. At the same time, if it is to prove durable and effective as a framing device and policy framework, the green economy concept should be revised to present a more equitable, defensible, and systemic strategy for change. It is in these terms that its influence as a long-term conceptual innovation in environmental policy could ultimately be realized.

Notes

1. For more on the Global Green Growth Institute, see http://gggi.org.

2. For more on the Green Economy Coalition, see http://www.greeneconomycoalition.org.

References

Alexander, Samuel. 2012. "Planned Economic Contraction: The Emerging Case for Degrowth." *Environmental Politics* 21 (3): 349–368. doi:10.1080/09644016.2012.671569.

Boston, Jonathan. 2011. "Biophysical Limits and Green Growth." *Policy Quarterly* 7 (4): 34–43.

Brand, Ulrich. 2012. "Green Economy—The Next Oxymoron?" *Gaia* 21 (1): 28–32.

Bratt, Leon C., and Rudolf de Groot. 2012. "The Ecosystem Services Agenda: Bridging the Worlds of Natural Science and Economics, Conservation and Development, and Public and Private Policy." *Ecosystem Services* 1:4–15. doi:10.1016/j.ecoser.2012.07.011.

Chong, Dennis. 2015. "Exploring Public Conflict and Consensus on the Climate." In *Changing Climate Policies: US Policies and Civic Action*, ed. Yael Wolinsky-Nahmias, 110–145. Washington, DC: CQ Press.

Costanza, Robert. 1989. "What Is Ecological Economics?" *Ecological Economics* 1 (1): 1–7. doi:10.1016/0921-8009(89)90020-7.

Daly, Herman E. 1991. *Steady-State Economics*. 2nd ed. Washington, DC: Island Press.

Daly, Herman E. 1998. "Sustainable Growth: An Impossibility Theorem." In *Debating the Earth: The Environmental Politics Reader*, ed. John S. Dryzek and David Scholsberg, 285–289. New York: Oxford University Press.

Dellink, Rob, Martijn Bennis, and Harmen Verbruggen. 1999. "Sustainable Economic Structures." *Ecological Economics* 29 (1): 141–159. doi:10.1016/s0921-8009(98)00061-5.

Dryzek, John S. 2013. *The Politics of the Earth: Environmental Discourses*. 3rd ed. New York: Oxford University Press.

Farley, Joshua. 2012. "Ecosystem Services: The Economic Debate." *Ecosystem Services* 1:40–49. doi:10.1016/j.ecoser.2012.07.002.

Federal Coordination of Internationalismus. 2012. *Ten Theses of a Critique of the Green Economy*. Accessed June 3, 2012, http://rio20.net/en/documentos/ten-theses-of-a-critique-of-the-green-economy.

Federal Democratic Republic of Ethiopia. 2011. *Ethiopia's Climate Resilient Green Economy Strategy*. Accessed December 23, 2016, http://www.greengrowthknowledge.org/resource/ethiopia%E2%80%99s-climate-resilient-green-economy-green-economy-strategy.

Fiorino, Daniel J. 2011. "Explaining National Environmental Performance: Approaches, Evidence, and Implications." *Policy Sciences* 44 (4): 367–389. doi:10.1007/s11077-011-9140-8.

Fiorino, Daniel J. 2014. "The Green Economy: Mythical or Meaningful?" *Policy Quarterly* 10 (1): 26–34.

Fitzgerald, Joan. 2010. *Emerald Cities: Urban Sustainability and Economic Development.* New York: Oxford University Press.

Gibbs, David. 2000. "Ecological Modernization, Regional Economic Development, and Regional Development Agencies." *Geoforum* 31 (1): 9–19. doi:10.1016/s0016-7185(99)00040-8.

Graham, Carol. 2011. *The Pursuit of Happiness: An Economy of Well-Being* . Washington, DC: Brookings.

Guber, Deborah Lynn, and Christopher J. Bosso. 2013. "Issue Framing, Agenda Setting, and Environmental Discourse." In *The Oxford Handbook of U.S. Environmental Policy*, ed. Sheldon Kamieniecki and Michael E. Kraft, 437–460. New York: Oxford University Press.

Hajer, Maarten. 1996. "Ecological Modernisation as Cultural Politics." In *Risk, Environment, and Modernity*, ed. Scott Lash, Bronislaw Szerszynski, and Brian Wynne, 246–268. London: Sage.

Hawken, Paul, Amory Lovins, and L. Hunter Lovins. 1999. *Natural Capitalism: Creating the Next Industrial Revolution.* Boston: Little, Brown.

Holland, Tim G., Garry D. Peterson, and Andrew Gonzalez. 2009. "A Cross-National Analysis of How Economic Inequality Predicts Biodiversity Loss." *Conservation Biology* 23 (5): 1304–1313. doi:10.1111/j.1523-1739.2009.01207.x.

Jackson, Tim. 2011. *Prosperity without Growth: Economics for a Finite Planet.* New York: Earthscan.

Jänicke, Martin. 1996. "The Political System's Capacity for Environmental Policy." In *National Environmental Policies: A Comparative Study of Capacity-Building,* ed. Martin Jänicke and Helmut Weidner, 1–24. New York: Springer.

Jänicke, Martin. 2008. "Ecological Modernization: New Perspectives." *Journal of Cleaner Production* 16 (5): 557–565. doi:10.1016/j.jclepro.2007.02.011.

Jänicke, Martin. 2012. "Green Growth: From a Growing Eco-Industry to Economic Sustainability." *Energy Policy* 48:12–21. doi:10.1016/j.enpol.2012.04.045.

Jones, Van. 2008. *The Green Collar Economy: How One Solution Can Fix Our Two Biggest Problems* . New York: HarperCollins.

Khor, Martin. 2011. *Risks and Uses of the Green Economy Concept in the Context of Sustainable Development, Poverty, and Equity*. Geneva: South Centre.

Kubiszewski, Ida, Robert Costanza, Carol Franco, Philip Lawn, John Talberth, Tim Jackson, and Camille Aylmer. 2013. "Beyond GDP: Measuring and Achieving Global Genuine Progress." *Ecological Economics* 93:57–68. doi:10.1016/j.ecolecon.2013.04.019.

Lundqvist, Lennart J. 2000. "Capacity Building or Social Construction: Explaining Sweden's Shift toward Ecological Modernization." *Geoforum* 31 (1): 21–32.

Magnani, Elisabetta. 2000. "The Environmental Kuznets Curve, Environmental Protection Policy, and Income Distribution." *Ecological Economics* 32 (3): 431–443. doi:10.1016/s0016-7185(99)00041-x.

Matthews, James A. 2012. "Green Growth Strategies: Korean Initiatives." *Futures* 44 (8): 761–769. doi:10.1016/j.futures.2012.06.002.

McCauley, Douglas J. 2006. "Selling Out on Nature." *Nature* 443 (September 7): 27–28. doi: 10.1038/443027a.

Meadows, Donella, Jørgen Randers, and Dennis Meadows. 2004. *Limits to Growth: The 30-Year Update*. White River Junction, VT: Chelsea Green.

Millennium Ecosystem Assessment. 2005. *Living beyond Our Means: Natural Assets and Human Well-Being*. Accessed December 23, 2016, http://www.millenniumassessment.org/documents/document.429.aspx.pdf.

Mol, Arthur P. J., and Gert Spaargaren. 2000. "Ecological Modernisation Theory in Debate." *Environmental Politics* 9 (1): 17–49.

Murphy, Joseph. 2000. "Ecological Modernisation." *Geoforum* 31 (1): 1–8. doi:10.1016/s0016-7185(99)00039-1.

Ophuls, William. 1977. *Ecology and the Politics of Scarcity: Prologue to a Political Theory of the Steady State*. San Francisco: W. H. Freeman.

Organisation for Economic Co-operation and Development (OECD). 2009. *Green Growth: Overcoming the Crisis and Beyond*. Accessed December 23, 2016, http://www.oecd.org/env/43176103.pdf.

Organisation for Economic Co-operation and Development (OECD). 2011. *Towards Green Growth*. Accessed December 23, 2016, http://www.oecd.org/greengrowth/towards-green-growth-9789264111318-en.htm.

Pearce, David, and Edward B. Barbier. 2000. *Blueprint for a Sustainable Economy*. London: Earthscan.

Pearce, David, Anil Markandya, and Edward Barbier. 1989. *Blueprint for a Green Economy*. London: Earthscan.

Pollin, Robert, Heidi Garrett-Peltier, James Heintz, and Bracken Hendricks. 2014. "Green Growth: A U.S. Program for Controlling Climate Change and Expanding Job Opportunities." Center for American Progress. Accessed September 18, 2014, https://www.americanprogress.org/issues/green/report/2014/09/18/96404/green-growth.

Porter, Michael E., and Claas van der Linde. 1995. "Toward a New Conception of the Environment-Competitiveness Relationship." *Journal of Economic Perspectives* 9 (4): 97–118. doi:10.1257/jep.9.4.97.

Portney, Kent. 2013. *Taking Sustainable Cities Seriously: Economic Development, Environment, and Quality of Life in American Cities*. 2nd ed. Cambridge, MA: MIT Press.

Scruggs, Lyle. 2003. *Sustaining Abundance: Environmental Performance in Industrial Democracies*. Cambridge: Cambridge University Press.

UN Environment Program (UNEP). 2011. *Towards a Green Economy: Pathways to Sustainable Development and Poverty Eradication—A Synthesis for Policy Makers*. Accessed December 23, 2016, https://sustainabledevelopment.un.org/index.php?page=view&type=400&nr=126&menu=35.

United Nations. 2010. "Objectives and Themes." Accessed December 23, 2016, https://sustainabledevelopment.un.org/rio20/objectivethemes.

US Environmental Protection Agency. 2013. *The Importance of Water to the United States Economy: Synthesis Report*. Accessed December 23, 2016, http://bafuture.org/sites/default/files/key-topics/attachments/Importance-of-water-synthesis-report.pdf.

Uslaner, Eric. 2002. *The Moral Foundations of Trust*. Cambridge: Cambridge University Press.

van den Bergh, Jeroen C.J.M. 2001. "Ecological Economics: Themes, Approaches, and Differences with Environmental Economics." *Regional Environmental Change* 2 (1): 13–23. doi:10.1007/s101130000020.

van den Bergh, Jeroen C.J.M. 2011. "Environment versus Growth—A Criticism of 'Degrowth' and a Plea for 'A-Growth.'" *Ecological Economics* 70 (5): 881–890. doi:10.1016/j.ecolecon.2010.09.035.

Victor, Peter. 2008. *Managing without Growth: Slower by Design, Not Disaster*. Cheltenham, UK: Edward Elgar.

Weidner, Helmut. 2002. "Capacity-Building for Ecological Modernization: Lessons from Cross-National Research." *American Behavioral Scientist* 45 (9): 1340–1368. doi:10.1177/0002764202045009004.

Wilkinson, Richard, and Kate Pickett. 2009. *The Spirit Level: Why Equality Is Better for Everyone*. London: Penguin.

World Bank. 2012. *Inclusive Green Growth: The Pathway to Sustainable Development*. Accessed December 23, 2016, http://siteresources.worldbank.org/EXTSDNET/Resources/Inclusive_Green_Growth_May_2012.pdf.

World Business Council for Sustainable Development. 2012. *Vision 2050: The New Agenda for Business*. Accessed December 23, 2016, http://www.wbcsd.org/Overview/About-us/Vision2050/Resources/Vision-2050-The-new-agenda-for-business.

13 Sustainable Consumption: An Important but Ambiguous Concept

Philip J. Vergragt

Sustainable consumption is an emerging normative concept gaining traction at present in policy circles. It mainly refers to the environmental problems related to material consumption patterns in affluent societies, but also applies to inequities in consumption patterns between rich and poor. It is discussed in emerging economies that are often on a pathway toward affluent consumption patterns. It is a concept that recognizes consumption is part of a complex system, and that a shift toward sustainable consumption patterns requires systemic changes in the economy, technology, governance, culture, lifestyles, values, and institutions. It marks a shift in focus from the supply side (sustainable production) to the demand side.

The origins of sustainable consumption go back to the UN Earth Summit of 1992, where "unsustainable patterns of production and consumption" were identified as a main cause of environmental deterioration. It was taken up subsequently by academic research, which until then had criticized consumerism from a social rather than environmental perspective. Sustainable consumption is closely related to sustainable production; the difference is that while sustainable production refers to sustainable technological innovations, sustainable consumption concentrates more on behavior, lifestyles, and well-being.

A wide range of academic disciplines is researching possibilities for transitions to sustainable consumption patterns and systems, ranging from economics (ecological, steady state, and degrowth) to happiness and well-being research. The concept has gained traction in the European Union and municipal policy circles in Europe, in parts of the United States, and elsewhere. It is presently associated with localism and the sharing economy, but there is also a strong tendency to analyze the issue at the macro level.

The concept suffers from tensions and ambiguities, however. These are related to scale (from individual to institutional and macro), equity (overconsumption in affluent societies versus underconsumption in poor communities), technology (technology optimism versus sufficiency and consuming differently), and between weak and strong versions. The last contrast refers to the belief in business and government circles that sustainable consumption can be reached without deep changes in institutions versus those who criticize the economic growth paradigm and the capitalist system itself. Finally, there are tensions among those who see the issue as mainly an economic one (steady state and degrowth), those who consider it essentially a cultural problem, and those who view it as both.

This chapter will first explore the origins of the concept in UN policy circles, and analyze the conceptual developments in academic research and how it merged with earlier research traditions criticizing consumer society. It will then look at the present significance of the concept in policy circles, mainly focusing on the United Nations, business, and regional and local policy making. Next it discusses ambiguities and tensions inherent in the concept, and the ways these are addressed in research and practice. Finally, it will examine the possible future development and significance of the concept.

Origins in UN Policy Circles and Definition of the Concept

The concept sustainable consumption first emerged in the political discourse after the UN Conference on Environment and Development, or Earth Summit, in Rio de Janeiro in 1992. In Agenda 21, it was stated that "the major cause of the continued deterioration of the global environment is the unsustainable pattern of consumption and production, particularly in industrialized countries, which is a matter of grave concern, aggravating poverty and imbalances" (United Nations 1992).

Agenda 21 contained a chapter on "Changing Consumption Habits," which focused on addressing unsustainable patterns of production and consumption and it argued that action was needed to meet the following broad objectives: "to promote patterns of consumption and production that reduce environmental stress and will meet the basic needs of humanity; to develop a better understanding of the role of consumption and how to bring about more sustainable consumption patterns." It developed an action agenda for management, research, policies, and strategies (United Nations 1992, chapter 4).

In Agenda 21 the term *sustainable consumption* is used for the first time in a widely read document. It subsequently appeared in the academic discourse. A first and widely quoted definition is from the Oslo Symposium of 1994: "the use of goods and services that respond to basic needs and bring a better quality of life, while minimizing the use of natural resources, toxic materials and emissions of waste and pollutants over the life cycle, so as not to jeopardize the needs of future generations" (International Institute for Sustainable Development 1995). To quote further:

> Sustainable consumption is an umbrella term that brings together a number of key issues, such as meeting needs, enhancing the quality of life, improving resource efficiency, increasing the use of renewable energy sources, minimizing waste, taking a life cycle perspective and taking into account the equity dimension. (ibid.)

More recently, Joachim H. Spangenberg (2014, 62) proposed a new and more encompassing definition, based on "an overall restructuring of the social, economic, and institutional fabric of societies and institutions, of production, allocation, and consumption patterns." Sustainable consumption is "the ability to lead a dignified life, maintaining or enhancing quality of life despite shrinking resource availability."

Nevertheless, in the first ten years after Rio, action was limited, except for the already-mentioned Oslo Symposium of 1994. The Organisation for Economic Co-operation and Development (1997) wrote a major report on sustainable consumption, and the UN Development Program in 1998 mentioned it in its human development report. In 2002, the International Coalition for Sustainable Production and Consumption (2002), a global coalition of nongovernmental organizations, wrote a progress report for the World Summit in Johannesburg. The title of this report, *Waiting for Delivery*, summarizes the main conclusion: although there was a lot of talking, little progress has been made toward implementation, and the world is fast moving in the wrong direction of unsustainable development. In 2002 at the World Summit on Sustainable Development, governments formally agreed that "poverty eradication and changing unsustainable patterns of production and consumption are overarching objectives of sustainable development and an essential requirement for promoting environmental protection" (United Nations 2002).

In 2003, the United Nations launched the Marrakech process with the aim of bringing together the expertise and leadership needed to develop a ten-year framework of programs on sustainable consumption and production (10YFP). The Marrakech process mainly consisted of a series of task forces led by individual countries. These were sustainable lifestyles (focusing on sustainable consumption patterns related to lifestyles and culture), sustainable products (to raise awareness on product policy and ecodesign), cooperation with Africa, and sustainable public procurement (to promote understanding of the issue, exchange experience, identify best practices, and develop links between governments, nongovernmental organizations, and other actors). These task forces were later followed by others: education for sustainable consumption, sustainable buildings and construction, and sustainable tourism. (UN Environment Program 2014a). The Marrakech process has not been evaluated by independent research. The general opinion is that the results have been mixed at best. The individual task forces have resulted in important outcomes, but the dissemination of those results has so far hardly translated into concrete policies. The most visible outcome was the creation, at the Rio summit in 2012, of the 10YFP (UN Environment Program 2014b). It took ten years to establish this program.

Evolution of the Concept

As mentioned above, the Oslo Symposium of 1994 provided the first academic definition of sustainable consumption. Research projects related to sustainable consumption emerged in that period, but were not yet labeled as such. For instance, the EU-funded project Strategies for the Sustainable Household (1998–2000) developed methodologies for future visions for sustainable consumption practices and methods of backcasting how to get there (Green and Vergragt 2002), but did not use the term sustainable consumption. The term first emerged in a report on sustainable consumption and globalization (Fuchs and Lorek 2000). Subsequently, Maurie Cohen and Joseph Murphy (2001) use the term in the title of their book *Exploring Sustainable Consumption*. In chapter 6 of this book, David Goodman and Michael Goodman (2001) make a distinction between a reorientation toward more environmentally friendly production technologies and consumption patterns versus alternatives to the capitalist economy. These two positions are consonant with what has later been called "weak" and "strong" sustainable consumption (Lorek and Fuchs 2011).

Scholars recognized in that period that addressing individual consumers, attitudes, behaviors, and lifestyles alone could not change unsustainable consumption practices. Consumers are often locked in by decisions made earlier in life, such as living in the suburbs and thus become dependent on multiple cars per family, or buying a large house that requires heating, cooling, and "stuff" (Sanne 2002). Consumers are "socially embedded"—meaning that individual choices are heavily influenced by contextual social forces like advertisements and social media, and structural features like convenience. Consumers are also "distanced"—meaning that they have no direct access or knowledge about the production processes of consumables that pollute the environment (Princen et al. 2002). Social practice theory emerged, arguing that instead of taking the consumer as the starting point, social practices like clothing, housing, food provision, travel, sport, and leisure should be taken as objects for analysis (Spaargaren 2003; Shove 2003).

More recently, the approach to influencing individual consumer behavior has seen a revival, stimulated by the emergence of the "nudging" concept (Thaler and Sunstein 2008; Sunstein 2015). This idea suggests that "choice architecture" may guide and enable consumers to make choices automatically. Nudges do not try to change value systems or provide additional information; instead they focus on enabling behaviors and private decisions. A prime example is to place green products in prominent places in the supermarket.

In economics, the neoclassic tradition has been challenged by, among others, ecological and steady-state economics (Daly 1990), and more recently, new economy and solidarity economy. In neoclassic economics, emphasis has been mostly on production. Consumption is treated as a "utility function," which is revealed by people's willingness to pay a certain price for goods. Ecological economics recognizes that the economy is a subsystem of a larger ecological system, and that there are thus boundaries to material growth. A steady-state economy is loosely defined as "a truly green economy. It aims for stable population and stable consumption of energy and materials at sustainable levels" (Center for the Advancement of the Steady State Economy 2015). Parallel with this discussion is the critique of GDP as an indicator of (economic) well-being. Since World War

II, politicians have adopted GDP not only as an indicator for the state of the economy but also for well-being in general. This notion has been criticized by many, and has resulted in a quest for alternative indicators of well-being. The most influential attempt has been in a report by Joseph Stiglitz, Amartya Sen, and Jean-Paul Fitoussi (2009).

Sustainable consumption research often addresses "systems of provision" like food, housing, or transportation. In the EU-funded Sustainable Consumption Research Exchanges project, these systems have been analyzed from the perspectives of business studies, design, consumer research, and system innovations, and published in five influential books (Tukker, Emmert, et al. 2008; Tukker, Charter, et al. 2008; Geerken and Borup 2009; Tischner et al. 2010; Lahlou 2011). A more sophisticated conceptual approach is developed in the "Transition Management" research tradition, also known as the "Multi-Level Perspective" (Geels and Schot 2007). In this research tradition, the above systems of provision can be viewed as "sociotechnical regimes," which are stable complex systems. In this approach, bottom-up initiatives in technology and social practices may constitute "niches" in which experimentation into alternatives to the dominant regime takes place. The bottom-up pressures provided by niche experiments, in combination with the top-down pressures from the highest "landscape" level, may destabilize the sociotechnical regime, which then could be modified or even replaced by a different regime. A historic illustration is the replacement of coal by oil and then natural gas for heating homes. A more contemporary example would be the possible replacement of the fossil fuel–based energy regime by a low-carbon regime based on renewables. Consumption practices are very much part of the dominant regimes, and alternative consumption practices in niches may help to challenge a dominant regime.

What happens in niches is that producers and consumers jointly develop new technologies, products, services, and practices that are more sustainable from environmental and social perspectives. Creativity, stakeholder participation, and learning are essential in such processes. Halina S. Brown and Philip J. Vergragt (2008) have coined the term *bounded sociotechnical experiment* and emphasized the collective higher-order learning processes that take place in, for instance, the design of a high-performance, energy-neutral building. Similarly, Gill Seyfang and Adrian Smith (2007) coined the term *grassroots innovations* to depict similar localized, multistakeholder processes to collectively develop sustainable solutions and thus change consumption patterns (Seyfang 2009).

These conceptual approaches were used in a Sustainable Consumption Research and Action Initiative workshop in Princeton, New Jersey, in 2011 in which the research traditions of sociotechnical transitions (Geels and Schot 2007), social practice theories (Shove 2003; Spaargaren 2013), and new economics (Harris 2013; Røpke 2013) were brought together. This workshop, which became the book *Innovations in Sustainable Consumption*, constructed bridges between the macrolevel analysis of new economics and microlevel analyses of social practices, and resulted in an emerging theory of change (Cohen et al. 2013).

Economic factors, policy, technological innovation, infrastructure, business forces, marketing and advertisements, personal needs, social values, and norms all drive unsustainable consumption patterns (Mont and Power 2009). More recently, the concept of *power* has entered the discourse around sustainable consumption. Doris Fuchs (2013a, 2013b) conceptualizes three forms of power: *instrumental power*, such as lobbying or campaign finance; *structural material power*, which predetermines decision and nondecision making through the shaping of actors' behavioral options; and *structural ideational power*, which reveals how policy problems, actors, interests, and solutions are constituted and defined before decision making commences.

Many authors have written about alternatives to unsustainable consumption practices and how to achieve change. In the 1990s, Duane Elgin (1993) developed the concept of *voluntary simplicity*, which includes the principles of frugal consumption, ecological awareness, and personal growth. In his book's newer edition, Elgin also includes the more recent trend toward "downshifting." The origins of voluntary simplicity go back to the work of E. F. Schumacher (*Small Is Beautiful*) and Mahatma Gandhi, and the concept contrasts the worldviews of the industrial and ecological eras. Tom Princen coined the term *sufficiency* as a challenge to the dominance of *efficiency* in modern society. While he acknowledges that we also need efficiency, Princen (2005) argues that more is not always better, and that in an environmentally constrained world, it may be more ethical to consume less and be content with sufficiency, instead of striving to consume as much as possible and then limit the negative side effects of this consumption by technological efficiency measures. This assertion clearly reflects the normative connotation of sustainable consumption.

Critique of Consumerism

Scholars and historians of the modern consumer society commonly date its advent from the first two decades of the twentieth century (Brown and Vergragt 2016). The provident remark made in 1929 by Charles Kettering (1929, 543), director of research at General Motors, has been widely quoted by various authors: "The key to economic prosperity is the organized creation of dissatisfaction. ... If everybody was satisfied, nobody would want to buy anything." Criticism of consumer society goes back to the mid-nineteenth century when Karl Marx (1867) developed the concept *commodity fetishism*, which means that the social relations of labor are obscured by consumer goods that carry other kinds of symbolic value for their users. He did not yet use the word *consumption*, because in the nineteenth century this famously meant "a progressive wasting away of the body, especially from pulmonary tuberculosis." The modern concept of consumption first appeared in the late nineteenth century in the writings of Thorstein Veblen (1899), who criticized "conspicuous consumption" by the rich, meaning ostentatiously displaying wealth while ignoring the poverty of the rest of the population. Émile Durkheim and Max Weber ([1905] 1992) pointed to the centrality of consumer goods when writing about the growing importance of them to social life in the late nineteenth century.

Much of the criticism of consumerism has revolved around its manipulation of human desires. Theodor W. Adorno and Max Horkheimer's ([1944] 2002) essay "The Culture Industry" contends that consumption of the easy pleasures of popular culture, made available by the mass communications media, renders people docile and content, no matter how difficult their economic circumstances. The inherent danger of the culture industry is the cultivation of false psychological needs that can only be met and satisfied by the products of capitalism. In contrast, authentic psychological needs relate to freedom, creativity, and genuine happiness. The US sociologist David Riesman's (et al. [1950] 2001) landmark book *The Lonely Crowd* sets the foundation for how sociologists study how people seek validation and community through material consumption, by looking to and molding themselves in the image of those immediately around them.

In his famous book *The Hidden Persuaders*, Vance Packard (1957) described the use of consumer motivational research and other psychological techniques by advertisers to manipulate expectations and induce desire for products. John Kenneth Galbraith (1958, 126) criticized neoclassic economics and the emphasis on "utility" rather than needs; he maintained that "if the individual's wants are to be urgent, they must be original with himself. They cannot be urgent if they must be contrived for him. And above all, they must not be contrived by the process of production by which they are satisfied. For this means that the whole case for the urgency of production, based on the urgency of wants, falls to the ground."

In the 1960s, the counterculture movement criticized consumerism and was inspired by the famous book *One-Dimensional Man* by Herbert Marcuse (1964), who described Western societies as awash in consumer solutions that are meant to solve one's problems, and thus provide market answers for what are actually political, cultural, and social problems. More modern critiques such as that of Robert Skidelsky and Edward Skidelsky (2012) place much of the responsibility for the current insatiable pursuit of material goods and money on neoclassic economic science. By replacing the concept of value with utility, and avarice and greed with self-interest, economics eliminated the controls—long recognized in religious and moral teachings—on the human tendency for excess. These ideas also inspired David Graeber (2013), who argues that the real objective of neoclassic economics was not free market capitalism but rather the suppression of dissent and creativity, which would henceforth be channeled into consumerism.

Since the 1970s, many sociologists have embraced French social theorist Jean Baudrillard's ([1970] 1998) ideas about the symbolic currency of consumer goods, and take seriously his claim that seeing consumption as a universal element of the human condition obscures the class politics in which it rests. Similarly, Pierre Bourdieu's (1984) research and theorizing on the differentiation between consumer goods, and how these both reflect and reproduce cultural, class, and educational differences and hierarchies, is a cornerstone of today's sociology of consumption (Allen and Anderson 1994).

Building on these early criticisms of consumer society, and supported by empirical data, Juliet Schor (1998), in *The Overspent American: Why We Want What We Don't Need*, criticizes the work-and-spend cycle, and asks why so many Americans are trapped in a cycle of working longer hours in order to spend more money on things that they do not need and that do not make them happy. She calls this the "upward creep of desire" and develops nine principles on how to address it. Schor moves beyond the individualistic approach toward the need for "coordinated intervention" through strong government policies, including taxation.

Recent Developments

Societal developments since 2008 have considerably changed the context for the evolution of the sustainable consumption concept. The great recession of 2008 and subsequent years undermined general confidence in progress and increased consumption, and has squeezed the middle class and its expectation of ever-increasing purchasing power. In the United States, massive unemployment, foreclosures of homes, and evaporation of retirement funds have eroded the American dream along with the expectation that the children of the middle class will enjoy greater prosperity than their parents. The Occupy movement put the inequity of the 1 percent rich and the remaining 99 percent squarely on the societal and political agenda (Wilkinson and Pickett 2009; Vergragt 2013; Graeber 2013; Piketty 2013). The emergence of the degrowth movement, first in southern Europe, and more recently in North America, has challenged the economic growth paradigm and existing notions of well-being through material consumption (Schneider, Kallis, et al. 2010; Schneider, Martinez-Alier, et al. 2011; Sekulova et al. 2013).

Sylvia Lorek and Doris Fuchs (2011) make an explicit connection between "strong sustainable consumption" and degrowth. The notion of the solidarity economy (Miller 2010) is based on increasing the quality of life through not-for-profit endeavors, and is critical of exploitation under capitalist economics and the corporate, shareholder-dominated economy. It has placed economic democracy and alternative forms of business ownership on the political agenda (Kelly 2012). Environmentalists have challenged the narrow focus of the environmental movement (Speth 2012), and argued for broadening it with social and cultural issues. More recently, the new economy movement has formulated novel principles that are consonant with sustainable consumption (Goodwin et al. 2011). The peer-to-peer or sharing economy calls for sustainable consumption practices like sharing. Grassroots innovations (Seyfang and Smith 2007; Seyfang 2009) and transition towns movements (Hopkins 2008) are reframing consumers as producers (for locally produced food, renewable energy, and repair services), and thus "prosumers" (Ritzer and Jurgenson 2010).

Another conceptual development is the reemergence of needs theory in research. Manfred A. Max-Neef (1991) developed a needs theory in which he made a distinction between basic needs, which are universal, and satisfiers, which are highly contextual and locally different. This theory was expanded by Tim Jackson (2006), who, following Baudrillard ([1970] 1998), argues that products not only satisfy needs; they also have important symbolic meaning, signifying status, personal identity, and belonging. As such, consumption has a crucial cultural meaning. Sustainable consumption discourse has incorporated the concepts of a good life, well-being, and happiness as a path toward societal flourishing (Easterlin 2003; Layard 2011). What is remarkable is the consistency and stability of certain basic determinants of happiness across different countries and cultures: a stable marriage, good health, community and friendships, social trust, and autonomy. Another consistent observation is that people judge the emotional value of their *material* wealth in comparison to others. Once basic subsistence needs are met, it is of greater significance to have more than others in an absolute sense. A distinction is made between the Benthamite perspective where well-being is understood as an emotional state of pleasure/contentment/joy, and the Aristotelian perspective that focuses on satisfaction arising from evaluating one's life, and emphasizes autonomy, search for meaning, spirituality, commitment, and ethical behavior as well as gaining respect, status, and a sense of achievement (Max-Neef 1991; Nussbaum 2011; Jackson 2009; Di Giulio et al. 2012; Jackson and Victor 2013).

Thus, a number of new societal developments frame the issues around sustainable consumption in a rather different way. Many of them do not explicitly invoke the notion of sustainability, but instead define framings around solidarity, community, equity, well-being, democracy, and self-reliance. Two aspects are particularly important in these new social movements: localism and community. Both seem to be a reaction to globalization as well as what Marxists used to call alienation due to globalization, distancing, and the pervasive influence of technologies in modern society. On the other hand, the Internet and especially social media play a critical role in the emergence of the new economy, and especially the peer-to-peer and sharing economy. The emerging threat of climate change is also important, mostly reinforcing local adaptation as opposed to global mitigation.

Cohen (2013) has analyzed the emerging signs of a possible postconsumerist culture, with millennials moving back into the cities, renouncing their car, and refusing to obtain drivers licenses while living in smaller spaces. Brown and Vergragt (2016) have hypothesized that change may come from a redefinition of well-being among the millennial generation, spurred by the Internet and social media as well as economic constraints.

In summary, sustainable consumption is rooted in both nineteenth-century criticisms of the consumer society and twenty-first-century understandings of environmental issues related to consumption patterns and reevaluations of what constitutes the good life. An understanding is emerging of what ultimately drives (over)consumption, but it still begs the question articulated by Bill Rees (2010): Is it part of the human condition that we will always consume as much as we can? Would it be possible to achieve a healthy economy without excessive material consumption? Are the problems of unsustainable consumption inherent in the system of capitalist economy and adherence to an inherently unsustainable growth ideology, or can they be solved within the present capitalist system?

Significance of the Concept in Policy and Business Circles

The European Union was among the first jurisdictions to develop policies on sustainable consumption. In the early years, these policies were mainly aimed at cleaner production and green products in the so-called Integrated Product Policy (European Commission 2001). EU policies subsequently moved toward sustainable production and consumption in a policy document called the "Sustainable Consumption and Production Industrial Policy Action Plan" (European Commission 2008). More recently, the "Roadmap to a Resource Efficient Europe" outlines how Europe's economy can be transformed into a sustainable one by 2050. It proposes ways to increase resource productivity as well as decouple economic growth from resource use and its environmental impacts (European Commission 2011). In June 2013, the European Parliament and European Council agreed on the seventh EU Environment Action Program to 2020, titled Living Well, within the Limits of Our Planet (European Commission 2012). One of the plan's priority objectives is "to turn the EU into a resource-efficient, green and competitive low-carbon economy," which requires that "the overall environmental impact of production and consumption is reduced, in particular in the food, housing and mobility sectors." The European Union also funded the SPREAD (2013) project, which envisions sustainable lifestyles and strategies to attain them (Mont et al. 2014). Yet the European Union seems to be moving away from the concept of sustainable consumption, and adopting new language such as the *circular economy* (Ellen MacArthur Foundation 2017) and *social innovation* (Hubert 2010), both of which appear to be less threatening to existing interests.

The United States does not have a formal policy on sustainable consumption. There is even a reluctance to use this concept in policy circles at the federal level. The US and Canadian governments organized two workshops on sustainable production and consumption in the context of the Marrakech process in 2008 in Washington, DC, and in 2011 in Ottawa. These workshops were well attended by policy makers, business and civil society representatives, and academics (UN Environment Program 2008, 2011). Sustainable consumption is somewhat better established at the level of some states in the United States and provinces in Canada, mainly on the west coast.

In 2010, the North American Roundtable on Sustainable Production and Consumption (2014) was formed as a multistakeholder platform. Since its start, it has attracted many participants from business and government, in addition to early civil society and academic participation. It brings the dialogue on the concept to federal policy makers.

In many regions of the world, cities have taken the lead in the quest for a transformation toward "ecocities" or "sustainable cities" (ICLEI 2015). It is interesting that on the US West Coast (Oregon and California), but also in other places, sustainable consumption has become a framework for municipal policies. Many cities are working to make their local economies more resilient and sustainable while building new forms of prosperity. Sustainable consumption and production provides a meaningful framework to advance this work as well as promote a broader societal shift. In comparison to suburbs and rural areas, urban centers in North America have higher population densities, smaller houses, lower levels of automobile use, more concentration of diverse activities, and the political capacity to implement innovative experiments with respect to personal mobility, public spaces, land use, and resource flows. This gives local governments the unique potential to shape opportunities for sustainable lifestyles and the local economies to support them.

For many cities, interest in sustainable consumption and production emerged from local climate action planning, but there are other dimensions that deserve exploration and analysis:

- *Structure of local economies*: sustainable consumption provides local businesses with new models and opportunities in a way that complements broader economic development strategies. Research suggests new ways for local businesses to meet these needs through a focus on providing services rather than products.
- *Social equity considerations*: enhancing the range of affordable alternatives supplies families of all income levels with ways to save money and meet their needs, freeing up time and resources for the things that really matter, such as time with family and friends, access to nature and recreation, community volunteerism, building memories, and acquiring new skills.
- *Social capital and cohesion*: as cities build toward more compact, cohesive, and livable communities, this urban form aligns with the collaborative nature of sustainable consumption. People living in close proximity have more opportunities to share idle resources like cars, sports equipment, or yard tools. Compact, cohesive communities also offer a platform for launching small-scale commercial ventures for sharing skills like equipment repair, clothing alternations, and hairstyling.

The Sustainable Consumption Research and Action Initiative together with the Urban Sustainable Directors Network organized a workshop in October 2014 to explore these questions further. This workshop resulted in the "Eugene Memo," a mobilizing document that describes the relevance of sustainable consumption for city policy makers (City of Eugene et al. 2015).

Business

At the global level, the World Business Council for Sustainable Development (WBCSD) has been an important player with respect to sustainable development. Since the 1990s, the WBCSD has published influential reports reflecting a progressive business vision on sustainable development issues. These reports represent the interests of mostly large companies, but also try to influence the business community with long-term visions of the role of business in a sustainable society.

In 2008, the WBCSD (2008) published *Sustainable Consumption: Facts and Trends*. In this report, the WBCSD acknowledges the impacts of global consumption on the earth's ecosystems and that human well-being does not necessarily rely on high levels of consumption. It notes that consumers are increasingly concerned about environmental and social issues, although this often does not translate into behavioral change. The role of business is then described along three dimensions: innovation to create sustainable products and services, choice influencing through marketing, and choice editing by removing unsustainable products from the marketplace. They call for dialogue with other stakeholders and among business to define sustainable products and lifestyles as well as to formulate actionable responses.

Consumers have become more sensitive to the reputation of companies, which has created pressure on companies to improve their record on ecological and social issues. The WBCSD acknowledges that consumer lifestyles should change and business needs to play a leadership role in fostering more sustainable consumer choices for more sustainable lifestyles.

The WBCSD (2010) also published *Vision 2050: A New Agenda for Business*, which presents a vision and pathway based on backcasting methodology (Vergragt and Quist 2011). Building on this, the WBCSD (2011) published *A Vision for Sustainable Consumption: Innovation, Collaboration, and the Management of Choice*. This vision is made up of five key elements: better products and services; enlightened consumers with more awareness and motivations to avoid negative impacts; maximized total value, which means both classic utility and environmental and social benefits; new indicators for success (meaning a suite of indicators beyond GDP); and a cohesive and responsive marketplace, which means a constant dialogue and information exchange among all actors in the value chain. The WBCSD acknowledges that to realize this vision, there will have to be increased collaboration and information exchange between stakeholders, deeper understanding of consumer behaviors, more use of technology for information sharing, an evolution in business models, and efforts to reinforce trust between social actors.

These reports by the WBCSD reflect the process through which they were developed, which was in close collaboration with the research community and civil society. They present a daring vision, at least from a business perspective. Notably absent in these reports is an appreciation of government's role in regulation and policies. They imply that business, in collaboration with civil society, will be able to implement the vision and pathways developed in these reports. They suggest that government regulation and policies are not only unnecessary but also possibly counterproductive. Implicitly, this reinforces the neoclassic ideology of free markets. The WBCSD does not question this free market ideology or the paradox of perpetual growth in a finite world. Notably, none of the initiatives undertaken by the WBCSD question economic growth or mass consumption; thus they are at best consistent with weak sustainable consumption. It is unclear how much influence these reports have on business behavior.

In a recent paper, Nancy Bocken and her colleagues (2014) developed a typology of sustainable business models. One of their archetypes is called "encouraging sufficiency," and one of its aspects is addressing a broader range of stakeholders. This archetype challenges advertising and overconsumption. Marjorie Kelly (2012) has described a "generative economy" where different business ownership models are prevalent.

Recent Developments on the Global UN Level

Two recent developments are important. The first, established at the Rio+20 conference in 2012, is the 10YFP. This initiative is under way and so far has established five programs: consumer information, sustainable lifestyles and education, sustainable public procurement, sustainable buildings and construction, and sustainable tourism, including ecotourism (UNEP 2014b).

The other crucial development is the establishment by the United Nations in 2015 of "Sustainable Development Goals." The twelfth goal is to "ensure sustainable consumption and production patterns" (United Nations 2015). Under this goal, twelve targets have been formulated, of which the first is to "implement the 10-Year Framework of Programmes on sustainable consumption and production (10YFP)." Other goals include sustainable management, efficient resource use, halving per capita global food waste, the sound management of chemicals and reduction of waste, sustainable reporting, sustainable public procurement, information provision and awareness raising, supporting developing countries in their scientific and technological capacities, sustainable tourism, and reforming taxes and subsidies. The implementation and financing of strategies and policies to achieve these goals are unclear. Now that these goals are formally accepted, though, the concept of sustainable consumption will remain anchored in the global political agenda of sustainable development.

Ambiguities, Tensions, and Possible Future Developments

The wider adoption of the concept of sustainable consumption has been hampered by its internal contradictions and ambiguities along with its wide scope. The first ambiguity is captured by the weak versus strong sustainable consumption metaphor. The strong version, mainly endorsed by academics and activists, criticizes consumerist culture and the present economic system while arguing that systemic change is necessary. This version also asserts that well-being above a certain minimum level is not dependent on material consumption and that economic indicators should go "beyond GDP." In contrast, many mainstream actors strive to reform government policies on all levels to incorporate elements of sustainable consumption or try to encourage business to take on the task; this often results in weak forms of sustainable consumption.

A second ambiguity is that in many cases, the consumer is addressed as an individual through efforts to seek to influence behavior through information, incentives, and nudging. This approach largely disregards how individuals are embedded in a social and physical infrastructure that makes individual changes difficult or impossible. Adding to this challenge, consumers are known to seek status, identity, and recognition through the consumption of material goods, and this complicates individual actions toward sustainable consumption. All this is reinforced by the advertising industry, which addresses consumers as individuals and is firmly established in consumer society.

A third problem with the concept is that many communities and countries are consuming below a sufficiency level, and strive to enhance their well-being through increased material consumption. This makes it harder to communicate the concept to a wide audience, especially because of widespread poverty in affluent societies. It is difficult to communicate sustainable consumption when many people cannot make ends meet. The conceptual challenge is finding a way to enhance well-being without "overshoot," so that individuals and communities reaching a sustainable minimum level of consumption do not seek to consume beyond that. This has so far proven challenging; consider, for instance, the emergent consumerist middle class in China.

A fourth problem is that communicating a message of "reduction" does not work well, so various different framings have been tried, like "consuming differently," "sufficiency," "well-being," and many others. Still, the drive to buy material goods seems to be deeply ingrained in people's consciousness, and the alternatives are difficult to communicate beyond a small elite of sustainability advocates. Even in affluent communities, many people feel they need to work more, earn more, and spend more to feel satisfaction.

Finally, there is the issue of how to quantify and measure consumption levels as well as establish an operationalization of sustainable consumption. Because the concept is so all encompassing—including equity issues as well as criticism of economic growth and lifestyles—establishing a norm of what constitutes sustainable is virtually impossible. Concepts like carbon and ecological footprints and tons of carbon dioxide reduction only capture aspects of sustainable consumption, while neglecting equity and cultural and lifestyle aspects.

These ambiguities and tensions have spurred researchers, activists, and policy makers to create more appealing concepts than sustainable consumption, such as resource efficiency, the green economy, green growth, and the circular economy. Many of these concepts, however, do not obviously relate to individual consumption and thus lose one of the strong aspects of sustainable consumption, which refers to both individual *and* collective consumption.

So in the future, two developments seem to be possible. The first is that the concept sustainable consumption will disappear because it will succumb to its internal contradictions and ambiguities, and be replaced by the more socially and politically acceptable alternatives mentioned above. The other possibility is that the concept survives and gains more strength in the policy domain. This could mirror a development similar to that of the concept sustainable development itself, which was also considered ambiguous and ill-defined, and was declared dead many times before it surged in popularity and influence in recent years.

If sustainable consumption survives as a concept, will it encourage more effective policies and actions for the creation of more sustainable consumption patterns? The barriers are formidable: entrenched interests and power relationships; the dominant culture of consumerism and consumer sovereignty; the advertising industry; the economic growth paradigm and the prevalence of GDP as an indicator of well-being; the distancing of environmental pollution from its sources; the lack of awareness by consumers; the lock-in of unsustainable life situations; and the list can go on and on. Vigorous government policies appear to be necessary, but only would be possible if backed up by a strong social movement. And such a social movement is probably only possible if triggered by major disasters (such as severe climate change) or a fundamental cultural transformation. Such cultural transformations have happened before: the broad acceptance of civil rights following the civil rights movement, the end of smoking nicotine, the broad acceptance of gay marriage, and others. So they are not impossible. Also needed are leaders with a strong and appealing vision as well as a public that is ready to make deep changes in lifestyles because it recognizes the limits of the present system of overconsumption.

References

Adorno, Theodor W., and Max Horkheimer. (1944) 2002. *Dialectic of Enlightenment*. Ed. Gunzelin Schmid Noerr. Trans. Edmund Jephcott. Stanford, CA: Stanford University Press.

Allen, Douglas E., and Paul F. Anderson. 1994. "Consumption and Social Stratification: Bourdieu's Distinction." In *NA—Advances in Consumer Research*, ed. Chris T. Allen and Deborah Roedder John, 70–74. Vol. 21. Provo, UT: Association for Consumer Research.

B Corporation. 2015. "What Are B Corps?" Accessed February 25, 2015, http://www.bcorporation.net/what-are-b-corps.

Baudrillard, Jean. (1970) 1998. *The Consumer Society: Myths and Structures*. London: SAGE Publications.

Bocken, Nancy M. P., Samuel William Short, P. Rana, and S. Evans. 2014. "A Literature and Practice Review to Develop Sustainable Business Model Archetypes." *Journal of Cleaner Production* 65:42–56. doi:10.1016/j.jclepro.2013.11.039.

Bourdieu, Pierre. 1984. *Distinction: A Social Critique of the Judgment of Taste*. Cambridge, MA: Harvard University Press.

Brown, Halina S., and Philip J. Vergragt. 2008. "Bounded Socio-Technical Experiments as Agents of Systemic Change: The Case of a Zero-Energy Residential Building." *Technological Forecasting and Social Change* 75 (1): 107–130. doi:10.1016/j.techfore.2006.05.014.

Brown, Halina S., and Philip J. Vergragt. 2016. "From Consumerism to Wellbeing: Towards a Cultural Transition?" *Journal of Cleaner Production* 132:308–317. doi:10.1016/j.jclepro.2015.04.107.

Center for the Advancement of the Steady State Economy. 2015. "What Is a Steady State Economy?" Accessed February 25, 2015, http://steadystate.org/wp-content/uploads/CASSE_Brief_SSE.pdf.

City of Eugene, Urban Sustainability Directors Network, and Sustainable Consumption Research and Action Initiative. 2015. "Eugene Memo: The Role of Cities in Advancing Sustainable Consumption." Accessed March 7, 2015, http://scorai.org/wp-content/uploads/wordpress/Eugene-Memo-Cities-Sust-Consumption-FINAL.pdf.

Cohen, Maurie. 2013. "Collective Dissonance: The Transition to Post Consumerism." *Futures* 52:42–51. doi:10.1016/j.futures.2013.07.001.

Cohen, Maurie, Halina S. Brown, and Philip J. Vergragt, eds. 2013. *Innovations in Sustainable Consumption: New Economics, Socio-Technical Transitions, and Social Practices*. Northampton, MA: Edward Elgar.

Cohen, Maurie, and Joseph Murphy, eds. 2001. *Exploring Sustainable Consumption: Environmental Policy and the Social Sciences*. Bingley, UK: Emerald Group Publishing.

Daly, Herman E. 1990. "Toward Some Operational Principles of Sustainable Development." *Ecological Economics* 2:1-6. doi:10.1016/0921-8009(90)90010-r.

Di Giulio, Antonietta, Bettina Brohmann, Jens Clausen, Rico Defila, Doris Fuchs, Ruth Kaufmann-Hayoz, and Andreas Koch. 2012. "Needs and Consumption: A Conceptual System and Its Meaning in the Context of Sustainability." In *The Nature of Sustainable Consumption and How to Achieve It*, ed. Rico Defila, Antonietta Di Giulio, and Ruth Kaufmann-Hayoz, 45–66. Munich: Oekom.

Easterlin, Richard. 2003. "Explaining Happiness." *Proceedings of the National Academy of Sciences of the United States of America* 100 (19): 11176–11183. doi:10.1073/pnas.1633144100.

Elgin, Duane. 1993. *Voluntary Simplicity: Toward a Way of Life That Is Outwardly Simple, Inwardly Rich*. New York: William Morrow and Company.

Ellen MacArthur Foundation. 2017. Accessed April 17, 2017, https://www.ellenmacarthurfoundation.org.

European Commission. 2001. "The Green Paper on Integrated Product Policy." Accessed January 29, 2014, http://ec.europa.eu/environment/ipp/2001developments.htm.

European Commission. 2008. "European Sustainable Consumption and Production Policies." Accessed January 29, 2014, http://ec.europa.eu/environment/eussd/escp_en.htm.

European Commission. 2011. "The Roadmap to a Resource Efficient Europe." Accessed January 29, 2014, http://ec.europa.eu/environment/resource_efficiency/about/roadmap/index_en.htm.

European Commission. 2012. "Living Well, within the Limits of Our Planet." Accessed January 29, 2014, http://ec.europa.eu/environment/pubs/pdf/factsheets/7eap/en.pdf.

Fuchs, Doris. 2013a. "Sustainable Consumption." In *The Handbook of Global Climate and Environmental Policy*, ed. Robert Falkner, 215–230. Hoboken, NJ: Wiley-Blackwell.

Fuchs, Doris. 2013b. "Theorizing the Power of Global Companies." In *The Handbook of Global Companies*, ed. John Mikler, 77–95. Hoboken, NJ: Wiley-Blackwell.

Fuchs, Doris A., and Sylvia Lorek. 2000. "An Inquiry into the Impact of Globalization on the Potential for 'Sustainable Consumption' in Households." Report presented at the Workshop on Sustainable Household Consumption: Impacts, Goals, and Indicators for Energy-Use, Transport and Food, November. Accessed September 9, 2014, http://doc.utwente.nl/67163/1/Fuchs00inquiry.pdf.

Galbraith, John Kenneth. 1958. *The Affluent Society*. Boston: Houghton Mifflin.

Geels, Frank, and Johan Schot. 2007. "Typology of Sociotechnical Transition Pathways." *Research Policy* 36:399–417. doi:10.1016/j.respol.2007.01.003.

Geerken, Theo, and Mads Borup, eds. 2009. *System Innovation for Sustainability 2: Case Studies in Sustainable Consumption: Mobility*. Sheffield, UK: Greenleaf Publishing.

Goodman, David, and Michael Goodman. 2001. "Sustaining Foods: Organic Consumption and the Socio-Ecological Imaginary." In *Exploring Sustainable Consumption: Environmental Policy and the Social Sciences*, ed. Maurie Cohen and Johan Murphy, 97–119. Bingley, UK: Emerald Publishing Group.

Goodwin, Neva, Richard Rosen, and Allen White. 2011. "Principles for a New Economy." Accessed January 28, 2014, http://ase.tufts.edu/gdae/advancing_theory/principlesforaneweconomy.pdf.

Graeber, David. 2013. "A Practical Utopian's Guide to the Coming Collapse." *Baffler* 22:23–35. doi:10.1162/bflr_a_00129.

Green, Ken, and Philip Vergragt. 2002. "Towards Sustainable Households: A Methodology for Developing Sustainable Technological and Social Innovations." *Futures* 34:381–400. doi:10.1016/s0016-3287(01)00066-0.

Harris, Jonathan. 2013. "The Macroeconomics of Development without Throughput Growth." In *Innovations in Sustainable Consumption: New Economics, Socio-Technical Transitions, and Social Practices*, ed. Maurie Cohen, Halina S. Brown and Philip J. Vergragt, 31–47. Northampton, MA: Edward Elgar.

Hopkins, Rob. 2008. *The Transition Handbook: From Oil Dependency to Local Resilience*. Cambridge, UK: Green Books.

Hubert, A., ed. 2010. "Empowering People, Driving Change: Social Innovation in the European Union." Brussels: BEPA–Bureau of European Policy Adviser.

ICLEI. 2015. "ICLEI: Local Governments for Sustainability." Accessed February 25, 2015, http://www.iclei.org.

International Coalition for Sustainable Production and Consumption. 2002. *Waiting for Delivery*. Accessed September 2, 2012, http://icspac.net/files/WaitingForDeliveryReport.pdf.

International Institute for Sustainable Development. 1995. "The Imperative of Sustainable Production and Consumption." *Oslo Roundtable on Sustainable Production and Consumption*. Accessed January 21, 2014, http://www.iisd.ca/consume/oslo004.html.

Jackson, Tim. 2006. *The Earthscan Reader in Sustainable Consumption*. Abingdon, UK: Earthscan.

Jackson, Tim. 2009. *Prosperity without Growth: Economics for a Finite Planet*. Abingdon, UK: Earthscan.

Jackson, Tim, and Peter Victor. 2013. *Green Economy at Community Scale*. Toronto: Metcalf Foundation.

Kelly, Marjorie. 2012. *Owning Our Future: The Emerging Ownership Revolution: Journeys to a Generative Economy*. San Francisco: Berrett Koehler Publishers.

Kettering, Charles F. 1929. "Keep the Consumer Dissatisfied." *Nation's Business* 17 (1): 30–31, 79. http://www.wwnorton.com/college/history/archive/resources/documents/ch27_02.htm Accessed January 28, 2014.

Lahlou, Saadi, ed. 2011. *System Innovation for Sustainability 4: Case Studies in Sustainable Consumption: Energy Use and the Built Environment*. Sheffield, UK: Greenleaf Publishing.

Layard, Richard. 2011. *Happiness: Lessons from a New Science*. London: Penguin Books.

Lorek, Sylvia, and Doris A. Fuchs. 2011. "Strong Sustainable Consumption Governance: Precondition for a Degrowth Path?" *Journal of Cleaner Production* 38:36–43. doi:10.1016/j.jclepro.2011.08.008.

Marcuse, Herbert. 1964. *One-Dimensional Man: Studies in the Ideology of Advanced Industrial Society*. Boston: Beacon.

Marx, Karl. 1867. *Capital: Critique of Political Economy*. Hamburg: Verlag von Otto Meissner.

Max-Neef, Manfred A. 1991. *Human Scale Development: Conception, Application, and Further Reflections*. New York: Apex Press.

Miller, Ethan. 2010. "Solidarity Economy: Key Concepts and Issues." In *Solidarity Economy I: Building Alternatives for People and Planet*, ed. Emily Kawano, Tom Masterson, and Jonathan Teller-Ellsberg, 25–42. Amherst, MA: Center for Popular Economics.

Mont, Oksana, Aleksi Neuvonen, and Satu Lähteenoja. 2014. "Sustainable Lifestyles 2050: Stakeholder Visions, Emerging Practices, and Future Research." *Journal of Cleaner Production* 63:24–32. doi:10.1016/j.jclepro.2013.09.007.

Mont, Oksana, and Kate Power. 2009. "Understanding Factors That Shape Consumption." ETC/SCP Working Paper No 1/2013. Accessed February 5, 2014, http://scp.eionet.europa.eu/publications/wp2013_1/wp/wp2013_1.

North American Roundtable on Sustainable Production and Consumption. 2014. "Advancing Sustainable Production and Consumption in North America." Accessed September 16, 2014. http://narspac.weebly.com.

Nussbaum, Martha. 2011. *Creating Capabilities: The Human Development Approach*. Cambridge, MA: Belknap Press.

Organisation for Economic Co-operation and Development. 1997. *Sustainable Consumption and Production*. Paris: Organisation for Economic Co-operation and Development.

Packard, Vance. 1957. *The Hidden Persuaders*. New York: David McKay Company.

Piketty, Thomas. 2013. *Capital in the Twenty-First Century*. Cambridge, MA: Harvard University Press.

Princen, Tom. 2005. *The Logic of Sufficiency*. Cambridge, MA: MIT Press.

Princen, Tom, Michael Maniates, and Ken Conca, eds. 2002. *Confronting Consumption*. Cambridge, MA: MIT Press.

Rees, William. 2010. "What's Blocking Sustainability? Human Nature, Cognition, and Denial." *Sustainability, Science, Practice, and Policy* 6 (2). Accessed December 23, 2016, http://www.gci.org.uk/Documents/BlockingSustainability(Final0910).pdf.

Riesman, David, Nathan Glazer, and Reuel Denney. (1950) 2001. *The Lonely Crowd: A Study of the Changing American Character* . New Haven, CT: Yale University Press.

Ritzer, George, and Nathan Jurgenson. 2010. "Production, Consumption, Prosumption: The Nature of Capitalism in the Age of the Digital 'Prosumer.'" *Journal of Consumer Culture* 10 (1): 13–36. doi:10.1177/1469540509354673.

Røpke, Inge. 2013. "Ecological Macroeconomics: Implications for the Role of Consumer-Citizens." In *Innovations in Sustainable Consumption: New Economics, Socio-Technical Transitions, and Social Practices,* ed. Maurie Cohen, Halina S. Brown and Philip J. Vergragt, 48–64. Northampton, MA: Edward Elgar.

Sanne, Christer. 2002. "Willing Consumers—or Locked-In? Policies for a Sustainable Consumption." *Ecological Economics* 42 (1–2): 273–287. doi:10.1016/s0921-8009(02)00086-1.

Schneider, François, Giorgos Kallis, and Joan Martinez-Alier. 2010. "Crisis or Opportunity? Economic Degrowth for Social Equity and Ecological Sustainability." Special issue, *Journal of Cleaner Production* 18 (6): 511–518. doi:10.1016/j.jclepro.2010.01.014.

Schneider, François, Joan Martinez-Alier, and Giorgos Kallis. 2011. "Sustainable Degrowth." *Journal of Industrial Ecology* 15:654–656. doi:10.1111/j.1530-9290.2011.00388.x.

Schor, Juliet B. 1998. *The Overspent American: Why We Want What We Don't Need.* New York: Harper Perennial.

Sekulova, Filka, Giorgos Kallis, Beatriz Rodriguez-Labajos, and François Schneider. 2013. "Degrowth: From Theory to Practice." *Journal of Cleaner Production* 38:1–6. doi:10.1016/j.jclepro.2012.06.022.

Seyfang, Gill. 2009. *The New Economics of Sustainable Consumption.* New York: Palgrave Macmillan.

Seyfang, Gill, and Adrian Smith. 2007. "Grassroots Innovations for Sustainable Development: Towards a New Research and Policy Agenda." *Environmental Politics* 16 (4): 584–603. doi:10.1080/09644010701419121.

Shove, Elizabeth. 2003. *Comfort, Cleanliness, and Convenience: The Social Organization of Normalcy.* Oxford: Berg.

Skidelsky, Robert, and Edward Skidelsky. 2012. *How Much Is Enough: Money and the Good Life.* New York: Other Press.

Spaargaren, Gert. 2003. "Sustainable Consumption: A Theoretical and Environmental Policy Perspective." *Society and Natural Resources* 16 (8): 687–701. doi:10.1080/08941920309192.

Spaargaren, Gert. 2013. "The Cultural Dimension of Sustainable Consumption Practices: An Exploration in Theory and Policy." In *Innovations in Sustainable Consumption: New Economics, Socio-Technical Transitions, and Social Practices*, ed. Maurie Cohen, Halina S. Brown, and Philip J. Vergragt, 229–251. Northampton, MA: Edward Elgar.

Spangenberg, Joachim H. 2004. "The Society, Its Products, and the Environmental Role of Consumption." In *The Ecological Economics of Consumption*, ed. Lucia Reisch and Inge Røpke. Northampton, MA: Edward Elgar.

Spangenberg, Joachim H. 2014. "Institutional Change for Strong Sustainable Consumption: Sustainable Consumption and the Degrowth Economy." *Sustainability, Science, Practice, and Policy* 10 (1): 62–77. Accessed December 23, 2016, https://sspp.proquest.com/institutional-change-for-strong-sustainable-consumption-sustainable-consumption-and-the-degrowth-e587130f2666#.e9sxcolcb.

Speth, James Gustave. 2012. *America the Possible: Manifesto for a New Economy*. New Haven, CT: Yale University Press.

SPREAD. 2013. "The SPREAD Sustainable Lifestyles 2050 Project." Accessed February 5, 2014, http://www.sustainable-lifestyles.eu.

Stiglitz, Joseph, Amartya Sen, and Jean-Paul Fitoussi. 2009. *Report by the Commission on the Measurement of Economic Performance and Social Progress*. London: New Press.

Sunstein, Cass R. 2015. "Behavioural Economics, Consumption, and Environmental Protection." In *Handbook of Research on Sustainable Consumption*, ed. Lucia Reisch and John Thøgersen, 313–327. Northampton, MA: Edward Elgar Publishing.

Thaler, Richard H., and Cass R. Sunstein. 2008. *Nudge: Improving Decisions about Health, Wealth, and Happiness*. New Haven, CT: Yale University Press.

Tischner, Ursula, Eivind Sto, Unni Kjaerness, and Arnold Tukker. 2010. *System Innovation for Sustainability 3: Case Studies in Sustainable Consumption: Food and Agriculture*. Sheffield, UK: Greenleaf Publishing.

Tukker, Arnold, Martin Charter, Carlo Vezzoli, Eivind Sto, and Maj Munch Andersen, eds. 2008. *System Innovation for Sustainability 1: Perspectives on Radical Change to Sustainable Consumption and Production*. Sheffield, UK: Greenleaf Publishing.

Tukker, Arnold, Sophie Emmert, Martin Charter, Carlo Vezzoli, Eivind Sto, Maj Munch Andersen, Theo Geerken, et al. 2008. "Fostering Change to Sustainable Consumption and Production: An Evidence-Based View." *Journal of Cleaner Production* 16 (11): 1218–1225. doi:10.1016/j.jclepro.2007.08.015.

United Nations. 1992. Agenda 21. Accessed September 2, 2014, https://sustainable development.un.org/content/documents/Agenda21.pdf.

United Nations. 2002. *Report of the World Summit on Sustainable Development*. Johannesburg, South Africa. A/CONF.199/20. Accessed January 22, 2014, http://www.unmillenniumproject.org/documents/131302_wssd_report_reissued.pdf.

United Nations. 2015. "Sustainable Development Goals." Accessed December 21, 2015, http://www.un.org/sustainabledevelopment/sustainable-development-goals.

UN Environment Program. 2008. *North American Workshop on Sustainable Consumption and Production (SCP) and Green Building: 2008 Workshop Report*. Accessed January 31, 2014, http://www.rona.unep.org/documents/scp/part_4/Final_Meeting_report_Washington_DC.pdf.

UN Environment Program. 2011. *North American Workshop on Sustainable Consumption and Production (SCP) and Green Building: 2011 Workshop Report*. Accessed January 31, 2014, https://scpgreenbuild.files.wordpress.com/2011/04/10-april-n-am-scp-workshop-report.pdf.

UN Environment Program. 2014a. "About the Marrakech Process: Towards a Global Framework for Action on Sustainable Consumption and Production." UNEP Sustainable Consumption and Production Branch. Accessed September 17, 2014, http://www.unep.fr/scp/marrakech/about.htm.

UN Environment Program. 2014b. "What Are the 10YFP Programmes?" Accessed February 21, 2015, http://www.unep.org/10yfp/Programmes/Whatarethe10YFPProgrammes/tabid/106264/Default.aspx.

Veblen, Thorstein. 1899. *Theory of the Leisure Class*. Accessed January 28, 2014, http://www.gutenberg.org/files/833/833-h/833-h.htm.

Vergragt, Philip J. 2013. "A Possible Way Out of the Combined Economic-Sustainability Crisis." *Environmental Innovation and Societal Transitions* 6:123–125. doi:10.1016/j.eist.2012.10.007.

Vergragt, Philip, Lewis Akenji, and Paul Dewick. 2014. "Sustainable production, consumption, and livelihoods: Global and regional research perspectives." *Journal of Cleaner Production* 63:1–12. doi:10.1016/j.jclepro.2013.09.028.

Vergragt, Philip J., and Jaco Quist. 2011. "Backcasting for Sustainability: Introduction to a Special Issue." *Technological Forecasting and Social Change* 78 (5): 747–755. doi:10.1016/j.techfore.2011.03.010.

Weber, Max. (1905) 1992. *The Protestant Ethic and the Spirit of Capitalism*. Translated by Talcott Parsons. London: Routledge.

Wilkinson, Richard, and Kate Pickett. 2009. *The Sprit Level: Why Greater Equality Makes Societies Stronger*. New York: Bloomsbury Press.

World Business Council for Sustainable Development (WBCSD). 2008. *Sustainable Consumption: Facts and Trends*. Accessed September 3, 2014, http://www.wbcsd.org/contentwbc/download/479/5182.

World Business Council for Sustainable Development (WBCSD). 2010. *Vision 2050: The New Agenda for Business*. Accessed December 26, 2016, https://archive.epa.gov/wastes/conserve/tools/stewardship/web/pdf/vision2050.pdf.

World Business Council for Sustainable Development (WBCSD). 2011. *A Vision for Sustainable Consumption: Innovation, Collaboration, and the Management of Choice*. Accessed December 26, 2016, http://wbcsdservers.org/wbcsdpublications/cd_files/datas/business-solutions/consumption/pdf/AVisionForSustainableConsumption.pdf.

14 Conceptual Innovation and the Future of Environmental Policy

James Meadowcroft and Daniel J. Fiorino

This volume has drawn attention to the importance of conceptual innovation in the development of environmental policy. Following two introductory chapters, readers were presented with a set of individual concept studies, each of which traced the career of an important environmental policy concept, examined its ideational and practical ramifications, and considered the face it presents to the future. This chapter will synthesize themes that emerged from the preceding discussion and provide some concluding reflections. It focuses on two broad issues. First, it considers the light these individual studies shed on understanding processes of conceptual innovation in environmental policy. Second, it examines what these cases (and attention to concepts and conceptual innovation more generally) can tell us about the underlying structure and evolution of the environmental policy domain. Before turning to these two issues, however, we start with a brief review of key findings related to the eleven concepts explored in the preceding chapters and the place they occupy in environmental policy argument.

Highlights from the Conceptual Narratives

Each of the chapter authors presented a high-level but nuanced analysis of the career of a specific environmental concept. Central elements of these individual narratives are highlighted below.

The environment emerged in political and policy discourse in the final third of the twentieth century. It was applied to surroundings (built and natural) important for human health and economic well-being, but that were being damaged by human activity. The threatened environment constituted an object for care, with citizen activists demanding action and governments gradually incorporating environmental protection into their core mandates, although to varying degrees. Today the environment is taken for granted as a natural focus for political argument, government regulation, and public expenditure. The concept is anthropocentric (*the* environment is the *human* environment), but emphasizes the interdependence between society and its biophysical context. According to James Meadowcroft, the idea embodies continuing tensions over the extent to which it is understood as "natural" or a product of human activity, something separated from or partly constitutive of human identity, a threatened victim or a site of human flourishing and creativity. Environment is the metaconcept that provides the ideational foundation for the entire sphere of environmental policy. Four other concepts examined in this book draw explicitly on this notion: environmental assessment, environmental risk, environmental justice, and environmental security.

Environmental assessment has come to refer to a specific policy instrument: a formal procedure for examining projects (but also policies and programs) to evaluate potential impacts. The idea was first institutionalized in the 1969 US National Environmental Policy Act, obliging government departments to analyze and make available to the public the environmental implications of proposed actions as well as consider alternatives. As Richard N. L. Andrews shows in his chapter, it was originally intended as an "action-forcing" measure to ensure that policy internalized environmental concern. Yet the US courts curtailed the substantive ambitions of the instrument, and environmental assessment spread internationally in a tamer, largely procedural form. Such assessments are now a routine element of the practice of environmental governance, not only in the United States but around the world, too.

Environmental risk is a mesoscale concept associated with the evaluation and mitigation of hazards to human health and the environment. The concept came to the fore in the late 1970s and 1980s, initially appealing to US regulators as a way to professionalize and depoliticize the management of environmental problems. Causal models and probabilistic assessments could link the anticipated frequency of an adverse outcome with potential impacts to provide a risk assessment that could be used to ground decisions over priorities and (when integrated with cost-benefit analysis) identify a proportionate policy response. While environmental risk assessment, management, and communication remain staples of contemporary environmental governance in both the public and private sectors, there are many practical difficulties. Michael E. Kraft points to the limitations of applying the concept in the context of pervasive uncertainty and limited knowledge of ecosystem processes. He further suggests that in responding to global environmental problems such as climate change or biodiversity loss, the idea should be linked to broader concepts such as sustainable development.

Environmental justice has served primarily as an insurgent concept, invoked by policy outsiders to mobilize opposition to the inequitable treatment of disadvantaged groups with respect to the distribution of environmental harms and benefits. The concept emerged in the United States, where it linked environmental issues to established movements fighting racial and class inequalities. Grassroots organizations concerned with exposures to toxic chemicals, air and water pollution, and urban decay invoked the claim to environmental justice to ground demands for change. The idea is linked to a series of allied concepts related to environmental entitlements and distribution, including environmental rights, environmental equity, ecojustice, climate justice, and climate equity. Karen Baehler argues that the concept of environmental justice has been relatively successful in identifying problems and mobilizing community-based protest, but has achieved only limited success with respect to the construction and implementation of solutions in the policy sphere.

Environmental security links two established policy spheres—suggesting that environmental problems can constitute security threats while offering security as a lens through which to view environmental challenges. Initially proposed by environmentalists worried that ecosystem degradation menaced societal well-being and survival, in the post–Cold War context it captured the attention of security specialists looking to define emerging threats to international stability. As ideas of climate disruption have been taken more seriously (including threats to food supplies and civilian infrastructure along with the possibilities of climate refugees, regime instability, or even "resource wars"), military planners have also been drawn into the discussion. Although the concept dramatizes issues such as climate change, making them intelligible to core state concerns with order and stability, Johannes Stripple worries that the "violent imaginary" associated with the security discourse—that presents the environment as a source of tension, conflict, suffering, and war—can be counterproductive, hardening the environmental sphere rather than greening the security domain.

Sustainable development points to an alternative development trajectory that takes account of environmental limits, and promotes inter- and intragenerational equity. Propelled to international attention by the 1987 Brundtland Report, the concept was officially endorsed by world leaders at the 1992 Rio Earth Summit. It embodies the core insight that environmental and development problems are irrevocably intertwined. Human societies are already pressing against ecological limits; meeting the urgent development needs of the poor will inevitably intensify such pressures. Discussing the complexities associated with the concept and the many criticisms to which it has been subjected, Oluf Langhelle argues that the problems it raises will remain at the heart of international politics for the foreseeable future.

Biodiversity denotes variation in the living world—within species, between species, and among ecosystems. Coined in the mid-1980s by biologists deeply concerned about extinction rates driven by human activity, the idea diffused rapidly within the scientific and policy communities. It was embedded institutionally in the 1992 Convention on Biodiversity, and is now at the center of debates over human impacts on the biosphere and societal dependence on natural systems. Biodiversity has become closely meshed with ideas of resilience, ecosystem services, natural capital, and adaptive management. Yrjö Haila contends that government agencies charged with nature protection have successfully invoked the urgency of protecting biodiversity to expand conservation initiatives. Yet he worries that the concept's comprehensiveness and ambiguity are also weaknesses. In particular, the political and policy objectives loaded onto the concept (that posit, for example, immediate and comprehensive prevention of further ecosystem loss) appear impossibly broad. Today almost every human activity threatens to erode some dimension of biodiversity, and more attention should be paid to socioecological interdependence and the forms of biodiversity that are essential to, and/or compatible with, human flourishing.

The concept of *critical loads* refers to the maximum pollutant burden an ecosystem can tolerate without suffering serious degradation. It emerged from efforts to establish a scientific foundation for policy action to manage acidification driven by air pollutant deposition. The concept was closely linked to the development of integrated assessment models, which allowed scientists and negotiators to appreciate the environmental impacts and economic costs of various courses of action. It played a central role in the negotiation of the revised (1999) sulfur protocol of the UN Economic Commission for Europe's Convention on Long-Range Transboundary Air Pollution (CLRTAP). Karin Bäckstrand suggests critical loads has been so successful because of productive ambiguity that allows it to be simultaneously understood as identifying critical thresholds, carrying out risk assessment, and generating cost-effective policy responses. Yet she also points to challenges going forward, including the disturbing reality that air pollution remains a significant cause of mortality and other health effects within the UN Economic Commission for Europe's region.

Adaptive management is a mesolevel concept applied to the governance of landscape and natural resource systems under conditions of uncertainty. It stresses structured learning about ecosystems and the continuous adjustment of management strategies as knowledge accumulates. The emergence of the concept in the late 1970s and 1980s was linked to changing scientific understandings of ecosystem evolution—from perceptions of linear movement toward climatic communities at equilibrium, to more dynamic and regenerative models with multiple equilibriums and great uncertainty. It offered a critique of the failure of traditional resource management practices based on reductionist assumptions along with optimizing output of one or a few target species. Taken up by natural resource management agencies and nongovernmental organizations in many countries, adaptive management has also been associated with the involvement of stakeholders in interactive management processes that are meant to contribute to social and ecosystem resilience. Judith A. Layzer and Alexis Schulman leave open the question of whether adaptive management is most suitable for well-bounded resource systems or large landscape-scale problems. Yet they maintain that in many respects, it is the only game in town.

The *green economy* promises increasing levels of human well-being while protecting ecosystems. The essential idea is that environment and economy can be mutually supportive: protecting the environment is good for the economy while sound economic development promotes environmental integrity. A green economy would be resource efficient and low carbon; it would safeguard critical environmental assets while creating jobs as well as promoting social welfare and equity. Although the concept has a long genesis, it is relatively new to the world stage, having been actively promoted by international organizations such as UN Environment Program (UNEP) and the World Bank in the run up to the Rio+20 event in 2012. It is linked to concepts such as ecosystem services, natural capital, biodiversity, resource efficiency, and the decoupling of environmental burdens from economic growth. Controversies around the concept involve the relative importance accorded to social objectives, and extent to which it adequately problematizes dominant understanding of economic growth and environmental limits. According to Daniel J. Fiorino, the idea could be strengthened by a greater emphasis on economic equity, nonmaterial contributions to welfare, and the preservation of irreplaceable ecological assets and services.

Sustainable consumption draws attention to the massive expansion of material consumption associated with modern lifestyles, attendant waste and destruction of environmental resources, and pervasive inequalities between the consumption of rich and poor. "Unsustainable patterns of production and consumption" were criticized at the 1992 Rio Earth Summit, and twenty years later the attainment of "sustainable consumption and production patterns" was defined as the twelfth of the sixteen United Nations' "Sustainable Development Goals." In the interim, sustainable consumption has had an uneven record, struggling to find a home in the policy world, although it has represented a more active site for academic research and social movement intervention. The idea draws strength from long-standing critiques of materialistic and consumerist society, linking this to contemporary concerns with the environment and global inequalities. Philip J. Vergragt identifies a series of tensions and ambiguities associated with the concept, including whether to focus on individual consumption choices or the social structuring of consumption, and the relative emphasis accorded to economic or cultural critique. He notes the opposition between "strong" and "weak variants," with the former linked to a radical critique of the existing economic system and cultural values, while the latter concentrates on more pragmatic business or consumer-led adjustments to contemporary lifestyle impacts. Such tensions, together with difficulty the concept has experienced in generating popular enthusiasm or penetrating policy circles, lead him to wonder about its long-term prospects.

Understanding Conceptual Innovation in the Environmental Domain

The first chapter of this volume raised a number of questions about conceptual innovation in the environmental policy realm, including the circumstances in which innovation occurs and its practical ramifications. It introduced a simple model of the process that highlighted the shaping of an idea as a policy-relevant concept, its uptake and institutional embedding, competition among conceptual variants, and selective pressures that influence a concept's evolution. And it suggested three factors that were particularly important for the successful introduction of novel ideas into the policy world: the response to a perceived need; the ability to appeal to a variety of constituencies; and some affinity with dominant ideas and practices. We will now return to these issues.

The Innovation Process

With respect to the *process* of innovation, casting an idea in policy-relevant terms is critical to its successful launch. The use of *environment* to denote surroundings and highlight organism/context interdependence was well established in the first half of the twentieth century. But it was bringing the human environment into focus, stressing the threats from human activities, and pointing to potentially serious consequences for health and welfare (as well as for the natural world itself) that transformed the idea in a way that allowed it to anchor political and policy actions. Similarly, critical loads provided a mechanism to link regions generating and receiving air pollution, using a quantitative and science-based assessment process that left room for political interpretation and negotiation. In each case, we have seen how the emerging concept was shaped deliberately to be operative in the policy world. Often it is possible to identify the specific individuals or groups that played the decisive role in such policy-relevant conceptual design, and/or the meetings or publications that initially propagated the new idea (see, for example, chapter 9 on biodiversity). Yet sometimes it is difficult to find a single point of origin, as the policy launch can involve a series of contributions spread over time. Here the concept of the environment serves as an illustration.

Conceptual innovation is in one sense always unique: each process brings forth a novel idea, and only detailed examination can determine the circumstances of its birth and subsequent evolution. On the other hand, common strategies are used to make new ideas from old. Among those we have seen here are *up-framing*, where a larger idea integrates elements of a series of preexisting discrete ideas to create a concept with broader significance that lifts and shifts existing perceptions. Environment, biodiversity, and sustainable development are examples here. *Decomposing* is in some sense the opposite move, where distinct aspects are teased out from a broader concept. Thus climate change comes to embrace "climate mitigation" and "climate adaptation," or sustainable development is revealed to contain "three pillars." In chapter 2, we discussed the process of generating new concepts by coupling environment (or environmental) with other terms. In some cases, the result is not just to bring out the environmental dimension of a particular problem or practice but also to create a new lens with which to view environmental policy itself. This is the case with environmental risk, environmental justice, and environmental security.

The institutional embedding of the concepts examined here has taken place in different ways and to varying degrees. Sustainable development is now rooted in the UN system, EU governance mechanisms, and the national institutions of many countries. Critical loads has been built into the CLRTAP process. In contrast, the institutionalization of the green economy is more recent and fragile. The concept can be found in the work programs of some international organizations (particularly UNEP) and policy frameworks in a selection of national governments (South Korea). But its staying power remains to be established. Of the other concepts studied here, environmental justice, sustainable consumption, and environmental security are among those less firmly anchored in the official policy world. Yet in each case, there has been sufficient embedding to allow the concept to play some role in policy argument, and not just in general societal debate.

Competition among conceptual variants is ubiquitous, and apparent to some extent in each of the narratives presented in earlier chapters. In the case of environmental security, there are tensions between understandings that privilege *human security* (for example, protecting the livelihoods of the most vulnerable populations from environmental crises), and mainstream *national security* approaches that emphasize conflict and threats to order. Alternative perspectives on sustainable consumption may prioritize the social structuring of consumption or the consumption decisions of individuals. Consider also ongoing arguments over the meaning of *the Anthropocene* or *sustainability science*. Although institutional embedding performs an important role in stabilizing dominant understandings, alternative perspectives on what a concept "really" implies can play out over decades.

At the outset we referred to selective pressures that influence a concept's development. Here we had in mind factors rooted in the ideational—but also practical political and economic—contexts that enable, constrain, or shape the uptake of certain configurations of meaning and practice. There are pressures that encourage innovation to take some routes and not others. So one can think in terms of a "selection environment" within which a particular concept makes its career, or a changing "landscape" within which an idea evolves. Either way, the point is that environmental concepts are shaped by struggles among actors who themselves are anchored in broader contexts of established norms and expectations, institutional hierarchies and entitlements, and social practices and conflicts. Thus the idea of adaptive management focused initially on experimental approaches to resource governance grounded in more dynamic conceptions of ecosystem development. But there were powerful pressures to also include the stakeholder-interactive dimension as a core feature, due to the institutional contexts where adaptive management was taken up and the evolution of political contestation. Moreover, truly experimental designs have proven challenging to implement (due in part to the resistance of risk-adverse managers and environmentalists), so there has been a tendency to adopt a looser understanding of what the concept entails. More generally, there are always pressures to clip the wings of novel ideas and sand off edges that rub against accepted ways of doing things. In the early days of environmental assessment in the United States, for instance, actors in business and government attempted to reduce the practice to a legal formality. Political struggles culminating in important court decisions ultimately defined a largely procedural norm that was far from trivial, but that fell short of the original ambition.

Turning to the question of factors that facilitate a novel concept's acceptance into the policy domain, meeting a perceived need is the most obvious. To catch on, a concept must appear to fill some void. It articulates something actors believe needs to be said, but that previously could not be expressed so directly or explicitly. In the environmental policy realm, this typically relates to the identification of novel problems and solutions, with the new concept promising to make issues more *intelligible* and *tractable*. Adaptive management presents an alternative to traditional resource governance: it filled a gap that became manifest as the critique of established resource management theory and practice took shape. The green economy provides a vision of an economy that avoids the environmentally destructive practices of today, but represents a recognizable continuity with existing institutions. One can think of such ideas as "filling a gap" in the conceptual field, so long as it is clear that the critique of existing ideas and the articulation of an alternative perspective is what opens up the gap that the novel concept fills.

Another feature of concepts that become established in the environmental policy sphere is the capacity to appeal to multiple constituencies. Environmental policy involves interactions among different sorts of actors—elected officials, civil servants, the media, scientists, businesses, and civil society organizations. This was true even in the days of closed policy communities. Innovations that stick must speak to multiple groups. All the concepts we have examined possess this property to some extent. Constructive ambiguity helps in this regard: a certain interpretative flexibility allows a wider range of actors to weave the new concept into perspectives they wish to articulate. Another contributory element is the advantage of resonance with everyday language. Paradoxical as it may seem, conceptual novelties can gain traction by drawing on the familiar, invoking terms that are *already* in common use, but combining them and shifting their meanings in new directions. Although more esoteric or technically demanding expressions appear in specialized disciplines and practices (scientific discourse, legal documents, detailed administrative regulations, etc.), broader policy concepts typically employ relatively familiar elements, even if the conceptual construct is ultimately rather complex. Think of "adaptive" and "management," "environment" and "security," "critical" and "load," "precaution" and "principle," "triple" and "bottom line," and so on.

Finally, there is the issue of a concept's "fit" with broader practices, structures, and policy discourse. This links to the idea of "selection pressures" discussed above. It helps explain the difficulties that concepts such as sustainable consumption or environmental justice have experienced in catching on in official circles. They sit somewhat awkwardly with dominant political and economic structures and perspectives. Sustainable consumption, for example, raises issues about the shaping of individual and collective needs, the continued expansion of material consumption, equity in living standards, and the legitimate role of the state in steering consumption decisions that can challenge economic and political orthodoxy regarding "consumer sovereignty," corporate power, and the growth economy. The tensions are even clearer in relation to the idea of *degrowth*, which has been essentially shut out of the policy circuit. The importance of fit is also demonstrated by the largely additive nature of the set of concepts—particularly the mesolevel analytic or management concepts—invoked in the modern environmental policy sphere. Novel concepts that have successfully entered the policy realm have not typically been presented as outright substitutes for established concepts; instead, they have been integrated as complements that can coexist with previously institutionally embedded understandings.

The Context and Consequences of Innovation

So this leads us back to the general issues with which we began. What gives rise to conceptual innovation in the environmental policy realm? Who innovates? In what context? And what are the consequences? At its core, conceptual innovation is driven by dissatisfaction with existing policy orientations, processes, and outcomes. Established ways of seeing and doing are criticized, and new concepts reflect these critiques and come to embody alternative approaches. Each of the concepts considered in this volume was advanced by actors engaging with policy problems and trying to effect some form of adjustment. At the outset we have frustration with existing arrangements, a diagnosis of perceived political and policy failure, and an impulse to bring something new to the discussion. So while conceptual innovation in environmental policy is in one sense a movement in the realm of ideas, it is driven by a perception of practical consequences. New ideas gain traction because they appear to offer advantages in managing practical political and policy dilemmas related to the environment.

To say innovation in conceptual categories is a response to practical problems is not to deny that those who promote innovation may be animated by diverse motives, including the desire to further individual careers or engage in bureaucratic turf wars. Nor is the suggestion that the goal of those promoting novel environmental concepts is always the more perfect protection of the natural world. What one can say is that the process is driven by attempts to redefine issues so as to make problems more "manageable." Actors are looking for new ways to approach issues, communicate with broader audiences, handle political conflicts, and reorient policy. But this leaves open the question of exactly how matters are to be rendered manageable and to what end.

Who, then, are the innovators? Although many individuals and groups are involved in the broad process, those responsible for the initial construction of a novel policy concept (or subsequent major readjustments) and early institutional embedding appear particularly significant. The studies in this volume highlight the role of scientists, government officials, the staff of international organizations, politicians, and activists. Yet it is not clear that any one of these groups is systematically more important than the others. Interactions involving issue experts from multiple professional backgrounds appear typical, with their specific roles varying in relation to particular concepts. In other words, the "expertise" of those most closely involved in conceptual innovation is drawn from multiple realms of social experience (science, public management, political action, international negotiations, etc.).

This multipartite contribution links to the contexts in which innovation occurs, which often involve attempts to wrestle with long-running policy problems, sometimes in settings where different kinds of expertise can interact, where tension and conflict among alternative perspectives and interests are manifest, but there is pressure to achieve some form of closure or output. Sustainable development emerged from the process associated with the UN Brundtland Commission; critical loads gelled in the CLRTAP negotiations; environmental justice emerged from interaction among movement activists and sympathetic academic and legal scholars.

How do these observations relate to the categories frequently invoked in the policy literature, which in relation to change variously underscore the significance of "advocacy coalitions" (Sabatier 1988), "epistemic communities" (Haas 1989), "policy networks" (Marsh and Rhodes 1992), and "discourse coalitions" (Hajer 1997)? The straightforward answer seems to be: not directly. Each of these sorts of groups could be said to contribute to developing and promoting novel concepts, or rethinking existing ideas. But it is unclear that any of these theoretical constructs sheds definitive light on the processes examined here. This, however, is a topic that should be explored in further research on conceptual innovation, especially in understanding their effects on policy processes and outcomes.

Lastly, there is the matter of practical implications: To what extent does conceptual innovation matter for policy outputs and ultimately policy outcomes? Our sample of concepts admittedly was skewed from the outset toward those that have had high visibility in the environmental policy realm. And they were deliberately selected to represent different kinds of concepts (for example, metaconcepts and mesolevel analytic and management concepts) as they are invoked in different levels of policy debate (general argument, expert forums, etc.). That said, most of the case studies illustrated quite specific ways in which the shift in conceptual categories was associated with a change in policy orientation and practical outcomes. Actors embraced new ideas to make policy-relevant arguments in order to secure a shift in the way the environment was actually governed. And in fact things did then come to be done differently.

At the broadest level, the environment came to anchor a new field of policy making that drew together existing activities, but over time substantially increased the control states exercise over activities deemed to have an environmental bearing. Environmental assessment introduced a new environment-focused practice of mandated and systematic ex ante reviews of impacts from projects, policies, or programs. Critical loads grounded the formulation of quantitative international air pollution abatement programs. And so on. Rather than saying conceptual innovation caused these practical changes, it is perhaps more accurate to assert that conceptual innovation was an essential element of the process of reflection, political struggle, and policy adjustment that led to altered governance practices and ultimately changed policy outcomes.

Conceptual Change and the Evolution of Environmental Policy

Finally, we return to the more general discussion of the longer-term evolution of the conceptual field that was introduced at the end of chapter 2. There we noted that over time, the range of concepts invoked in environmental policy has broadened dramatically. A greater diversity of concepts are now being applied to manage a vastly more complex array of issues. Moreover, these concepts reflect a deeper appreciation of the societal entanglement of environmental problems, which are now understood to reach across economic sectors, involve producers and consumers, and be deeply entwined with core economic and technological practices. Concepts have also come to articulate a deeper understanding of the scale of social adjustment and policy intervention that will be required to manage environmental issues. Another observation was that environmental concepts have been marked by more general political/ideological changes, especially phenomena such as the increased skepticism toward the state's capacity to implement solutions to societal problems, globalization, and a more active involvement of business corporations and civil society organizations in international governance. And despite continued change and innovation, there is also evidence of substantial conceptual continuity. Many of the ideas raised at the original 1972 Stockholm Conference on the Human Environment continue to echo down through subsequent decades. This continuity has been cemented by the early embedding of concepts in governance routines and law even as new ideas continue to come to the fore. But it also reflects underlying structural circumstances driving environmental issues onto the policy agenda, which has remained relatively constant over more than half a century. To some degree, these issues are reflected in key themes that reoccur across the conceptual landscape. Four such crosscutting themes that can be seen in our inquiry are science and policy, environmental limits, economy and environment, and environmental equity.

Science and Policy

In a sense, all the critical organizing concepts in the environmental realm trade, at least indirectly, on findings from the natural sciences. Evidence from modern science is crucial for identifying problems, establishing causal linkages, and implementing solutions. But for some concepts the connection is more direct. Problem-defining concepts, such as biodiversity loss or climate change, are directly underpinned by science. Their legitimacy rests on scientific credentials: research is systematically extending knowledge about the physical and social processes involved, and scientists have been active in arguing that these issues should be of concern to policy makers. Some mesolevel analytic and management concepts articulate a still more explicit vision of the science-policy interaction. Among the concepts examined in detail here, this is particularly evident for environmental assessment, environmental risk, critical loads, and adaptive management.

Scientific knowledge is assumed to inform the environmental assessment process evaluating the ex ante impacts of planned interventions. But such assessments do not typically drive the extension of scientific knowledge; instead, established understandings are taken "off the shelf" to complete individual evaluations. Early formulations of environmental risk articulated a rather-unproblematic separation between science (risk assessment) and policy decisions (risk management). With time, the interpenetration of normative and scientific elements in risk governance has been more widely recognized. Critical loads offers science-based analysis of ecosystem pressures and vulnerabilities to underpin transnational air pollution governance. Yet the concept also allows officials substantial interpretative flexibility to determine the scope and pace of remedial action—a factor that has been central to its appeal. Adaptive management acknowledges uncertainties inherent in attempts to manage ecosystems, and suggests that "experiments" and formal modeling—two pillars of the scientific endeavor—should be at the heart of resource governance. Yet in practice, these characteristics have been substantially diluted. Within the wider canvas of environmental policy, other concepts that speak to science policy interaction include the precautionary principle, ecosystem services, and planetary boundaries. (Table 14.1 presents a summary of the thematic focus of selected environmental concepts.)

Environmental Limits

Another concern reflected across the policy field is the idea of *limits*—that there are limits to the environment's capacity to absorb the impacts of human activity without adverse consequences, and that to prevent such harm, societies must set limits to environmentally damaging behavior. Among the concepts examined in this volume, limits are most explicitly articulated in the notion of critical loads. As Bäckstrand (this volume) writes, the core idea is that "policies should be based on *ecosystem tolerance levels* ... so as to avoid long-term, significant change in the structure and function of ecosystems." Science can help determine key thresholds so that policy can be designed to roll back impacts into safer territory. The authors of *Our Common Future* defined limits as one of the two key ideas contained in sustainable development. As Langhelle (this volume) explained, these limits are "both flexible and absolute." They are flexible, because technology and human ingenuity can frequently overcome resource constraints or reduce environmentally destructive practices. But they are absolute, because irreversible damage to critical ecosystems (such as the effects of dangerous climate change) can erode the foundations for human development. Yet as

Table 14.1
Thematic Focus of Selected Environmental Concepts

Theme	Concepts examined here	Additional established policy concepts	Other concepts
Science/policy	Environmental risk Critical loads Adaptive management Environmental assessment	Precautionary approach/ principle Ecosystem services	Sustainability science
Limits	Critical loads Sustainable development Biodiversity Sustainable consumption Environmental security (Adaptive management) (Green economy)	Climate change Precautionary approach/ principle	Planetary boundaries Ecological footprint
Economy/ environment	Sustainable development Green economy Sustainable consumption (Adaptive management)	Decoupling Ecosystem services Natural capital Polluter pays principle Ecoefficiency Environmental policy integration	Circular economy
Equity, distribution	Sustainable development Environmental justice Sustainable consumption (Green economy)	Common but differentiated responsibilities Polluter pays principle	Climate justice

the concept of sustainable development has been deployed over the past few decades, the idea of absolute limits has been eclipsed, with the image of a "balance" among economic, social, and environmental goals more often finding favor (Meadowcroft 2013).

Understandings of limits are implicit in other concepts examined here. Biodiversity was formulated because of worries over unprecedented rates of species extinction. So ideas about irreversible loss and critical thresholds lie close to its core. Sustainable consumption engages with limits through questions about the long-term viability of profligate modern lifestyles, the critique of overconsumption, and a determination to avoid ecological overshoot. Concern that the existing economy—with its dependence on fossil fuels, wasteful patterns of resource use, and systematic destruction of ecological systems—faces real environmental constraints ground the call for a green economy, even if its advocates are often unclear precisely how these are to be drawn.

In the wider universe of environmental policy, the concept of *planetary boundaries*, with its aspirations to define a "safe operating space" for humanity, is perhaps the most eye-catching incarnation of limits since the 1970s' "limits to growth" debate. But it is the discussion of anthropogenic climate change that recently has contributed most to bringing environmental limits into focus. Evidence shows human intervention is already transforming the climate system, and this leads to questions about the implications for human society and the biosphere. Hence the political formulation of the two-degree limit for warming by the end of this century along with other limits-related concepts such as carbon budgets (Committee on Climate Change 2010), unburnable carbon (Jakob and Hilaire 2015), and so on.

Economy and Environment

The relationship between *economy and environment* forms a third broad locus of conceptual engagement. At some point in their practical application, virtually every environmental policy concept is influenced by or impinges on economic issues. Environmental assessments and environmental risk, for example, often factor into broader economic decision making. Moreover, concepts may evolve in ways that increase the relevance of economic issues; consider the progressive association of ecosystem valuation with the concept of biodiversity (Christie et al. 2006). On the other hand, the economic dimension is not *formally* part of the core preoccupations of environmental assessment, environmental risk, or biodiversity.

The three concepts examined here that have the most to say about these issues are sustainable development, the green economy, and sustainable consumption. Sustainable development tried to sidestep the perceived zero-sum opposition between economic growth and environmental protection by shifting the focus to a process of "development" that included progress in health, education, science and culture, and so on—dimensions that could in principle grow indefinitely. And it emphasized changing the "quality of growth" by paying greater attention to equity and environmental limits. The green economy takes economy and environment as the critical nexus that will determine the likelihood that the economic growth aspirations of all countries will be reconciled within the limits of local, regional, and global ecosystems. Sustainable consumption is another concept in which the economy-environmental relationship is central. Here the spotlight is turned on the problematic link between expanded material consumption and the enhancement of genuine individual and social welfare. It typically has a sharper critical edge than the dominant understanding of the green economy, raising questions about established economic structures, individual consumption choices, and authentic human needs. Concepts from the wider environmental policy lexicon that revolve around the economy-environment linkage include cost-benefit analysis, natural capital, ecosystem services (especially payment for ecosystem services), and the circular economy (European Commission 2015).

Environment and Equity

Finally, ideas about *equity* and the distribution of environmental harms and goods percolate across the policy field. In our subset of policy concepts, environmental justice takes this on most directly, deploring the inequities that impose disproportionate environmental burdens on vulnerable groups. But equity also plays a central role in the two sustainability concepts—sustainable development and sustainable consumption—with an implication that production and consumption policies in developed countries need to be tempered to enable economic progress in poor and developing ones. It also makes an appearance in some manifestations of the green economy—hence UNEP's (2012) "inclusive green economy."

Except for the more recent attention given to social and community issues, social equity and justice play a small role in environmental assessment. Nor is social equity particularly central to concepts like adaptive management, critical loads, or biodiversity, although (as is the case for many other environmental policy concepts) in the course of their application they can be drawn into equity-linked controversies. Across the broader field, the equity theme is especially present in the concepts of *common but differentiated responsibilities*, *climate equity*, and the *polluter pays principle*.

That many of the most prominent environmental policy concepts engage with these basic themes points to underlying features of the environmental policy realm. The preoccupation with science and policy reflects tension about the place of scientific expertise in political argument and policy design. It underlines the contrasting rationalities of science as a knowledge-building enterprise, policy as a practical administrative realm, and politics as an interest-mediating venture. It also reflects the irreducible role of science in defining environmental problems and finding solutions, even as it simultaneously enables the generation of new or more expansive problems. For science as a knowledge system is embedded in societal relationships, and linked inextricably to the development and commercialization of technologies that may well have future environmental impacts.

That limits feature so prominently in environmental argument (even if they are articulated in a form that avoids connotations of "ultimate limits") is rooted in the tensions between the physical characteristics of ecosystems and the continually expanding reach of human civilization (population, appropriation of land, resources, and technological potency). Societies negotiate such limits, adopting novel technologies and social practices, exploiting new resources, adjusting expectations and ways of life, and actively managing socioecological interactions (Meadowcroft 2013). While local or regional limits preoccupied policy makers in the early days of the modern environmental era, and many of the most damaging environmental impacts continue to be linked to limits at the local and regional levels, it is global limits that have increasingly come to the fore. That a serious climate policy actually implies a rapid and complete shift away from greenhouse gas–emitting fossil energy that has powered centuries of economic development is just beginning to percolate into the core of government. Although biodiversity loss is not "global" in quite the same way (it is happening across the world, but ecosystems are not integrated in the same way as the one atmosphere), it is also grounded in the continuing expansion of the human presence.

The economy/environment issue remains a staple of everyday environmental policy making. Environmental measures typically have costs that must be paid by someone (governments, citizens, or businesses). There are frequently concerns over the impact of more stringent policy on the macroeconomy and competitiveness. And for poorer countries there is the challenge of playing economic catch-up in a more environmentally constrained world. Although arguments are typically fought out in the language of the public good, powerful economic interests can mobilize resources to resist change that would on balance benefit much wider but more diffuse publics. Moreover, modern polities are structured on the expectation of steady economic growth, and the political and social consequences when this stalls for any prolonged period of time are ugly. Yet existing growth pathways typically result in greater aggregate environmental burdens. Faced with this tension, it is hardly surprising that policy makers have been attracted by the possibility of reconciling economic growth with forms of economic activity that meet societal needs without undermining critical environmental assets, as is promised in concepts like sustainable development and green economy.

Then there are the issues of justice and distribution. There are two transcendent realities here. The first is that society is already an ongoing concern, with established entitlements and expectations. Any newly recognized environmental problem will be embedded in an existing set of practices, routines, and property relationships, and brings with it a corresponding set of distributional challenges. New environmental issues may shuffle existing interests by bringing into alliance or opposition different sets of constituencies, but political struggles always play out in the context of existing power relations and inequalities. The second reality is the split between developed and developing countries that (although rendered more complex and nuanced in recent decades with the rapid growth of countries like China) remains a fundamental feature of the landscape within which international environmental policy is made.

One of the more important developments over recent decades has been the perception of the drawing together of issues—across the environmental field, but more important, between environmental issues and other major societal problems. Conceptually, this has been manifest in the rise of integrative ideas such as sustainable development and the green economy yet also *resilience, ecosystem services,* and so on. Again, climate change has played a major role here, with its mitigation and adaptation dimensions as well as its links to energy, forestry and agriculture, the built environment, national security, and so forth. Whatever the ultimate assessment of the notion of *environmental policy integration,* the entwining of environmental issues with broader political questions has become ever more complex in practice. Environmental policy debates collide and intersect with other debates about the role of the state in societal steering, the appropriate place of markets, public participation in decision making, economic inequality, welfare and social policy, and so on.

In keeping with a growing appreciation of the scale of change required to sustain existing social institutions, broader ideas relating to "long-term strategies," "transitions," "transformations," and "pathways" are also coming to the fore (Scoones et al. 2015; Wiseman et al. 2013). Yet as environmental policy seeks to expand outward to engage with a broader range of issues, it simultaneously appears trapped within a larger political matrix that constrains the direction in which it may evolve.

* * *

At the outset we said that this volume was an exploratory venture, intended to open up terrain as well as encourage researchers to place greater attention on concepts and conceptual innovation in the environmental realm. The analysis has been focused at a relatively broad level, reflecting on the nature of innovation processes, the longer-term evolution of the conceptual field, and a number of policy-relevant environmental concepts. There are opportunities to extend this sort of work in many fruitful ways. It would be possible to conduct more detailed analysis of conceptual change in specific policy and implementation contexts—for example, examining concepts or groups of concepts over quite-bounded periods of time, the work of specific agencies, or programs at the local, national, or international level. Such efforts could track the evolution of arguments along with the integration of conceptual and practical change, bringing out more clearly the ways different understanding compete in specific controversies. In would also be possible to map conceptual innovation in specific areas of environmental policy (water governance, waste management, agriculture, energy, and so on) as well as the interactions among them.

The focus here has been on *policy concepts*, defined rather loosely as those that have actually been taken up and been integrated into the idiom of governance. But one could look at environmental concepts that have been important in broader societal debate (even if they have not necessarily found home in policy institutions), such as degrowth or environmental space. Another promising field is a more systematic examination of the evolving categories used in social science to understand environmental issues (consider ecological modernization, transition studies, environmental governmentality, green democracy, etc.). What does this tell us about the evolution of social science and the environmental problematic? There is scope in this broad research agenda to employ a variety of methods ranging from ethnographic studies to quantitative document analysis or social network analysis. Another question relates to comparisons between environmental policy and other spheres. Our assumption here has been that in recent decades, environmental policy has been one of the most dynamic and innovative areas of policy and social contestation—that conceptual innovation has been more marked than in comparable fields such as social welfare policy, economic policy, security policy, cultural policy, and forth. But perhaps this is little more than a prejudice—because it is after all our special field of study. Here we were not able to develop a methodology to assess this assumption, but it is one we would like to encourage others to take up.

A focus on conceptual innovation brings agency to the forefront—the agency of individuals and groups as they struggle to formulate new ideas, incorporate them into their activities, and realize political and policy goals. But it also highlights the structural contexts, both ideational and institutional, within which this innovation takes place. Conceptual reimagining is part of the process of political and policy change, and the character of that reimagining influences how debates and actions play out. If ideas do matter, as the analysis in this book suggests, conceptual innovations deserve to be studied for what they tell us about the past as well as what they portend for the future.

References

Christie, Mike, Nick Hanley, John Warren, Kevin Murphy, Robert Wright, and Tony Hyde. 2006. "Valuing the Diversity of Biodiversity." *Ecological Economics* 58 (2): 304–317.

Committee on Climate Change. 2010. *The Fourth Carbon Budget: Reducing Emissions through the 2020s*. Accessed December 27, 2016, https://www.theccc.org.uk/publication/the-fourth-carbon-budget-reducing-emissions-through-the-2020s-2.

European Commission. 2015. *Closing the Loop—An EU Action Plan for the Circular Economy*. Communication from the Commission to the European Parliament, the Council, the European Economic and Social Committee, and the Committee of the Regions, COM/2015/614. Accessed December 27, 2016, http://eur-lex.europa.eu/legal-content/EN/TXT/?uri=CELEX%3A52015DC0614.

Haas, Peter. 1989. "Do Regimes Matter? Epistemic Communities and Mediterranean Pollution Control." *International Organization* 43 (3): 377–403.

Hajer, Maarten. 1997. *The Politics of Environmental Discourse*. Oxford: Oxford University Press.

Jakob, Michael, and Jerome Hilaire. 2015. "Climate Science: Unburnable Fossil-Fuel Reserves." *Nature* 517:150–152.

Marsh, David, and Rod Rhodes. 1992. *Policy Networks in British Government.* Oxford: Oxford University Press.

Meadowcroft, James. 2013. "Reaching the Limits? Developed Country Engagement with Sustainable Development in a Challenging Conjuncture." *Environment and Planning C: Government and Policy* 31 (6): 988–1002.

Sabatier, Paul. 1988. "An Advocacy Coalition Framework of Policy Change and the Role of Policy-Oriented Learning Therein." *Policy Sciences* 21:129–168.

Scoones, Ian, Melissa Leach, and Peter Newell. 2015. *The Politics of Green Transformations.* London: Earthscan.

UN Environment Program (UNEP). 2012. *Measuring Progress towards an Inclusive Green Economy.* Accessed December 27, 2016, http://web.unep.org/greeneconomy/sites/unep.org.greeneconomy/files/publications/measuring_progress_report.pdf.

Wiseman, John, Taegen Edwards, and Kate Luckins. 2013. "Post-Carbon Pathways: A Meta-analysis of 18 Large Scale Post Carbon Economy Transition Strategies." *Environmental Innovation and Societal Transitions* 8:76–93.

Notes on the Contributors

Richard N. L. Andrews is a professor emeritus of public policy and environmental sciences and engineering at the University of North Carolina at Chapel Hill. He was one of the first scholars to examine in depth the origins and consequences of the National Environmental Policy Act and its requirement for environmental impact assessments. Andrews is the author of *Environmental Policy and Administrative Change: The National Environmental Policy Act* (Lexington Books, 1976), *Managing the Environment, Managing Ourselves: A History of American Environmental Policy* (Yale University Press, 2nd ed., 2006), and many articles on US and comparative environmental policy.

Karin Bäckstrand is a professor in environmental social science at the Department of Political Science at Stockholm University. She has held positions as a Wallenberg Academy Fellow at the Massachusetts Institute of Technology and senior fellow at the University of Oxford. Her research revolves around global environmental politics, climate policy, the politics of expertise, and the legitimacy of environmental governance. Bäckstrand's most recent books are *Research Handbook on Climate Governance*, coedited with Eva Lövbrand (Edward Elgar, 2015), and *Rethinking the Green State: Environmental Governance toward Climate and Sustainability Transitions*, coedited with Annica Kronsell (Routledge, 2015).

Karen Baehler is a scholar in residence at American University's School of Public Affairs. She is author of *Value-Adding Policy Analysis and Advice*, with Claudia Scott (University of New South Wales Press, 2010), and editor of *Chinese Social Policy in a Time of Transition* (Oxford University Press, 2013), with Douglas Besharov, and the forthcoming *Improving Public Services*, with Douglas Besharov and Jacob Klerman

(Oxford University Press). Her current research interests relate to adaptive governance of water systems, including environmental justice dimensions.

Daniel J. Fiorino is director of the Center for Environmental Policy and a distinguished executive in residence in the School of Public Affairs at American University. Among his other books are *The New Environmental Regulation* (MIT Press, 2006), winner of the Brownlow Award of the National Academy of Public Administration; *Environmental Governance Reconsidered* (coedited with Robert F. Durant and Rosemary O'Leary, MIT Press, 2004); and the forthcoming *A Good Life on a Finite Earth: The Political Economy of Green Growth* (Oxford University Press). Before joining American University in 2009, he served in management and staff positions with the US Environmental Protection Agency.

Yrjö Haila is a professor of environmental policy (emeritus) at the University of Tampere. He was educated as an ecologist, and his research interests have centered on the nature-society interface and ecosocial dynamics, from several complementary perspectives. His books include *Humanity and Nature: Ecology, Science, and Society*, together with Richard Levins (Pluto Press, 1992), and *How Nature Speaks: The Dynamics of the Human Ecological Condition*, coedited with Chuck Dyke (Duke University Press, 2006), as well as several books in Finnish.

Michael E. Kraft is a professor of political science and public affairs emeritus at the University of Wisconsin at Green Bay. He is the author of *Environmental Policy and Politics* (Routledge, 7th ed., 2017); coauthor of *Coming Clean: Information Disclosure and Environmental Performance* (MIT Press, 2011); coauthor of *Public Policy: Politics, Analysis, and Alternatives* (CQ Press, 6th ed., 2017); and coeditor and contributing author of *Environmental Policy* (CQ Press, 10th ed., 2018), *The Oxford Handbook of U.S. Environmental Policy* (Oxford University Press, 2013), and *Toward Sustainable Communities: Transition and Transformations in Environmental Policy* (MIT Press, 2nd ed., 2009).

Oluf Langhelle is a professor of political science in the Department of Media, Culture, and Social Sciences at the University of Stavanger in Norway. His research has focused on the concept of sustainable development, strategies for sustainable development, environmental politics and policy, and transitions toward low-carbon societies, concentrating on carbon capture and storage (CCS) and other issues. He has coedited a number of books and published articles on various aspects of sustainable development.

Judith A. Layzer was a professor of environmental policy in the Department of Urban Studies and Planning at the Massachusetts Institute of Technology until her death in 2015. Her research focused on several aspects of US environmental politics, including the roles of science, values, and storytelling in environmental politics as well as the effectiveness of different approaches to environmental planning and management. Layzer was the author of *The Environmental Case*, now in its fourth edition (CQ Press, 2015), *Natural Experiments: Ecosystem-Based Management and the Environment* (MIT Pres, 2008), and *Open for Business: Conservative's Opposition to Environmental Regulation* (MIT Press, 2012).

James Meadowcroft is a professor in the Department of Political Science and the School of Public Policy and Administration at Carleton University in Ottawa. He holds a Canada Research Chair in Governance for Sustainable Development. Meadowcroft has written widely on environmental politics and policy, democratic participation and deliberative democracy, national sustainable development strategies, greening the state, and sociotechnical transitions. His recent work has focused on energy and the transition to a low-carbon society, and includes publications on CCS, smart grids, the development of Ontario's electricity system, the politics of sociotechnical transitions, and negative carbon emissions.

Alexis Schulman is a PhD candidate in the Department of Urban Studies and Planning at the Massachusetts Institute of Technology. She focuses on urban sustainability and environmental governance. Schulman is currently researching urban institutional change in the United States, and investigating why some cities, and not others, accomplish the difficult work of translating their sustainability objectives into enduring reforms.

Johannes Stripple is an associate professor of political science at Lund University in Sweden. He explores and develops "critical climate studies" at the intersection between social/political thought and emerging practices of a carbon-constrained and rapidly warming world. In his work, Stripple has traced the governance of climate change through a range of sites, from the United Nations to the everyday, from the economy and the urban to the low-carbon self. He is the editor of *Governing the*

Climate: New Approaches to Rationality, Power, and Politics (Cambridge University Press, 2014) and *Toward a Cultural Politics of Climate Change* (Cambridge University Press, 2016).

Philip J. Vergragt is an academic and activist, a fellow at the Tellus Institute, a research professor at Marsh Institute at Clark University, and a professor emeritus of technology assessment at Delft University of Technology. He has coauthored eighty scientific publications and four books. His research interests are: sustainable technological and social innovations in transportation, energy, and housing; grassroots innovations; sociotechnical transitions; sustainable consumption and production; sustainable cities; and assessment of emerging technologies. Vergragt is a founding board member of SCORAI, the North American (and European) Sustainable Consumption and Action Initiative, and holds a PhD in chemistry from the University of Leiden.

Index

American and Comparative Environmental Policy

Sheldon Kamieniecki and Michael E. Kraft, series editors

Russell J. Dalton, Paula Garb, Nicholas P. Lovrich, John C. Pierce, and John M. Whiteley, *Critical Masses: Citizens, Nuclear Weapons Production, and Environmental Destruction in the United States and Russia*

Daniel A. Mazmanian and Michael E. Kraft, editors, *Toward Sustainable Communities: Transition and Transformations in Environmental Policy*

Elizabeth R. DeSombre, *Domestic Sources of International Environmental Policy: Industry, Environmentalists, and US Power*

Kate O'Neill, *Waste Trading among Rich Nations: Building a New Theory of Environmental Regulation*

Joachim Blatter and Helen Ingram, editors, *Reflections on Water: New Approaches to Transboundary Conflicts and Cooperation*

Paul F. Steinberg, *Environmental Leadership in Developing Countries: Transnational Relations and Biodiversity Policy in Costa Rica and Bolivia*

Uday Desai, editor, *Environmental Politics and Policy in Industrialized Countries*

Kent Portney, *Taking Sustainable Cities Seriously: Economic Development, the Environment, and Quality of Life in American Cities*

Edward P. Weber, *Bringing Society Back In: Grassroots Ecosystem Management, Accountability, and Sustainable Communities*

Norman J. Vig and Michael G. Faure, editors, *Green Giants? Environmental Policies of the United States and the European Union*

Robert F. Durant, Daniel J. Fiorino, and Rosemary O'Leary, editors, *Environmental Governance Reconsidered: Challenges, Choices, and Opportunities*

Paul A. Sabatier, Will Focht, Mark Lubell, Zev Trachtenberg, Arnold Vedlitz, and Marty Matlock, editors, *Swimming Upstream: Collaborative Approaches to Watershed Management*

Sally K. Fairfax, Lauren Gwin, Mary Ann King, Leigh S. Raymond, and Laura Watt, *Buying Nature: The Limits of Land Acquisition as a Conservation Strategy, 1780–2004*

Steven Cohen, Sheldon Kamieniecki, and Matthew A. Cahn, *Strategic Planning in Environmental Regulation: A Policy Approach That Works*

Michael E. Kraft and Sheldon Kamieniecki, editors, *Business and Environmental Policy: Corporate Interests in the American Political System*

Joseph F. C. DiMento and Pamela Doughman, editors, *Climate Change: What It Means for Us, Our Children, and Our Grandchildren*

Christopher McGrory Klyza and David J. Sousa, *American Environmental Policy, 1990–2006: Beyond Gridlock*

John M. Whiteley, Helen Ingram, and Richard Perry, editors, *Water, Place, and Equity*

Judith A. Layzer, *Natural Experiments: Ecosystem-Based Management and the Environment*

Daniel A. Mazmanian and Michael E. Kraft, editors, *Toward Sustainable Communities: Transition and Transformations in Environmental Policy*, second edition

Henrik Selin and Stacy D. VanDeveer, editors, *Changing Climates in North American Politics: Institutions, Policymaking, and Multilevel Governance*

Megan Mullin, *Governing the Tap: Special District Governance and the New Local Politics of Water*

David M. Driesen, editor, *Economic Thought and US Climate Change Policy*

Kathryn Harrison and Lisa McIntosh Sundstrom, editors, *Global Commons, Domestic Decisions: The Comparative Politics of Climate Change*

William Ascher, Toddi Steelman, and Robert Healy, *Knowledge in the Environmental Policy Process: Re-Imagining the Boundaries of Science and Politics*

Michael E. Kraft, Mark Stephan, and Troy D. Abel, *Coming Clean: Information Disclosure and Environmental Performance*

Paul F. Steinberg and Stacy D. VanDeveer, editors, *Comparative Environmental Politics: Theory, Practice, and Prospects*

Judith A. Layzer, *Open for Business: Conservatives' Opposition to Environmental Regulation*

Kent Portney, *Taking Sustainable Cities Seriously: Economic Development, the Environment, and Quality of Life in American Cities*, second edition

Raul Lejano, Mrill Ingram, and Helen Ingram, *The Power of Narrative in Environmental Networks*

Christopher McGrory Klyza and David J. Sousa, *American Environmental Policy: Beyond Gridlock*, updated and expanded edition

Andreas Duit, editor, *State and Environment: The Comparative Study of Environmental Governance*

Joseph F. C. DiMento and Pamela Doughman, editors, *Climate Change: What It Means for Us, Our Children, and Our Grandchildren*, second edition

David M. Konisky, editor, *Failed Promises: Evaluating the Federal Government's Response to Environmental Justice*

Leigh Raymond, *Reclaiming the Atmospheric Commons: Explaining Norm-Driven Policy Change*

Robert F. Durant, Daniel J. Fiorino, and Rosemary O'Leary, editors, *Environmental Governance Reconsidered: Challenges, Choices, and Opportunities*, second edition

James Meadowcroft and Daniel J. Fiorino, editors, *Conceptual Innovation in Environmental Policy*